REQUIREMENTS MODELING AND CODING An Object-Oriented Approach

REQUIREMENTS MODELING AND CODING An Object-Oriented Approach

Liping Liu

The University of Akron, USA

World Scientific

NEW JERSEY · LONDON · SINGAPORE · BEIJING · SHANGHAI · HONG KONG · TAIPEI · CHENNAI · TOKYO

Published by

World Scientific Publishing Europe Ltd.

57 Shelton Street, Covent Garden, London WC2H 9HE
Head office: 5 Toh Tuck Link, Singapore 596224
USA office: 27 Warren Street, Suite 401-402, Hackensack, NJ 07601

Library of Congress Cataloging-in-Publication Data
Names: Liu, Liping (Computer scientist), author.
Title: Requirements modeling and coding : an object-oriented approach /
  Liping Liu, The University of Akron, USA.
Description: London ; Hackensack, NJ : World Scientific Publishing Europe
  Ltd., [2020] | Includes bibliographical references and index.
Identifiers: LCCN 2020018498 | ISBN 9781786348821 (hardcover) |
  ISBN 9781786348876 (paperback) | ISBN 9781786348838 (ebook) |
  ISBN 9781786348845 (ebook other) | ISBN 9781786348920 (ebook other)
Subjects: LCSH: Object-oriented programming (Computer science) | Software engineering. |
  Computer software--Specifications. | Model-driven software architecture.
Classification: LCC QA76.64 .L58 2020 | DDC 005.1/17--dc23
LC record available at https://lccn.loc.gov/2020018498

British Library Cataloguing-in-Publication Data
A catalogue record for this book is available from the British Library.

For any available supplementary material, please visit
https://www.worldscientific.com/worldscibooks/10.1142/Q0260#t=suppl

Desk Editors: Ramya Gangadharan/Michael Beale/Shi Ying Koe

Typeset by Stallion Press
Email: enquiries@stallionpress.com

This book is dedicated to the memory of my mother, Shengju Zhen (1939–2017), who believed in and insisted on my education against all the odds during the Great Proletarian Cultural Revolution.

Preface

This book serves as a text for a capstone course on Systems Analysis and Design in Information Systems programs. It conceptualizes business objects and functions, develops business models and software architectures, and enriches the models and the architectures by storyboarding use cases along with user interface designs.

There are two obstacles in teaching object-oriented techniques in Information Systems programs in business schools. First, many professors were trained in structured methodologies, and/or have a native instinct of thinking in functions due to their business background. They tend to have a hard time converting their thinking into objects. Second, many Information Systems programs offer one course on basic programming principles but do not have the luxury to offer advanced object-oriented programming, which is a prerequisite to a better understanding of object-oriented models. Therefore, most schools still teach structured methodologies, although the software industry is primarily using object-oriented ones.

This book assumes little or no preparation in traditional structured methodologies and modern object-oriented programming languages. It bridges structured and object-oriented methodologies and turns existing knowledge in business data and functions into an asset, rather than a burden, in learning object-oriented analysis and modeling.

This book is not about software development processes or methodologies such as systems development life cycle, agile, scrum, DevOps, etc. Rather, it is concerned with how to faithfully model business requirements and how to effectively develop systems specifications. All development methodologies entail requirements analysis and modeling, and some of the modern ones put more emphasis on software delivery rather than the incremental commitment of models, documentations, and contracts. As the Manifesto for Agile Software Development states, "We want to restore a balance. We embrace modeling, but not in order to file some diagram in a dusty corporate repository. We embrace documentation, but not hundreds of pages of never-maintained and rarely-used tomes. We plan but recognize the limits of planning in a turbulent environment."

This book can serve any methodology but appears to work better with the agile. The book emphasizes the integration of programs and models with two goals in mind. First, it helps the reader to connect a modeling construct to code so that he or she appreciates how certain models are more useful than others to programmers. In over 20 years of teaching, I found that students and even instructors often create models that lack clarity to be interpreted and/or precision to be coded. Second, it helps to reduce the gap between the end users and programmers; whether to model a business function, a procedure, a business object, a use case, or a user interface, the book shows its conversion into testable code. Thus, testing and end-user involvement can be integrated into every stage of systems development.

This book is not a text on computer programming, and so it does not go in-depth into the nitty-gritty details. However, one of the key features of the book is to present requirements modeling and code expression in parallel for students to understand modeling concepts better and for professionals to reduce the gap between analysis and development. Instructors may choose to review the essential concepts and principles in an object-oriented programming language such as C#, Java, or C++ from day one. This book will use C# in all examples because C# bridges Visual Basic, Delphi, C, and Java very well and also possesses an advantage over others in prototyping graphical user interfaces (GUIs). Yet, the instructor may choose Java instead without any difficulty.

This book uses Unified Modeling Language (UML) for diagramming notations and IBM Rational Rhapsody as the modeling tool. Rhapsody is a visual development environment that software developers can use to create real-time or embedded systems. It is an integrated computer aided software engineering tool that uses graphical models to generate software applications in various languages including C, C++, Ada, Java, and C#. The reader may choose other similar tools such as Poseidon, Visual Paradigm, etc. The vendors of these tools typically provide free or low-cost educational licenses to instructors and students.

About the Author

Liping Liu is Professor of Management and Information Systems at The University of Akron. He received his Bachelor of Science in Applied Mathematics in 1986 from Huazhong University of Science and Technology, Bachelor of Engineering in River Dynamics in 1987 from Wuhan University, Master of Engineering in Systems Engineering in 1991 from Huazhong University of Science and Technology, and Ph.D. in Business in 1995 from the University of Kansas. His research interests are in the areas of Uncertainty Reasoning and Decision-Making in Artificial Intelligence, Electronic Business, Systems Analysis and Design, Technology Adoption, and Data Quality. Dr. Liu has published articles in *Decision Support Systems*; *European Journal of Operational Research*; *IEEE Transactions on System, Man, and Cybernetics*; *International Journal of Approximate Reasoning*; *Information and Management*; *Journal of Association for Information Systems*; *Journal of Optimization Theory and Applications*; *Journal of Risk and Uncertainty*; among others. Dr. Liu has made distinct contributions in such fields as Decision Theory, Artificial Intelligence, and Research Methodology. He provided the best axiomatization of the rank-and-sign utility function and proposed a theory of coarse utility that explains the St. Petersburg Paradox, Allais Paradox, and others better than any other utility functions. He developed a theory of linear belief functions and applied the concept to information integration in auditing, investment analysis, model combination, matrix computation, etc. He proposed the concept of predictive and mediating efficiencies to test the nomological validity of second- or higher-order measurement models. His theories of coarse utilities and linear belief functions are taught at the nation's top Ph.D. programs in such subjects as Accounting, Computer Science, Economics, Management, and Psychology. His concept of predictive and mediating efficiencies is one of two authoritative references on nomological networks, along with Cronbach's classic paper in 1955, by OMICS International.

Dr. Liu has served as a guest editor for *International Journal of Intelligent Systems*, a co-editor for *Classic Works on Dempster–Shafer Theory of Belief Functions*, and on the editorial boards of a few academic journals as well as on the program committees of many academic conferences.

Dr. Liu has strong practical and teaching interests in e-business systems design, development, and integration using advanced DBMS, CASE, and RAD tools. He has won several teaching awards. His consulting experience includes designing and developing a patient record management system, a payroll system, a course management system, and an e-travel agent, and providing corporate trainings on Oracle database administration, Oracle applications development, and object-oriented requirements analysis and modeling for large corporations.

During the years of his Ph.D. study at the University of Kansas, Dr. Liu was on the Dean's List every year. His GPA at graduation was 4.0. His dissertation has received the Best Dissertation Award.

During his undergraduate studies at Huazhong University of Science and Technology, Dr. Liu was as the Chief Editor for *College Mathematics* and *Journal of Undergraduate Academy* since his freshman year. He began to publish papers in the top academic journals such as *Journal of Systems Science & Mathematical Sciences*, *Journal of Huazhong University of Science & Technology*, etc. His publications won him both the first and second place in the 1986 Hubei Province Best Student Research Competition. He also won China's National Outstanding Research Achievement Award for his research on energy planning in 1987.

Dr. Liu was significantly publicized by *Changjiang Daily*, *Hubei Daily*, *Guangming Daily*, etc. for being the first person in China to pursue a double major simultaneously from two different colleges. Starting from his sophomore year, he attended both Wuhan University and Huazhong University of Science and Technology and earned two Bachelor degrees in three years. His second degree thesis led to the obtainment of an analytical solution to a long-lasting engineering problem related to the Three-Gorges Dam project.

Contents

Chapter 1

Introduction

The process of systems analysis is to chart a course to achieve a vision. This text teaches how to analyze and model business requirements that can then eventually be transformed into systems specifications for developing a computer-based information system that supports the vision. In the process, we will learn how to develop evolving artifacts that represent business requirements at one end and systems specifications at the other. The target readers of the book are students and professionals who intend to become or understand business analysts, whose primary role is to bridge the gap between programmers and business end-users.

In studying this text, the reader shall pay close attention to the following three streams of the course development: *System*, *Process*, and *Techniques*. This chapter introduces three streams. First, we will introduce the concept and the components of information systems and the typical roles assumed by a system analyst. Then, we will introduce the systems development life cycle and explain the deliverables of each phase as well as the techniques to be used to produce said deliverables.

Information Systems

A *system* is a set of interrelated and interacting elements that collaborate to accomplish a specific purpose. This is a generic concept because it applies to many other subjects of study such as biology, economics, and politics. A system has the following features: (1) each element has its own purpose, which serves the purpose of the entire system; (2) different elements are interdependent; (3) the purpose and function of the individual elements serve the purpose and functions of the entire system; and (4) the whole is greater than the sum of individual elements.

An information system is a set of computer hardware, software, database, and people that are integrated to provide a platform for transactional and decisional support. A typical information system includes elements as exampled in Table 1.

Computer hardware includes input and output devices, communication devices, central processing units (CPU), and data storage. The CPU acts as the brain of a computer and is essentially an electronic circuitry made of microswitches that use on/off states for 0s and 1s to perform basic arithmetic, logic, controlling, and input/output operations. Inputs and outputs include monitors or terminals, keyboards, pointing devices, printers, speakers and sound cards, video cards, scanners, etc. There are two broad categories of storage: random access memory (RAM) and permanent memory. The difference is that RAM is much faster than permanent memory (at least 10^7 faster), so almost all programs will store their temporary data in blocks of RAM — called *variables* — for fast access. Yet, RAM is temporary, and it will not survive a power shutdown. Permanent memory will stay for a long period of time despite power failures. Examples of this form include hard drive, floppy drive, CD, and DVD, etc. A hard drive

Table 1.   Information system constituents.

Hardware	• Input (e.g., keyboard, mouse, touchscreen, microphone) • Output (e.g., display, speakers) • Central processing units • Storage (random access memory and permanent memory like hard drive, CD, DVD, jump drive, etc.) • Communication devices (modem, network interface card, cable, hub, switch, router)
Software	• Operating systems (e.g., Windows, Mac OS, Android, iOS) • Databases • Business applications (e.g., forms and reports)
People	• End users • Programmers, developers, software engineers • Business analysts, systems analysts • Network engineers • Database administrators, systems administrators

stores 0s and 1s by magnetizing drive materials to different directions. A CD records 0s and 1s by creating small dots (pits) so that pits and lands reflect light differently.

Communication devices include modems, repeaters, bridges, routers, network interface cards, and cables. *Modem* is a portmanteau made of two words, modulation and demodulation, and these actions are responsible for the conversion between analog data and digital data (0s and 1s). *Cables* are responsible for sending raw electric or light signals representing 0s and 1s. Typical examples include *patch cord* made of unshielded twisted pairs of copper wires and fiber optical ones made of a shielded glass thread. *Network interface cards* are responsible for packing individual 0s and 1s into data packets called frames and controlling the error of transmission. Switches are used to create segments inside a network to improve its performance and security. Routers are responsible for connecting individual networks to form inter-networks, or internet.

People are an important part of the information system. The people involved include end users, programmers, and those who play the role of bridging these two groups or supporting them. The first set is systems analysts (or business analysts, business engineers, or systems engineers), who acts as the middleman between users and programmers; they facilitate communications between the two groups so that the users' wants can be translated into program specifications, according to which the program can then be developed. The second set includes the database administrators, system administrators, and network administrators, who support both end users and programmers in sharing data, system, and network resources. The third set is made up of support technicians who help troubleshoot hardware and software issues for other users.

This book will not teach how to manage people; the reader can take a course in management or psychology to learn how to design and develop effective organizations. This book will not teach how to analyze, model, develop, and manage hardware components; the reader should take courses in computer engineering or technology management to learn those aspects of an information system. This book will also not teach how to analyze and develop operating systems, which is usually taught in computer science. As far as this book is concerned, an information system contains *business applications* (or programs) that serve business end users for transactional or decisional support, with *databases* in the back end as the central repository of data resources for applications (Figure 1).

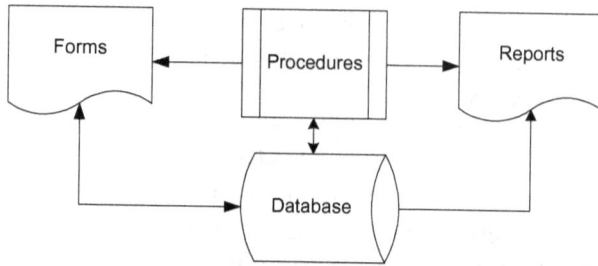

Figure 1.   The essential components of an information system.

Business applications

Forms and reports are the typical business applications. They are the interfaces between the system and the user, and so are often referred to as *user interfaces*. The difference between them is that a form is usually used for viewing and entering data whereas a report is used to display and summarize data.

Forms and reports often interact with the database through intermediate program modules called *procedures,* which run behind the scene. In other words, procedures are the connectors of the front-end forms/reports with the back-end databases. They collect data from forms and/or retrieve data from databases, process them, and finally write the result back to the database or display the result to the user. Procedures can be located with front-end applications (such as in client/server systems), stored on back-end databases (so-called *stored procedures*), or stored somewhere in between such as applications servers. Procedures usually implement business rules that are subject to change during daily business practice. Forms, reports, and procedures constitute business applications.

Business applications are developed using application development tools — computer programs that compile or interpret commands in a programming language. Well-known examples include Visual Studio, Eclipse, Power Builder, Oracle Developer, C++ Builder, Delphi, Dynasty, etc. These tools often embed compilers to compile or interpret one or more programming languages. For example, Eclipse supports Java development, while Visual Studio supports C#, Visual Basic, Visual C++, Fortran, etc. A generic program compiler or interpreter may be able to develop applications. Yet modern systems development works better with a more sophisticated tool possessing the feature of *rapid application development* (RAD) for quick prototyping and modular assembling because these tools include a library of components or program modules that are ready to be plugged into a project without re-inventing the wheel.

Databases

A database is simply a set of *records*, each of which is an array of observations made on one business object. The records must be connected based on a certain logical data model. For example, a relational database packs the records into tables or relations, and the records are then linked by sharing common columns or by duplicating a *primary key* into a *foreign key*. Another example is a network database that organizes the data into a network of records linked by pointers, a concept commonly seen in programming languages such as C++, COBOL, etc. An object-oriented database encapsulates both records and the program modules that process the records into higher-level units, called *objects*.

Databases are created and managed using database management systems (DBMS), computer programs that organize, validate, secure, and manipulate data. In other words, it is a program for us to build a new database, manage an existing database, and manipulate the data in the database.

In the old days, databases were developed and operated using a programming language such as COBOL. Now, as a standard, a relational DBMS speaks structured query language (SQL), and so any commands to create and access a database must be in SQL. The use of SQL has become extremely pervasive: you can use SQL to talk to a database interactively. You can also embed SQL commands into a program and have them talk to a database automatically.

Systems Analysts

To create a new system or improve existing ones, we need business end users, who understand what to program but do not know how to program, and programmers, who are the opposite. One may think it is enough to assemble these two groups of people into a project team, as some development methodologies suggest. The reality is that these two groups of people often speak different languages and have different interests. A typical programmer speaks Java or C# and likes to lock himself into a basement cell without interacting with people, especially those who do not speak programming languages.

Wherever there is a language and/or interest barrier, there must be a middleman who can overcome this barrier. Systems analysts are such middlemen. *Systems analysis and design* is essentially a process of bridging end users and programmers. It discovers and documents end user requirements, converts the requirements into programming specifications, and communicates these specifications to the programmers. In this sense, a system analyst is a communicator or interpreter between users and programmers.

In order to be an interpreter, one needs to speak two languages. To be an effective systems analyst, one needs to speak the business languages to talk to end-users as well as the programming languages to talk to programmers. Therefore, information systems programs are typically housed in business schools and have courses in business functional areas as well as in programming languages. Students take the programming courses not to become programmers, but rather, to learn how to talk to programmers, how to develop meaningful specifications, and how to prototype ideas to meet end user requirements.

The role of the interpreter is passive. In fact, the purpose of an information system is to support business needs. Thus, when creating or improving an information system, there is often a need to change how one can run a business with the new or improved systems. From this perspective, systems analysis is also a process of business re-engineering. It analyzes existing business processes and designs a new way to improve them. In this sense, a systems analyst is a process architect or engineer. An associated subtle role that a systems analyst plays is that of a politician. When re-engineering a business process, various stakeholders may be affected. Those who receive negative impacts, e.g., losing a comfortable job, may object to the change and/or the new design, regardless of how brilliant it is.

Structured Development Processes

A systems development process may be roughly divided into six stages: project proposal, analysis, design, development, implementation, and maintenance. Systems development life cycle (SDLC) is a methodology that emphasizes incremental commitment to a system development project. Each stage starts with the approval of an overseeing committee and finishes with promised deliverables for review by the committee. The waterfall methodology has been challenged by newer alternatives such as agile, scrum, DevOps, etc., which emphasize continuous delivery of software products rather than incremental commitment of deliverables. The agile methodology, for example, encourages software testing and end-user involvement in every stage of a development process, while DevOps combines development with information technology operations and tries to address the gap between development and implementation.

There is also an alternative interpretation of the SDLC that treats the SDLC not as a development methodology but rather a map of the process that converts user requirements into programming specifications. From this viewpoint, RAD tools can be employed to paint forms and reports as systems prototypes at a very early stage of the process. Joint application development (JAD) may also be employed to involve users in every stage of the SDLC.

Requirements analysis and modeling, as the first stage of the systems development process, is done to discover, document, and validate business requirements and communicate these precisely as system specifications to the systems developer. At this stage, models, or pictures with interconnected graphical symbols, are often used, since a picture is worth a thousand words.

Conceptually, a model is an abstract representation of realities. Realities are complex and multifaceted. When we model realities, we will ignore insignificant details and focus on essentials. We will also have to take into consideration a viewpoint from which to observe and describe them. We often observe different facets of and build different models for the same object when viewed from different viewpoints. Like in the old Indian tale, the descriptions of an elephant by six blind men were dramatically different. Therefore, to model an information system, we may need many different models, with each one describing one perspective of the system.

Figure 2 shows an exploded view of the central stages of a typical structured development process: analysis, design, and development along two parallel paths, with one leading to the implementation of a database (data path) and the other to the development of business applications (function path).

- **Data path:** Business Object Models — Logical Data Model — Databases.
- **Function path:** Business Process Models — Procedure Models — Business Applications.

Before we go into detail about each model or deliverable in Figure 2, we shall note that these two paths, despite proceeding in parallel, are not

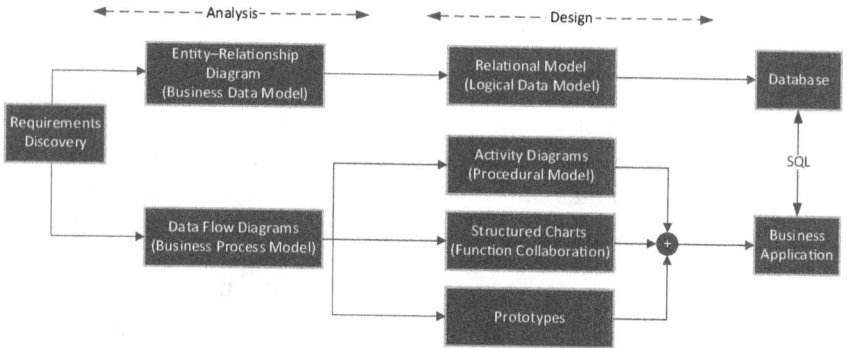

Figure 2. The structured systems development process.

independent because changing data requirements may alter functional requirements, and the identification of functions can bring up new data requirements. It must also be noted that in modern object-oriented analysis, which is what this textbook is about, the two paths are integrated so that data and functions are encapsulated into higher-level abstract units, called *objects*.

Requirements discovery

A new project is usually initiated by problems and opportunities. A problem with the legacy system may stimulate a bottom-up proposal from end-users to improve the system. Business opportunities and technology advancements may engender top-down planning for a new up-to-date system. In any case, a proposal must clearly identify the problems and opportunities. It must define the scope of the project, e.g., what business functions are to be included. It must have a forecast as regards the time frame and resource requirements. After the proposal is accepted, following studies of its financial and technology feasibilities, the analysis and design can be conducted in the two parallel paths.

Data path

Business object models, or conceptual data models, represent the user requirement on what data is needed for conducting business transactions and supporting future managerial decision-making in a structured way. Since data comes from observations on business objects, data models represent business objects such as things, events, concepts, people, etc., and the relationships between them that embody the requirements of data navigation and business rules. The typical database course provides comprehensive coverage on how to use entity–relationship diagrams as such models.

A conceptual data model, such as an entity–relationship diagram, is the language a business analyst uses to communicate with end users. This must then be converted into a logical data model — such as a relational model, a hierarchical model, or a network model — for computer programmers to understand database programming specifications. The conversion follows certain rules. For example, if the relational model is chosen, each entity set will be converted into a table and each attribute will be converted into a column of the table; for many-to-many relationships, we will add a junction table, which is made of the duplicated primary keys of both end tables. Also, if the relational model is chosen, certain normalization procedures must be applied to reduce redundancies and operational anomalies. The typical database course will introduce rules to convert an entity–relationship diagram into a relational model. It may also cover how to use a computer-aided software engineering (CASE) tool to automate the conversion process.

A logical data model represents programming specifications for a new database to be developed. In the olden days, you would have to ask a programmer to implement the design using COBOL. Now, with the availability of many commercial DBMS such as Oracle, Microsoft SQL Server, MySQL, and IBM DB2, database implementation is simplified into writing SQL commands. A special subset of SQL statements, called *data definition language*, is used for creating and altering database objects such as tables.

Function path

The function path starts with a process model that represents the functional requirements of a system. The emphasis of process modeling is on WHAT rather than HOW, i.e., what a user will do with the system and what functionalities the user would like the system to have. For example, when developing an online order system, usual functions include taking orders, billing customers, querying order status, handling returns, etc. A process model captures these processes and their relationships, such as workflows or data flows. A typical conceptual process model in the structured development is the *data flow diagrams* that graphically depict a network of processes connected by data inputs and outputs, called data flows.

A process model treats each process as a black box with inputs and outputs. It does not say what the box contains or how each function is performed, which is the realm of procedural modeling.

Like a business data model, business analysts use process models as a language to communicate with end-users who understand business processes. They must then translate process models into program specifications and talk to programmers. Unlike database specifications, however, there may be several aspects of application specifications. For example, in a structured methodology, there are three types of specifications: structured charts, procedural models, and application prototypes.

In structured development, one aspect of the program specification is the program structure, or how the code modules are commanding or being executed. The structural specifications are represented by structured models, like *structure charts*. A structure chart shows how functions work together or collaborate in a coordinated manner to achieve a higher-level functionality.

The second specification is a procedural model that opens each black box and details how each business process is performed logically and sequentially. A procedural model can be created using pseudocodes, structured English, activity diagrams (used to be called program flow charts), or even high-level scripting languages like Visual Basic, Oracle PL/SQL, Power Script, and JavaScript. The benefit of using pseudocode or structured English is that you do not need to be concerned with the constraints of strict programming syntax rules but focus on expressing the logics and sequences of a procedure. However, the business analysts familiar with a programming language may find it convenient and efficient to use actual code to express a procedural model. Instead of learning pseudocode or structured English and potentially enlarging the gap between modeling and coding, using a high-level programming language in procedural modeling can help achieve continuous delivery of software products.

Earlier, a program would run from start to finish without user intervention or with occasional stops for user inputs. It displayed nothing but a black box with plain lines of text outputs. The reader may have seen such so-called console programs in their first course on programming principles. To develop a console program, procedural models, one for each process, and a structured chart, would be sufficient program specifications.

However, if a program is executed with human user interventions, the last aspect of program specifications has to be user interfaces, like a form or report. The user interface is most likely a graphical one in a modern windowing environment or webpage. The collection of all linked user interfaces represents a prototype for the application to be designed. Note that a prototype is usually just a dummy framework to give end users the feel and look of the applications. It is not a working program because the code behind it may not exist or does not completely satisfy procedural requirements.

Prototypes, procedural models, and structural charts represent program specifications for an application to be developed by programmers. They are the language a business analyst will use to communicate with programmers. Often, there is mystery surrounding programmers who can speak the cryptic code language. In fact, the most difficult task in software development is to develop the specifications.

Data and function paths show the overall direction toward the completion of a development project. It spells out the deliverable of each phase and the sequential or parallel arrangement of related activities for project management. However, one should not misunderstand that it is possible to move straight ahead without coming back to make modifications on the deliverables of a previous stage. As a matter of fact, systems development tends to be an iterative or evolving process. For example, after communicating with programmers, you may realize that you need to go back to end-users and discuss some modifications on business data or process models. Methodologies such as the agile and Scrum tend to overemphasize this nature, thus calling for user involvement and code testing at every stage. What is essential in the development process, however, is to ensure that what is coded is exactly (or at least as close as possible) what a business wants. Due to misunderstandings or misspecifications, a programmer can deliver a donkey when the business wants a horse. There can be many factors contributing to these misunderstandings and misspecifications, and improper use of communication language is one of the most important. Regardless of the choice of a methodology, we cannot underemphasize the use of the right language to communicate with the right people. In addition, system analysts should attempt to understand the problem domain of a business and not make unjustified assumptions about its requirements. They must know how to protect themselves; if necessary, they should get end-users or project managers to sign off models and specifications before these are submitted, converted, and/or coded.

Object-Oriented Development Processes

A structured program is a collection of one or more reusable units, called *functions*. One of these is the main function, also named as `main()`, that acts the starting point of the whole program. Because of the separation of data and functions, each function can virtually do nothing without being fed the appropriate data inputs. Consequently, functional modules become highly dependent on each other and coupled into a highly complex web of functional calls and executions. A simple change in one function or data source can cause ripple effects in many related modules. Also, in order to achieve maximum reusability, each function tends to be a relatively small module. The reader may have seen functions like `upper()`, `lower()`, `trim()`, etc. These functions do nothing but change the data format or remove spaces from input data. Counting these types of functions, a typical medium- or large-sized system can easily consist of thousands of modules. Maintaining and managing such modules is a nightmare, if not impossible.

Because of these problems, structured development has given up its dominance or even existence, and the new object-oriented development has thus arisen. Object orientation advocates the encapsulation of both data and functions that process data into a higher-level reusable unit called *class*, which can then be used to create running instances, called *objects*. We will elaborate more on these concepts in later chapters. But for now, the reader can just imagine that things like windows, buttons, or menus are objects, and the code that create those objects are classes.

An object-oriented program is made of one or more reusable classes, one and only one of which must contain the `main()` function and act as the starting point of the program. This is the static view of a program. From a dynamic or running perspective, a program consists of one or more collaborating objects that are created by the classes. *Objects are data holders and behavior executors.* Instead of calling a function or issuing a command directly, an object is called to execute a function by *sending a message* to the object. Each object often contains many functions that are related to each other and responsible for the same area of concerns. Often, a task to be performed by the program can be delegated to one object specialized in the task. The object may need assistance from other objects to perform a portion of the task. It does so by sending messages to the other objects. Dynamically, a program is a network of objects that invokes each other through messages.

There are three benefits of the object-oriented approach. First, the number of data flows and functional calls (or equivalently messages) have been significantly reduced because data inputs to or outputs from a function are packaged inside the same object as the function. The following analogy may help to understand the point better. If we ask a boy how old he is, he does not need an input of his birth date and the current time to answer the question. However, if we ask a calculator the same question, the calculator will need to have the input data in order to answer the question.

Second, the number of program modules is dramatically reduced. Because each class can pack tens or hundreds of functions, the number of reusable program units is a lot lesser. Yet, all the functions can be still executed individually by sending a message to its housing object. Third, program modules become less coupled or dependent on each other. Although objects still need to collaborate to perform a large task, most objects are self-sufficient in performing a task.

What does object-oriented development mean for systems analysis and design? Since functions and data are no longer separate, the final product of analysis and design is not a set of interrelated functional modules. Rather, it is a set of classes that can be used to create functioning and collaborating objects. These are two specific implications. First, there is no need to model the data inputs and outputs of functions, and thus data flow diagrams are no longer useful for modeling functional requirements. Second, the dynamic view of a program is not function calls. Rather it is object creation and collaboration through messages. Thus, structured charts are no longer useful.

Regardless of structured or object-oriented development, we always need to capture and model data and functional requirements. However, since data and functions are now bundled into classes, we shall change our terminologies and use the terms classes and objects throughout the system development process. When modeling data requirements, we model business or domain objects. When modeling functional requirements, we model use cases, which are also classes.

Figure 3 shows the object-oriented systems development process and deliverables along the way. The requirements discovery stage is the same as for the structured development, including identifying problems and opportunities; setting up visions, goals, and objectives; and discovering the solutions and requirements that solve the problems, take advantage of the opportunities, and satisfy the goals and objectives. Then, in the analysis stage, we model two views of the system: static view and dynamic view. From the

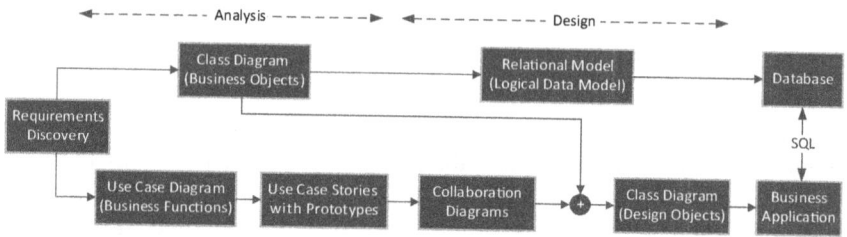

Figure 3. The object-oriented development process.

static point of view, we identify and model business objects that are business data and business function carriers and assess their relationships as governed by business rules. The deliverables are *class diagrams*. From the dynamic point of view, we model use cases and their associations with *actors*, which are groups of users that the use cases intend to serve. The deliverables are use case diagrams. We will then storyboard, or tell a story about, each use case by describing how the users interact with the system in a step-by-step manner to execute the use case. This is done using structured English with user interface prototypes. Structured English is used for procedural modeling in structured development, but it has two exceptions here. First, if a use case is performed by a human user, we will need to create one or more user interfaces. This is called prototyping, which is identical to the structured requirements analysis. Prototyping is often underemphasized, but it is a "must" for effective discovery of requirements and effective control of development risks. Second, what the system does during interactions with the users will have to be re-specified as the actions performed by one or more objects that constitute the system. Re-specification is modeled by either a communication diagram or a sequence diagram, from which one can derive the functions to be housed by each class. Adding derived functions to the classes in the initial class diagram, we arrive at an enriched class diagram at the design stage. This will become the system specifications to be communicated to programmers.

Data and functions are bundled into objects. Ideally, there should be an object-oriented database that can save object data to make objects persistent. Unfortunately, there is no commercial object-oriented DBMS that supports such a large amount of transactional data. In the conceivable future, we anticipate that organizations will continue using relational DBMS for business objects. Thus, we still need to convert a business object model into a logical data model as the database specification. The rules for such conversion are essentially the same as in the structured development described earlier.

Review Questions

1. What is an information system?
2. What is SDLC?
3. What are the key features of the structured systems analysis and design methodology?
4. What tools are needed to develop an information system?
5. What are the components of an information system in general, and in specific to be designed in this course?
6. What are the two most important roles played by systems analysts?
7. What is different between computer science and information systems as fields of study?
8. What language does a relational DBMS speak?
9. What models does a systems analyst use to communicate with end users? With programmers?
10. What is RAD? How is it different from a programming language?
11. What is different between a logical data model and a conceptual (business) data model?
12. What is different between a process model and a procedural model?

Exercises

1. Think about what tools you may need to build an information system to sell books on the Internet, like amazon.com.
2. Please describe the major activities during systems design and the deliverables of each activity (choose either the structured systems analysis and design methodology or the object-oriented methodology).
3. Write a short essay for a popular magazine to introduce information systems as a field of study and make sure you point out its differences from computer science.
4. Write a short ad for your employer on potential job openings in systems analyst positions and make sure you give a job definition in the ad.
5. Use diagrams to illustrate and explain the systems development life cycle. Make sure you list the deliverables, tools to be used, and the role of systems analysts at each stage of the SDLC.
6. Use diagrams to illustrate the components of an information system.

Chapter 2

A Review of Programming Principles

Introduction

Regardless of programming languages and development models, there are four basic types of instructions in any computer program that work around one central concept, called a *variable*. In programming, a variable is simply a memory block with a specific name, size, and location used to store temporary data. The first chapter mentioned the difference between random access memory (RAM) and permanent storage. In fact, RAM is important in computer programming: the locations or blocks in RAM are variables.

The four basic types of instructions are each concerned with how to create memory blocks, how to change values in the memory blocks, how to manipulate the values in the memory blocks, and how to view and save the values in the memory blocks.

Variable declaration is to create a memory block. The size of the memory block is determined by the type of the variable declared. Using C# or Java, we may declare the following types of variables: char (2 bytes) for holding a Unicode character like 'a', '\n' (new line character), '\t' (tab character), int (4 bytes) for holding whole numbers, double (8 bytes) for holding real numbers, and string for holding texts like "John Doe", "Ohio", "23" (a number but stored as a text). For example, we may declare two variables, `price` as a decimal and `quantity` as an integer, to store the price of a product and the quantity of the product, respectively, for computing the subtotal of the product in a purchase order.

Variable assignment is to put a value into a memory block, and it will replace the existing content of the memory block by a new value. For example, you may assign 6.98 to the `price` variable and 10 to the `quantity` variable so that those two memory blocks will now hold data 6.98 and 10, respectively. The new value can be obtained from a user through an input device, from a data file on a storage device, from a database, from a mail server, from a network server, or generated by a CPU operation. In C# or Java, the assignment is done by using the "=" sign, where the value on the right-hand side is assigned to the variable on the left-hand side. The following are some examples of variable assignments:

```
intHour = 23;
myFullName = "John Doe";
mySalary = 4500.98;
```

Variable manipulation is to use the values inside memory blocks to perform algorithmic or mathematical operations. For example, we may multiply the values stored in the `price` and `quantity` locations to compute the subtotal. Variable manipulation occupies the core position in any program and may consist of a series of simple additions, subtractions, multiplications, and divisions performed by the CPU. For example, the right-hand side of the following code increases `mySalary` by 5% and then uses the result to replace the existing content of `mySalary`:

```
mySalary = mySalary * 1.05;
```

Variable report is to present the data in memory locations to human users or to communicate the data to another device or a program. When it comes to human users, we are used to seeing data in a meaningful context in a familiar language, but data in memory blocks are simply binary strings of 0s and 1s. Thus, we need to transform and re-express the data in a format understandable to human users using an output device such as a screen, a data file, a database, a mail server, or another computer on the network for the user to see it or use it later. For example, we may use a message box to output the value of a variable as follows:

```
MessageBox.Show("my salary is " + mySalary);
```

which will display the words, made of two parts, using a dialogue box. Note that the operator + here is used to concatenate or combine multiple words into one. For example, "John" + "Doe" will result in "JohnDoe", "Price is " + "1.99" will result in "Price is 1.99", etc.

The four statements must be in a logical order. We need to create a memory block before assigning a value to the block. Only after the data are in memory blocks is it possible to manipulate or report the data. Note that when a variable is first declared, the current value in the memory location holds whatever is left by the prior execution of some programs, and so generally it is garbage. Thus, we need to put an initial value into it. This is called *initialization*. An initial value may or may not be useful data, but it must be of the right type. For example, you cannot put 1.00 into quantity because quantity is an integer variable but 1.00 is a decimal value.

Variable Declaration

Variables are named according to certain naming conventions and should avoid reserved words. For example, these are valid names: volume, length, mySalary, screenSize. Some invalid ones include new, int, private, public, class, because they use reserved words. class size, course#, student-ID are also invalid because they contain unpermitted characters. The following are example declarations:

```
char c, firstLetter, lastChar;
int intAge, intHours, intYear;
double mySalary, myWeight;
string myFirstName, myFullName, myStreet;
```

A variable can be simple, i.e., to hold only one item of data. It can be complex, i.e., to hold object(s) such as word documents, windows controls and events, and networks. In this case, the memory block holds a bunch of data rather than a single value. For example, the following code declares one variable to hold a point in time and another to hold a random number generator:

```
DateTime currentTime;
Random myRandomNumberGenerator;
```

The following creates an array or collection of memory blocks that holds a series of data of the same type:

```
double[] employeeSalaries;
string[] employeeNames;
string[,] names = new string[5,4];
```

In the following, we will review the declaration of three types of variables, from simple primitive types, to collections, and finally to custom types.

Primitive types

Five primitive types of variables are illustrated below: bool (for true or false values), int (for 32-bit integers), double (for 64-bit double precision decimal numbers), char, and string (String for Java).

```
double price; //declare first
price = 6.98; //initiate next

int quantity = 10;
//declare and initiate in the same time

bool isValid = true;

char letter = 'C';
//char values are inside single quotes

string username, password;
//declare multiple variables
username = "scott";
//string values are inside double quotes
```

Note that the string, or string of characters, is not really a primitive type, but there is no harm in treating it like one in C# programming.

Collection types

When we need to create a lot of variables of the same type, e.g., a list of product prices, a list of quantities, a list of state names, etc., we can use arrays and list types. Arrays are used if the number of values is known, while lists are used if the number of values may change over time in the code. In the following are illustrations of how to declare and initialize arrays and lists.

```
double[] prices = new double[3];
//declare first and then initiate
prices[0] = 6.98;
prices[1] = 1.25;
prices[2] = 69.90;

int[] quantities = {10, 200, 1};
//declare and initiate

List<double> lstPrices = new List<double>();
lstPrices.Add(6.98); //add list item
lstPrices.Add(1.25);
lstPrices.Add(69.90);
lstPrices.Remove(1.25); //remove list item
```

After we declare an array or list, we can access and manipulate the individual items in the collection using their location index, starting from 0, and then increasing values, such as 1, 2, and so on, until the last item index, as if they are usual primitive variables. For example, `prices[1]` will be the second price in the array `prices` and `listPrice[0]` will be the first price in the list `lstPrices`. The last item in `prices` has index `prices.count` − 1, and the last item in `lstPrices` has index `lstPrices.length` − 1 because the indices start from 0, not 1.

We may also use multi-dimensional arrays to store matrix or tabular data. For example, the following creates an array for a 3 × 2 table of observations:

```
double[,] obs = new double[3,2];
//declare first and initialize next
obs[0,0] = 5.7;
obs[0,1] = 130;
obs[1,0] = 6.2;
obs[1,1] = 145;
obs[2,0] = 3.9;
obs[2,1] = 120;

double[,] obs = {{5.7, 130}, {6.2, 145}, {3.9, 120}};
//declare and initiate
```

Custom types

If data are complex and cannot be stored in any predefined types of variables, we then create custom types first and then declare variables with these newly defined custom types. For example, if we want to create memory blocks to store data about each product, including its stock keeping unit (SKU) ID, price, and quantity, we will not be able to use either primitive types or collections because the data are not of the same type. SKU is a string, price is a double, and quantity is an int. In this situation, we can define our own custom types. The reader may realize later that this book is essentially about defining custom types. Here, we just want to point out its connection to the notion of variables.

The first custom type is enumeration, enum, which can be used to create a variable that has a fixed list of possible values to take on. For example, gender must be either male or female; course grade must be either A, B, C, D, or F; color must be Red, Green, Black, etc., and US state name must be OH, NY, MI, etc.

```
//custom enumeration definition
public enum Gender
{
        Male,
        Female
};
//variable declaration and initialization
Gender lisaGender;
lisaGender = Gender.Female;
```

The reader may see the keyword public in front of enum in the type definition. We will introduce this so-called access scope later, but for now the reader just need know that, without public, the custom type cannot be used to declare variables if the declarations are not located in the same place as the type definition in code.

The second custom type is a structure, struct, used for creating a collection of memory blocks of different kinds. For example, we can define a custom type Product to store values of each product.

```
//custom struct definition
 public struct Product
 {
             public string SKU;
             public double Price;
             public int Quantity;
 };
```

```
//variable declaration and initialization
```

```
 Product cup;
 cup.SKU = "1G-345-BE";
 cup.Price = 6.98;
 cup.Quantity = 10;
```

The third type of custom types, `class`, goes one step further than `struct`. It can create variables, called *objects*, that can store not only a collection of values of different types but also code to process the values. For example, `Product` class is defined in the following.

```
//custom class definition
```

```
Public class Product {
        private string sku;
        private double price;
        private int quantity;
        public double getSubTotal()
        //function to compute subtotal
        {
                double subtotal;
                subtotal = price * quantity;
                return subtotal;
        }
}
```

```
//declare variables of Product type
```

```
Product cup; //declare variable cup
cup = new Product();
//creating sub memory blocks for sku, price, etc.
Product milk = new Product();
//declare and create sub blocks
```

The reader may skip the section on the function to compute subtotal in the above code; we will get into this in later chapters on modeling and programing business functions.

To be able to manipulate the above variables cup and milk, we will need to assign values to sub-memory blocks for SKU, price, and quantity. This is usually done by constructors or *data access methods* such as get() and set() functions in Java and *properties* in C#. Note in the above code that Product() is a *constructor* for the class Product. A constructor always takes on the same name as a class, but a constructor is a function while a class is a type. A constructor is used to initialize sub-blocks of a collection variable. For example, Product() here will initialize the sub-blocks of cup and milk, including sku, price, quantity, and the function getSubTotal.

We will learn more about classes later in the book. For now, we use a few predefined custom classes in Visual Studio libraries, such as DateTime, Random, Pen, SolidBrush, File, OleDBConnection, StreamReader, etc., to appreciate how to manipulate object variables.

```
//objects handling date and time

DateTime myBD; //declare first and initialize next
myBD = new DateTime(1988, 8, 8);

DateTime myBD = new DateTime(1988, 8, 8);
//declare and initialize

//objects generating random numbers
Random myGenerator = new Random();
int myNumber = myGenerator.Next(0,100);

//objects for drawing

Pen myBluePen = new Pen(Color.Blue);
SolidBrush myRedBrush = new SolidBrush(Color.Red);

//objects for Oracle database connections

string sConn = "Provider=MSDAORA;Data
    Source=CBA12c;User ID=scott;Password=tiger";
OleDbConnection myConn= new OleDbConnection
    (sConn);
myConn.Open();
```

Code Structure

Where do we declare, initialize, and manipulate variables? Instructions in C# are organized into three layers: project (or namespace) layer, class layer, and function layer. Each project contains one or more classes, and each class contains one or more functions. Each layer is delimited by curly braces { ... }. Custom types such as enumerations, structures, and classes are defined inside a project or a class. Variables are declared inside a class or a function, but the assignments and manipulations of the variables must be inside functions. Figure 1 shows an example structure of three layers: inside `CodeExercise` project, there are two enumerations (`Day` and `Gender`), one structure (`Product`), and one class (`Universe`). Inside the `Universe` class, there is one function (`DoAll()`).

For initial programming exercises, the reader may use either Syntax Checker at ecourse.org/X/SyntaxChecker.aspx or Notepad ++ (open source program). However, to be able to do the demonstrations and exercises in later chapters, Microsoft Visual Studio is recommended. Here are the two steps to follow to start using Visual Studio.

Step 1: Open Visual Studio, then use File → New → Project menu to create a new project. Make sure to choose C# as the programming language and Windows Forms Application as a template (see Figure 2).

Step 2: Use Project → Add Class menu to create a new class, specify a class name, e.g., `Universe`, `MickyMouse`, `Customer`, `Order`, etc. (see Figure 3).

After the above two steps, you will see a code page as in Figure 4. Note that, before the namespace, there are a few lines of code, all starting with keyword "using", which provide default directives of built-in classes in the Visual Studio library. For now, do not bother to change any of them. Instead, we will restrict our code inside the namespace and class. First, add "public" in front of "class Universe" and add custom types anywhere in parallel to "class Universe {...}" layer. Figure 5 shows an example result.

For all other instructors, including variable declarations, assignments, and manipulations, let us create a function as in Figure 6 and then write all the code in this chapter inside the function *"public void DoAll () {...}"* as shown in Figure 1.

```csharp
namespace CodeExercise
{
    public struct Product
    {
        public string SKU;
        public double Price;
        public int Quantity;
    }

    public enum Day
    {
        Sunday, Monday, Tuesday, Wednesday, Thursday, Friday, Saturday
    }

    public enum Gender
    {
        Male, Female
    }

    public class Universe
    {
        public void DoAll()
        {
            double price; // declare first
            Product cup;
            cup.SKU = "1G-345-BE";
            cup.Price = 6.98;
            cup.Quantity = 10;
        }
    }
}
```

Figure 1.   C# three-layer code structure.

Operations

Computer programing involves four types of instructions in order: A variable must be declared before you can give a value to it; a variable must be initialized before you can use it for any operations that manipulate the variable. Table 1 lists the operations in C# language.

As in primary school arithmetic, operations involving variables follow the Parentheses, Exponents, Multiplication (*), Division (/), Addition, and Subtraction (PEMDAS) order, except that we have more operations than simple PEMDAS.

×

Create a new project

Search for templates (Alt+S)   🔎 ▾

Recent project templates

A list of your recently accessed templates will be
displayed here.

All languages   ▾   All platforms   ▾   All project types   ▾

⬜ **Blank Solution**
Create an empty solution containing no projects

Other

⬜ **Console App (.NET Framework)**
A project for creating a command-line application

C#   Windows   Console

⬜ **Windows Forms App (.NET Core)**
A project for creating an application with a Windows Forms (WinForms) user
interface

C#   Windows   Desktop

🌐 **ASP.NET Web Application (.NET Framework)**
Project templates for creating ASP.NET applications. You can create ASP.NET Web
Forms, MVC, or Web API applications and add many other features in ASP.NET.

C#   Windows   Cloud   Web

⬜ **Class Library (.NET Framework)**
A project for creating a C# class library (.dll)

Back   Next

Figure 2.   Create C# project.

Add New Item - CodeExercise   ?   ×

◢ Installed   Sort by: Default   ▾   ▦ ▤   Search (Ctrl+E)   🔎 ▾

◢ Visual C# Items   ⬜ Class   Visual C# Items   **Type:** Visual C# Items
    Code   An empty class definition
    Data   ●—○ Interface   Visual C# Items
    General
  ▷ Web   ▣ Form (Windows Forms)   Visual C# Items
    Windows Forms
    WPF   ⬜ User Control (Windows Forms)   Visual C# Items
    SQL Server
▷ Online   ⬜ Component Class   Visual C# Items

                             ⬜ User Control (WPF)   Visual C# Items

                             ⬜ About Box (Windows Forms)   Visual C# Items

                             ⬜ ADO.NET Entity Data Model   Visual C# Items

                             ⬜ Application Configuration File Visual C# Items

                             ⬜ Application Manifest File (Windows Only)   Visual C# Items

                             ⬜ Assembly Information File   Visual C# Items

                             ⬜ Bitmap File   Visual C# Items

                             ⬜ Code Analysis Rule Set   Visual C# Items

                             ⬜ Code File   Visual C# Items ▾

Name: Universe.cs

                                                                    Add   Cancel

Figure 3.   Add class to project.

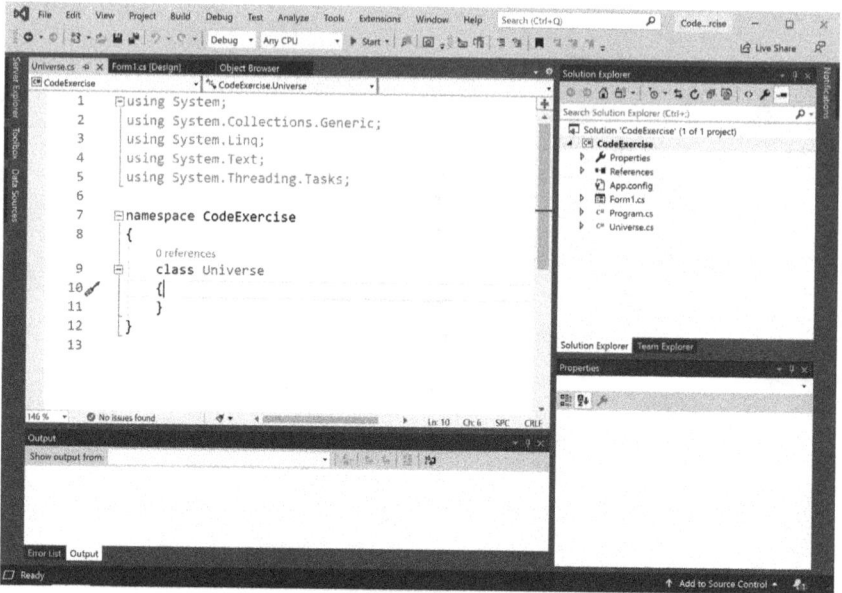

Figure 4.   Sample code page for Universe.

```
namespace CodeExercise
{
    public struct Product
    {
        public string SKU;
        public double Price;
        public int Quantity;
    }

    public enum Day
    {
        Sunday, Monday, Tuesday, Wednesday, Thursday, Friday, Saturday
    }

    public enum Gender
    {
        Male, Female
    }

    public class Universe
    {

    }
}
```

Figure 5.   Sample code page after adding custom types.

```
namespace CodeExercise
{
    public struct Product
    {
        public string SKU;
        public double Price;
        public int Quantity;
    }

    public enum Day
    {
        Sunday, Monday, Tuesday, Wednesday, Thursday, Friday, Saturday
    }

    public enum Gender
    {
        Male, Female
    }

    public class Universe
    {
        public void DoAll()
        {

        }
    }
}
```

Figure 6.   Sample code page after adding function DoAll().

A course in computer programing will cover the details of the operations, and here we will use a few examples to illustrate the meaning and
the order of some unfamiliar operators; PEMDAS operators are common
knowledge for all college students.

Example 1: What is 3/4 and 3.0/4? What is 3%4?

Integer division results in an integral quotient, and thus 3/4 is 0. The
remainder is obtained by % operator, and so 3% 4 is 3. When an operation involves data of mixed type, an implicit cast is done by the operation
that coverts all special value types to the most generalized type in the
operations, e.g., Boolean values to integers, integers to decimal numbers,
and numeric values to text values. For example, in 3.0/4, the integer 4
will be cast into 4.0, and thus, 3.0/4.0 will be 0.75. As another example,
3 + "10" will be "310" rather than 13.

Table 1.   Operations.

Operator	Description	Level	Associativity
[ ]	Access array element	1	Left to right
.	Access object member		
( )	Invoke a method		
++	Post-increment		
--	Post-decrement		
++	Pre-increment	2	Right to left
--	Pre-decrement		
+	Unary plus		
-	Unary minus		
!	Logical NOT		
~	Bitwise NOT		
( )	Cast	3	Right to left
new	Object creation		
*	Multiplicative	4	Left to right
/			
%			
+ -	Additive	5	Left to right
+	String concatenation		
<< >>	Shift	6	Left to right
>>>			
< <=	Relational	7	Left to right
> >=	Type comparison		
instanceof			
==	Equality	8	Left to right
!=			
&	Bitwise AND	9	Left to right
^	Bitwise XOR	10	Left to right
|	Bitwise OR	11	Left to right
&&	Conditional AND	12	Left to right
||	Conditional OR	13	Left to right
?:	Conditional	14	Right to left
= += -=	Assignment	15	Right to left
*= /= %=			
&= ^= |=			
<< = >>= >>>=			

Example 2: Evaluate the following based on the declaration "`int x = 3, y = 2;`":

- `"x + y = " + x + y`
- `"x + y = " + (x + y)`

Following the implicit cast and the order of operations, "`x + y = "
+ x + y` will be "`x + y = 32`", but "`x + y = " + (x + y)`
will be "`x + y = 5`" because in the second expression, `x + y` will be
performed before the implicit cast.

Example 3: Evaluate the following based on the declaration "`int x = 3, y = 2;`":

- `++x - y`
- `x++ - y`
- `x - --y`
- `x - y--`

This example is concerned about post- and pre-increments (++) or decrements (--). A pre-operator is used to increment or decrement the value of a variable before using it in an expression. In the pre-increment, a value is first incremented and then used inside the expression. A post-increment is just the opposite: a value is first used in the expression and then incremented. For example, in `++x - y`, `x` will become 4 before performing `x - y`. Thus, `++x - y` is 2. In contrast, in `x++ - y`, `x` will become 4 after `x - y`. Thus, `x++ - y` is 1.

Example 4: Evaluate the following operations following the declaration "`bool x = true; int y = 2;`":

- `!x`
- `y == 2`
- `y%2 == 1`
- `x && !(y == 2)`
- `x || (y == 1)`

This example is about how to use logical operations: ! for negation, &&
for AND, and || for OR. Thus, !x is false. Operator = is for assign-
ment, and == is for comparison. Thus, y == 2 will result in a logical
value, true, but y = 2 means to put value 2 into y's memory block. In
the expression y%2 == 1, should we compare 2 == 1 first or should
we do y%2 first? According to the data given in Table 1, we should do %
before ==. Thus, y%2 == 1 results in the logical value false.

Example 5: What is x in the following statements after the declaration
"int x = 4;"?

- x = x + 1;
- x += 2;
- x -= 2;
- x *= 2;
- x ^= 2;
- x /= 2;

This example illustrates how to use assignment operations, which are performed from right to left. In x = x + 1, x + 1 will be performed by
the assignment, and so x will be 5 after the assignment. +=, -=, *=, ^=,
and /= are just short-cut notations. For example, x += 2 stands for
x = x + 2, x ^= 2 for x = x^2, etc.

Controls

Operations may be repeatedly and/or contingently performed. The repetition is controlled by loops, and contingency is controlled by decision
controls. In the following, we will use examples to illustrate the use of
for-loop, while-loop, if-else, and switch controls.

Control *for-loop* is for definite steps of repeated execution of some
operations. The following two snippets are, respectively, for finding the
sum 1 + 3 + 5 + ... + 99, and the product 2*3*5*7*11*17*19.

```
int sum = 0;
for (int i = 1; i <= 100; i = i + 2) {
    sum = sum + i;
}
```

```
int product = 1;
int[] factors = {2, 3, 5, 7, 11, 13, 17, 19}
foreach (int i in factors) {
    product = product * i;
}
```

Note here that we use two different for-loops: one uses an index to control the steps of repetition and the other uses a list or an array to control the steps.

Also, note that, before each loop, we need to declare a variable to hold the result of the repeated operations. Here is a programming tip for initializing the result variable before the loop: *the variable should be initialized to 0 for repeated additions and to 1 for repeated multiplications.*

Control *while-loop* is for indefinite steps of repeated execution. For example, to find the corresponding ASCII code for an integer 9876224342, we need to repeatedly subtract the number by 128 until the result is between 0 and 128:

```
int i = 9876224342;
while (i > 127) {
    i = i - 128;
}
```

Yet another example, to find constant $e = \lim_{n \to \infty} \left(1 + \frac{1}{n}\right)^n = 1 + \frac{1}{1!} + \frac{1}{2!} + \cdots$, of course, we cannot perform additions infinitely. However, after many steps, the additional additions will not add any significant values to the result, and so we may stop, say when we reach the term $\frac{1}{i!}$, which is smaller than 0.000000001.

```
double x = 1, sum = 0;
int i = 0;
while (x > 0.000000001) {
    sum = sum + x;
    i++;

    //a definite loop for 1/i!
    x = 1;
    for (int k = 1; k <= i; k++) {
        x = x/k;
    }
}
```

Decision control `if-else` is a condition-based selective or contingent execution of operations. The following two snippets do the following: one tests if a random number is even or not and the other generates a text greeting based on the current system time via `DateTime.Now`.

```
//test a random number
bool isEven;
Random g = new Random();
int value = g.NextInt(1,100);
if (value % 2 == 0) {
    isEven = true;
}
else {
    isEven = false;
}

//generates a greeting message
string greeting = "";
int h = DateTime.Now.Hour;
if (h >= 8 && h < 12) {
    greeting = "Good Morning!";
}
else if (h >= 12 && h < 18) {
    greeting = "Good Afternoon!";
}
else {
    greeting = "Good Evening!";
}
```

To find the maximum or minimum value in a list, we can set a temporary result first and go through each list item using a loop to test it against the temporary one. Here is a programming tip for setting the temporary value: *the value should be initialized to the first element for finding the maximum or minimum element in a list.*

```
int[] values = {2, 3, 5, 7, 11, 13, 17, 19}
int max = values[0];
//initialize to the first list item
foreach (int v in values) {
    if (max < v) {
            max = v;
    }
}
```

The next example determines if 31123 is a prime number or not, i.e., can it be wholly divisible by any whole integer that is no greater than the square root of 31123.

```
int x = 31123;
int n = (int) Math.Sqrt(x); //explicit cast
bool isPrime = true;
//note to initialize the result to true
for (int k = 2; k < n; k++)
{
    if (x%k == 0) {
            isPrime = false;
            break;
    }
}
```

Note that, in the above case, we can discontinue the loop if at any step we find that 31123 is wholly divisible. Here is a programming tip for initializing the result variable in determining a truth value using a loop: *if the truth can be determined without going through the entire loop, the variable is initialized to false. Otherwise, it is initialized to true.* The following example tests if all the values in a given list are negative, and so the truth cannot be determined without going through the entire list.

```
double[] values = {2.1, 0.5, 5.0, 5.0, 1.1, 1.3, 1.7,
    -1.9}
bool positive = true;
foreach (double v in values) {
    if (v <= 0) {
            positive = false;
            break;
    }
}
```

Decision control `switch` is a case-based selective execution. It can shorten many nested if-else statements into a simple list of cases based on the value of a choice variable. For example, to convert a grade into a GPA point, the following snippet uses switch statement using grade value, which is a text such as A, B, C, etc.

```
double point;
switch (grade) {
    case "A":
            point = 4.0;
            break;
    case "B":
            point = 3.0;
            break;
    case "C":
            point = 2.0;
            break;
    case "D":
            point = 1.0;
            break;
    default:
            point = 0.0;
            break;
}
```

This example generates verbal feedback based on essay marks ranging from 1 to 5:

```
string verbal;
switch (mark) {
    case 5:
            verbal = "Excellent";
            break;
    case 4:
            verbal = "Very Good";
            break;
    case 3:
            verbal = "Good";
            break;
    case 2:
            verbal = "Fair";
            break;
    default:
            verbal = "Poor";
            break;
}
```

Finally, in the following is a more comprehensive example that maps the text "aabacdefgh" into a "phone number":

```
string x = "aabacdefgh";
x = x.ToUpper();
char[] chars = x.ToCharArray();
string pNumber = "";
int n;
foreach (char c in chars){
     switch (c)
     {
          case 'A':
          case 'B':
          case 'C':
               n = 2;
               break;
          case 'D':
          case 'E':
          case 'F':
               n = 3;
               break;
          case 'G':
          case 'H':
          case 'I':
               n = 4;
               break;
          default:
               n = 5;
               break;
     }
     pNumber += n;
}
```

Exercises

1. Create simple variables:
   a. Declare a variable to hold your county sales tax rate.
   b. Declare a variable to hold your name.
   c. Declare a variable to store true/false value on whether today is sunny or not.
   d. Declare a variable to hold the number of classes you are taking.
2. Create object variables:
   a. Create a random number generator and declare a variable to hold one of its generated value between 500 and 600.
   b. Declare a variable to hold your birth date.
   c. Declare a yellow brush object.

3. Create arrays:
   a. Declare an array to hold the names of pieces of a newly setup chess game.
   b. Get the names and heights of three of your friends. Declare an array to hold their names and another one to hold their heights.
   c. Declare an array to store data in the following matrix:

$$12.76 \quad 12.09 \quad 13.89$$
$$12.45 \quad 12.11 \quad 13.09$$

   d. Declare an array to store the birthdates of all your family members.
4. Create custom types and their variables:
   a. Define a struct type of memory block that can hold the dimension of rectangle — width and height — and then create a memory of the type to hold one rectangle.
   b. Define a struct type of point using x and y coordinates and then declare and initialize three-point variables.
   c. Define a struct of triangle using three points then declare and initialize one arbitrary triangle.
   d. Define an enumeration type for degree with possible values as either bachelor, master, or doctor and then declare and initialize two variables of degree type.
5. Create custom classes and objects:
   a. Define class Employee and then declare two employee objects to hold the data of your parents.
   b. Define a type of memory that holds a student's ID, name, major, and admission date. Then declare a variable to hold your own data.
   c. Define a type of memory block to hold a course's number, title, and credit hours. Then declare an array to hold all the courses you are taking (Hint: You can create an array of objects such as Employee[] employees = new Employee[3];).
6. Write a program that simulates throwing a dice three times giving random values from 1 to 6. The output should contain the values of the dice and the probability for this combination to occur.
7. What are the values of x, y, and z after the following code fragment?

```
int x = 5;
int y = 10;
int z = ++x * y-- + x++ - ++y + --x;
```

8. A year is a leap year if it is divisible by 4, but century years are not leap years unless they are divisible by 400. Add parentheses to the following expression to make the order of evaluation clearer.

```
year % 4 == 0 && year % 100 != 0 || year % 400 == 0
```

9. What does the following code fragment print?

```
Console.WriteLine(1 + 2 + "abc");
Console.WriteLine("abc" + 1 + 2);
Console.WriteLine("1 + 2 = " + 1 + 2);
Console.WriteLine("1 + 2 = " + (1 + 2));
```

10. Considering the following code segment, what is printed after executing the code segment?

```
List<Integer> list1 = new ArrayList<Integer>();
list1.add(new Integer(1));
list1.add(new Integer(2));
list1.add(new Integer(3));
list1.set(2, new Integer(4));
list1.add(2, new Integer(5));
list1.add(new Integer(6));
Console.WriteLine(list1);
```

11. Find the larger one of two numbers in variables a and b.
12. Given numbers in the variables a, b, and c, find the smallest number among them.
13. Write a code to determine if today is your birthday.
14. Given an array of decimal numbers, write a code to find out how many of the numbers are positive.
15. You assign a number between 1 and 100 to put into variable myGuess. Then you generate an integer between 1 and 100. Write a code to determine if your guess is correct.
16. Create an array that contains five words. Then write a code to reverse them.
17. Given coefficient values in the variables a, b, and c for quadratic equation $ax^2 + bx + c = 0$, find the real solutions.
18. Find the sum $1 + 4 + 7 + 10 + 13 + \cdots + 100$.
19. Find the sum $2.1 + 4.1 + 6.1 + \cdots + 200.1$.
20. Find the factorial $21!$.
21. Given real numbers in an array of observations, find the mean value.

22. Given real numbers in an array of observations, find the standard deviation of the numbers.

23. Given real numbers in an array of observations, find the maximum value.

24. Find the value of $1+\frac{1}{1!}+\frac{1}{2!}+\cdots+\frac{1}{500!}$.

25. Find an approximate value of $1+\frac{1}{3}+\frac{1}{3^2}+\frac{1}{3^3}+\cdots$ with the error of approximation less than 0.000000001.

26. Find the sum of $\frac{1}{1*2*3}+\frac{1}{2*3*4}+\frac{1}{3*4*5}+\cdots+\frac{1}{45*46*47}$.

27. Assume the interest rate is 7% and you save 200 each year. Find the total amount of money you will have in 20 years.

28. Assume the interest rate is 7% and you can pay back 50 each year for 50 years. How much money can you borrow today?

29. Write a code to determine if 876412313207 is a prime number or not.

30. Write a code to convert integer 9823432143243 into a binary number.

31. Write a code to determine an approximate value of π using the equation (error of approximation = 0.000000000000001):

$$\pi = 4\left(1-\frac{1}{3}+\frac{1}{5}-\frac{1}{7}+\frac{1}{9}-\frac{1}{11}+\cdots\right)$$

32. Suppose there are missing values in an array of observations. Write programs to move the numbers to the beginning contiguously so that there are no missing blocks that interrupt the numbers.

Chapter 3

Modeling Functions and Procedures

Introduction

A *business process* is a collection of one or more information activities by which data are transformed, stored, retrieved, modified, or distributed. It is an abstraction of physical data processing activities such as data retrieval, data modification, data insertion, data deletion, data transmission, or batch calculation. A business process may be also called a *business function*. Thus, these two terms are often used interchangeably.

Analysis involves decomposition and abstraction, and so does business process analysis. Here, the abstraction is about how to represent a business process. Decomposition means breaking down a large process into smaller chunks. It may also mean breaking down a process into non-breakable action units in a logical order, which is in the realm of procedure modeling. Therefore, in this chapter, we will study three aspects of process modeling: process decomposition, process representation, and procedure modeling.

Process analysis is an essential part of structured development. The modern object-oriented methods tend to downplay its importance, with the exception of procedural modeling, which has been reinstated to be an integral part of Unified Modeling Language (UML). This chapter introduces process decomposition and representation, aiming at enhancing the understanding of the object-oriented approach. First, the so-called use cases in the object-oriented method are nothing but a business process that delivers a tangible value to the user. Function decomposition will help understand functional requirements and identify use cases. Second, objects are nothing but a high-level encapsulation of both data and functions. To understand the concept of objects better, we first need to understand functions.

Capturing Function Requirements

A business process can be as large as the mission of an organization. It can be an ongoing, continuous action performed by people to further the mission of the organization, such as those involved in marketing, accounting, manufacturing, etc. It can be a low-level process, which supports a major function and *has time-oriented starting and ending points*, such as market analysis, promotional sales, annual auditing, etc. A process can be also as small as one data activity or action, such as changing the mailing address for a customer, creating a new account for a customer, etc. We sometimes differentiate actions from activities. The difference between them is that an action is atomic or not decomposable, whereas an activity may be decomposed into one or more smaller-scope actions or activities.

In general, systems analysis follows the divide-and-conquer approach to both requirements modeling and systems design. This is particularly true in process analysis. Large and complex processes are procedurally difficult to understand. Therefore, they are difficult, if not impossible, for systems analysts to specify and systems developers to implement. For such a process, the usual approach is to break it down into multiple smaller and simpler subprocesses so that each subprocess is easy to understand, specify, program, and test. We shall apply this procedure to any complex process until all processes are simple.

Besides vertical decomposition, which breaks down large and complex processes into small and simple ones, decomposition may be also done horizontally along the time line so that a long process is broken down into short ones. If a process consists of a long sequence of data activities that may be interrupted by physical activities, then it may be chopped into short subprocesses at the point of interruptions.

Regardless of whether it is vertical or horizontal, decomposition must satisfy the basic conservation principle as follows:

The Conservation Principle of Decomposition: Each process may be decomposed into two or more child processes. However, the function to be accomplished by the parent process should be accomplished by the sum of the subprocesses only, no more and no less.

Therefore, processes of different levels are better structured as a process hierarchy. On the top is the *overall business process* that is on a global scope and requires a long time to finish. At the next level are the *major business processes* that support the overall process. In the third level are the functions that support the major processes. In this manner each higher-level process is then progressively refined into more detailed processes. The refinement continues until there are processes that cannot be decomposed further. This process is called *process or function decomposition*.

The overall process is the mission of an organization or system. For example, the overall process of a manufacturing company is to design and produce its products, and the overall process for an inventory system is to manage inventories.

Major processes are often those that directly support the externals of the systems, such as end users and other connecting systems. They justify the mission of the system to be built or give direct tangible values to the users. So, operationally, the following processes usually qualify as major functions: (1) capturing data from external agents, (2) maintaining data storage, (3) generating and distributing data to externals, and (4) high-level descriptions of data transformation operations. Technically, each major function usually corresponds to a menu item on the main system menu. For example, the major processes for an inventory system shall include updating inventory, generating orders, and querying inventory levels. These processes interface with external agents and provide tangible values to them.

Theoretically, there is no definition on what is complex and what is not. Thus, it is not always obvious when the decomposition stops. In practice, however, a process is complex if one of the following conditions is true:

1. How the process is performed programmatically is still unknown or the code is perceived to be too long or too complex.
2. The process involves two or more data activities that are performed programmatically differently.
3. The process consists of two or more activities, which need not necessarily be performed at the same time or in certain times we may just need to perform a subset of them.

Let us elaborate the third point above. Here, code reuse is a major consideration for process decomposition. Even though a process is not complex programmatically, we may still need to decompose it just because one or more of the decomposed subprocesses can be re-used somewhere else. For example, suppose we are given a flow of payments, and we need a function to compute the net present value of the payments. The function is so simple that a few lines of code can do the work. However, it might be desirable to decompose it into two sub-processes: one that is specialized in checking and formatting any series of data into a standard array and the second that computes the net present value for any array of data. The advantage of doing this decomposition is that the two sub-processes created can be re-used in many other places.

Note that the reader should not be too concerned about the function decompositions beyond the identification of major functions; the use case storyboarding, to be covered later, will be a more elaborate alternative to function decomposition. The following shows a few classic systems and their function decompositions, primarily of major functions.

Food Order System: The overall process for a food order system is to process and maintain food orders, inventories, and payments. In this system, obviously we would like to have a process to handle food orders. Then we need a process to do accounting job and update goods sold data. We should also have a process to update inventories. Finally, there must be a process that generates reports for management. Among the four major functions for the food order system, except for the function that handles food orders, the other three functions should be simple enough to be implemented by one form, one batch program, or one report. They do not need to be decomposed. However, the function to handle food orders is still complex. As we can perceive, the function must take orders, transfer the orders to the kitchen, accept payments from customers, and generate receipts to give to the customers. In addition, order data may need to trigger other functions such as updating goods sold and inventories. Therefore, the function "Manage Food Order" is a complex process and needs to be exploded. Putting all the pieces together, we should have a function hierarchy as shown in Figure 1.

Student Registration System: A major process of a student registration system is to enroll students into courses. To this end, a few sub-processes must be performed. The first is to search for an available course to enroll, which may also include browsing all available courses. The second is to check for prerequisites, i.e., only if a student who has fulfilled

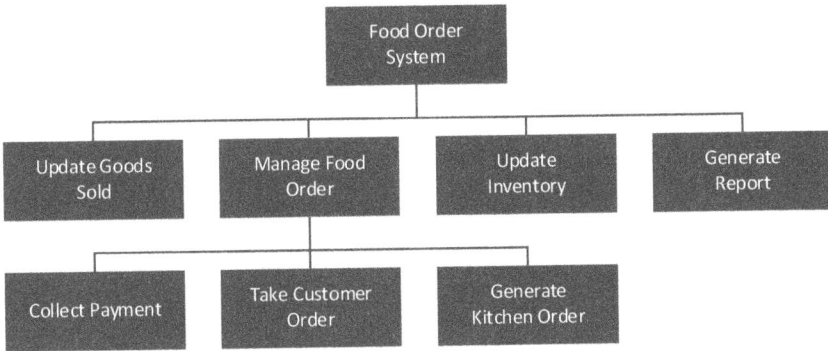

Figure 1.   Process hierarchy for the food order system.

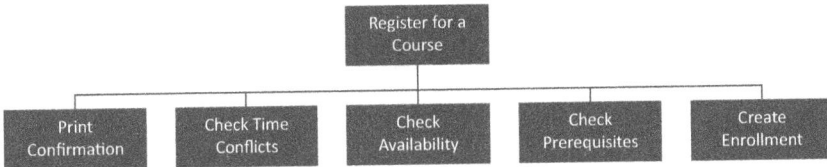

Figure 2.   Process hierarchy of enrollment.

the requirement by having taken required prerequisites can enroll in said courses. The third is to check for schedule conflicts: a student could not take a course if time conflicts with other courses already registered for. Finally, it will create actual enrollment by adding the student to the class roster and print a confirmation for the student. The process hierarchy is shown in Figure 2.

Student Admission System: The process of admitting new students is complex and involves a lot of data activities. However, at the high level, these activities may be organized into four major processes: manage applications, evaluate applications, handle acceptance affairs, and print management reports. These major processes need to be decomposed into more detailed and programmable sub-processes. For example, "manage applications" is too vague. It entails activities such as filling out applications, mailing applications, entering application data, obtaining test scores, and checking for materials completeness. Among these activities, some are purely physical, such as mailing applications, and thus outside our analysis. Some may be a hybrid of physical and data activities, such

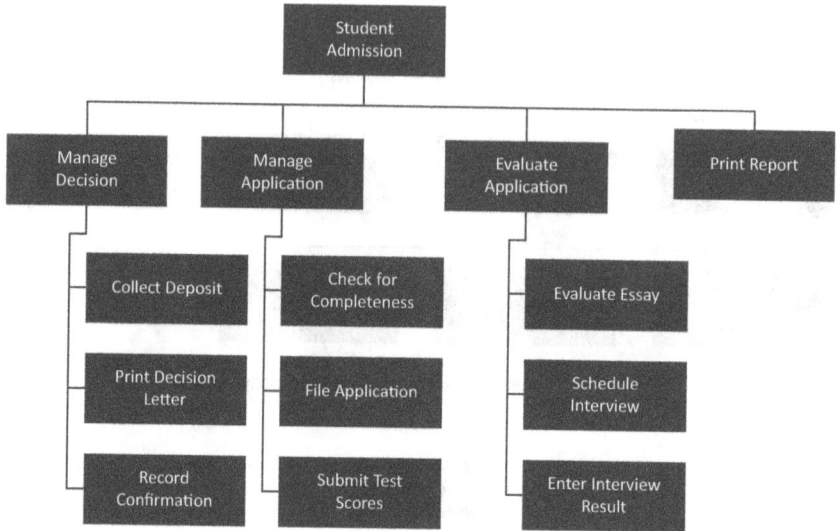

Figure 3.   Process hierarchy of student admission system.

as checking for materials completeness and obtaining test scores, and we will reframe them into programmable data processes. Figure 3 shows the process hierarchy. Here, we use "File Application" to denote the process of filling in applications by applicants or entering application data by clerks. We reframe "Obtain Test Scores" as "Submit Test Scores", that is, the process of entering and validating test scores.

Process Modeling

A process consists of many data activities, but externally it may be considered as a black box, or data processor, that takes inputs and generates outputs. Therefore, the external view of a process consists of data inputs, outputs, and a box for the process. The box has a name, which is usually a verb indicating the overall purpose of the process. For example, Figure 4 shows two simple functions: the first one is used to find the max of two numbers (Figure 4(a)), and the second one is used to validate user login (Figure 4(b)). They are simplified *data flow diagrams*.

The FindMax function is easy to imagine: if a > b, the max is a. Otherwise the max is b. Thus, the inputs are enough for the function to work. However, to validate a login, the user needs to enter a user name and

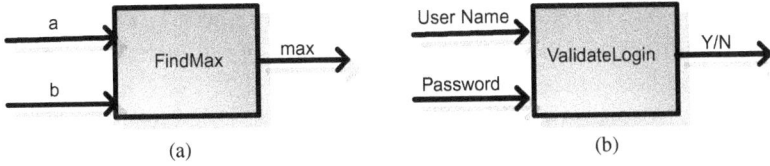

Figure 4.   Sample functions: (a) find max of two numbers and (b) validate user login.

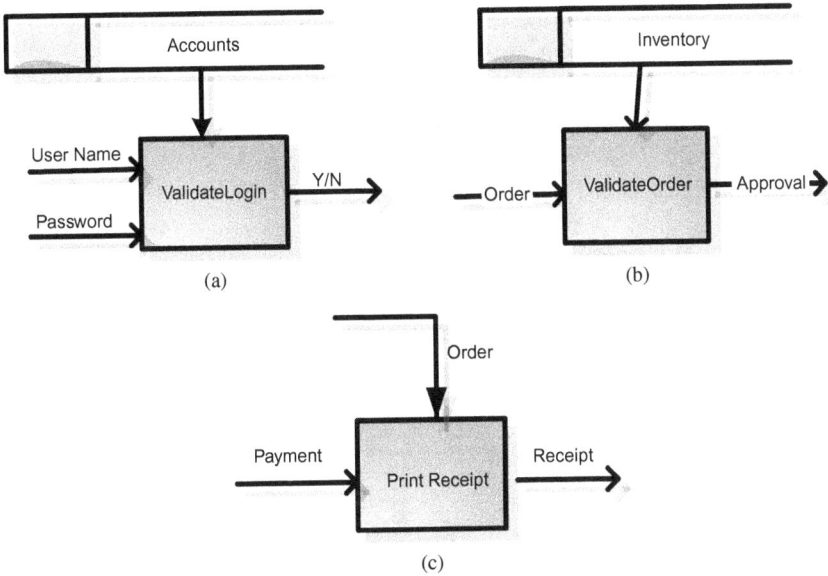

Figure 5.   Process representations: validate login (a), validate order (b), and print receipt (c).

a password, but the function must have access to a user account database or file, or a *data store*, to check if the user entry matches one of the accounts in the data store. Thus, a more elaborate data flow diagram shows not just inputs and outputs but also data sources or data destinations, including external agents such as users or other systems that interact with the process, data stores such as data files and reference books that keep the data at rest, and other processes that send data to or receive data from the current process. For example, the ValidateLogin process may be represented as in Figure 5(a). Similarly, Figures 5(b) and 5(c) are two processes used by an e-commerce store to validate an online order, i.e., to approve or reject orders and to print receipts. At a minimum, the `ValidateOrder` function will need to check for the availability of the ordered items, and so it requires both order and inventory data as inputs. The output is simply the approval decision of the order. The `PrintReceipt` function will need to have order and payment data as inputs and the printed receipt as the output.

In the following, we will show a few more processes in various systems, including patient appointment system, order system, and registration system. Figure 6 shows that to enroll a student into a course, the process needs to have data such as student ID, class ID, past student courses, and course catalog. The outputs generated include a new enrollment record in a data store and a confirmation and a billing invoice to be sent to the student. Figure 7 shows one of its sub-processes that does nothing but checking for prerequisites.

Figure 8 shows a process that makes appointments for patients. Of course, it will need an appointment request from a patient as well as existing schedule as inputs. It will print a confirmation to the patient and create a new appointment record, as represented by the output flow to the data store.

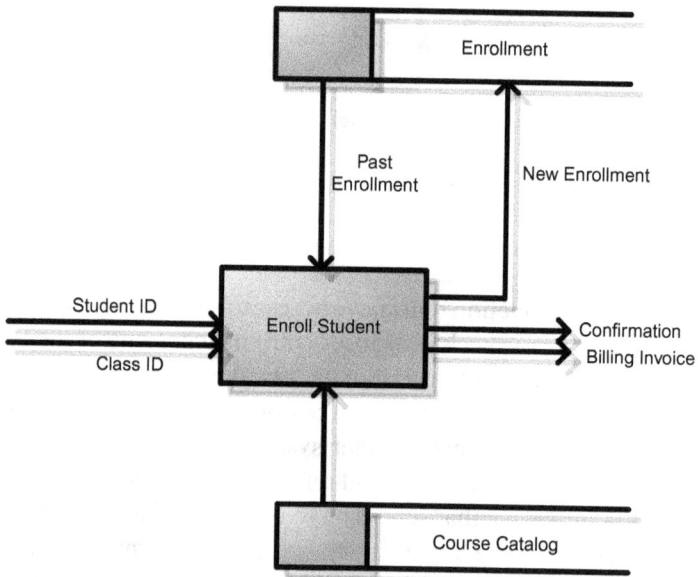

Figure 6.   Representation of the enrollment process.

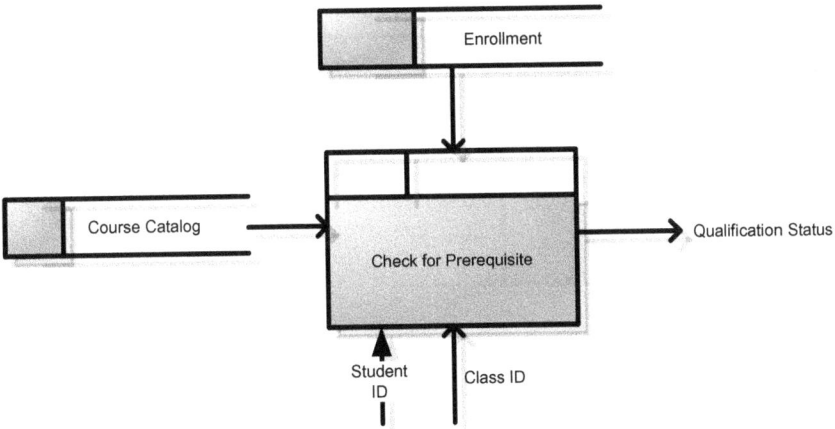

Figure 7.   Representation of check for prerequisites.

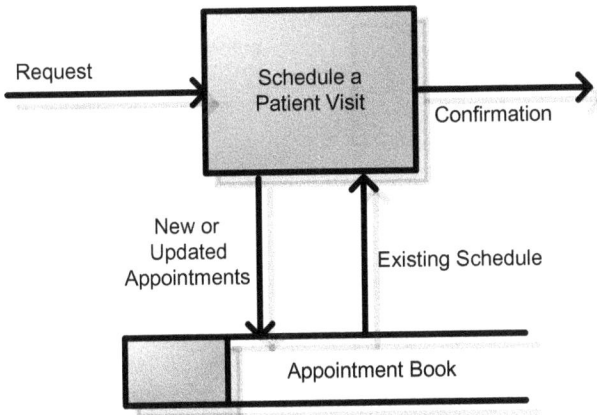

Figure 8.   Representation of schedule patient visits.

Figures 9 and 10 show two processes in a point of sale system. To receive a shipment, the user needs to have the invoice and the original order to check the quantity and identity of the shipped items. The process will also update the inventory and backlog data (if there is discrepancy). To check out items, the process will need to have a list of order items and a product catalog for pricing data. It will also need tax and payment information. The output may include a new order and a receipt.

There exist some syntax rules to follow to create data flow diagrams. For example, a function name must start with a verb, and there cannot be

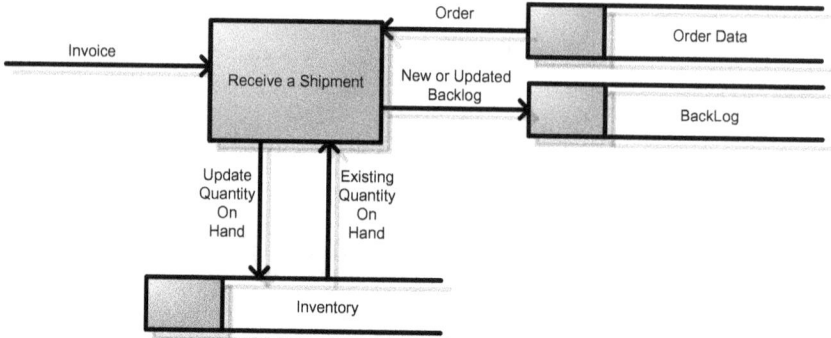

Figure 9.   Representation of receiving shipment.

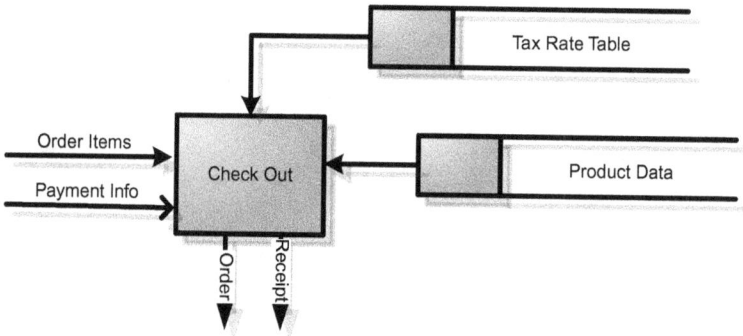

Figure 10.   Representation of checking out process.

a data flow directly connecting data stores and/or external entities without a process in the middle. The most important rule that has relevance to this book is the following so-called validity principle.

The Validity Principle of Process Representation: The basic principle regarding the validity of process representation is that *each process has enough data inputs for it to be properly performed, and it generates value-added outputs as the process name suggests.*

This is reasonable; a process is simply a data processor and will stop working without all data in place. The process must also generate value-added outputs. Imagine if a process does nothing but simply spit out what comes in? That process should be eliminated for the sake of efficiency. What if a process is supposed to validate the order as its name suggests but it prints a receipt? It does not do what it is supposed to do.

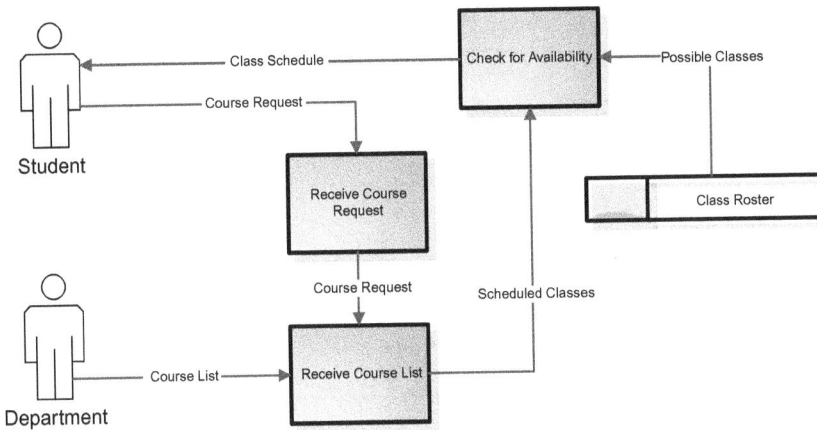

Figure 11.   Sample violations of the validity principle.

Figure 11 depicts a data flow diagram that shows the inputs and outputs of three processes performed in a student registration system, where students and departments are two external entities. We can point out the following violations of the validity principle in the process representation:

1. The process "Receive Course Request" appears to do nothing because outflow is identical to the inflow. Such functions should be removed to improve the efficiency of the system.
2. The process "Receive Course List" does not need the "Course Request" data flow in order to perform its function, as implied by its name.
3. To check for availability, it is necessary to have information about possible course offerings and the existing registration data. However, the process "Check for Availability" does not have past registration data to perform its function, as its name implies.
4. To the "Check for Availability" process, the "Scheduled Classes" input is overlapping with the "Possible Classes" input.
5. The output "Class Schedule" seems to be more than what the "Check for Availability" function can do. To make a class schedule for a student, you need to first have a course request from a student and then check for availability.

One possible correction is to rename the "Check for Availability" process as "Schedule Classes." The resulting diagram should be like the ones in Figure 12. Another data store, *Enrollment*, has been added to

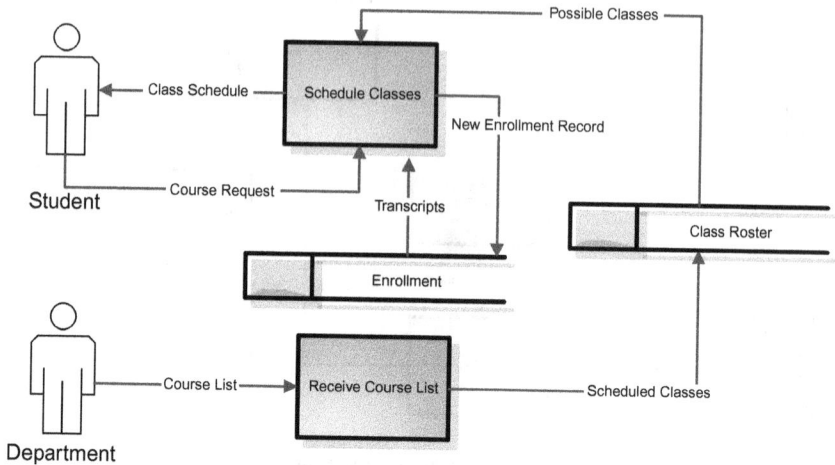

Figure 12.   A corrected diagram of Figure 11.

reflect the need to check the existing registration to be able to check for availability and fulfill course requests. After a class schedule is made for a student, the enrollment data must be updated accordingly.

Activity Diagrams

Process modeling focuses on what functions are required and treats each function as a black box. Procedure modeling exposes the black box inside out and represents the process internally.

Although process decomposition and representation are important conceptual building blocks toward the understanding of objects, they are not formal deliverables or artifacts in the object-oriented analysis and design. Procedural models are different, and they are used to describe use cases as well as object behaviors and collaborations. The formal procedure models include activity diagrams, structured English, state transition diagrams, communication diagrams, and sequence diagrams. In this chapter, we will use activity diagrams and leave other procedural modeling tools to be discussed in later chapters.

Activity diagrams were called *program flow charts* earlier and have now been reinstated as a part of UML 2.0 standard. An activity diagram is the best for describing how a function is performed sequentially and logically as a series of elementary or algorithmic actions and control flows. It is often used as a programmer's main point of understanding a process

Figure 13.   A simple activity diagram.

or brainstorming a strategy for problem solving. Figure 13 depicts a simple activity diagram showing a car wash process.

An activity diagram is read from top to bottom and left to right. It consists of two elements:

- **Activities or action states** represent the invocation of an operation, a step in a business process, or an entire business process. Activity is represented as a rectangle node except for the initial state represented as a solid circle and a final state represented as a buck eye. Both initial and final nodes are action states.
- **Transitions or threads** represent the flow of control from one activity to another through a link between the activities. A transition that involves an object is known as an object flow. The control flow is represented by a directed arrow.

To create an activity diagram in Rational Rhapsody, first we create a new project, then right click with the mouse on Package in the model browser to add a new package, and finally right click on Activity group to add a new activity. Note that all diagrams and diagram elements like activities are listed in groups in the model browser. Check View → Browser menu item if the model browser does not open by default.

More sophisticated activity diagrams include additional diagram elements such as the following:

- **Decision node** represents a transition where one action is followed by one of a few possible actions.
- **Merge node** is where two or more transitions resulting from a decision node are combined.
- **Fork** is a transition where one action is followed by two or more actions to be performed in parallel.
- **Join** is a point where two or more forked transitions re-join.
- **Swimlane** is a mechanism to group activities performed by the organizational units.

Decision nodes are used when we want to execute a different sequence of actions depending on a condition. For example, to determine the maximum between two numbers a and b, the action depends on whether $a > b$ or not. A decision node is represented by a diamond with one control flow in and multiple exclusive flows out. Each branchout control flow contains a *guard condition* written in brackets. Guard conditions determine which edge is taken after a decision node. The branchout flows join at a *merge* node, which marks the end of the conditional behavior that was started at the decision node. Merges are also shown with diamond-shaped nodes, but they have multiple incoming flows and one outgoing flow. Figure 14 shows an example to use both decision and merge nodes. Note that to create a guard condition in Rhapsody, we double click on the branchout flow to open its property dialogue and enter the condition there.

Example 1 (Validate User Login): To validate a user login, we need to check if a user name and a password match one of the account records in a login database. To be tolerant of entry errors but protect against hacking, we may restrict the number of entry errors to no more than three times. Figure 15 shows an activity diagram for the process.

First, we initialize the output variable, `valid`, on whether an entry is valid. Note that the validation process essentially corresponds to a while-loop in programming, during which if any step finds a match, the loop breaks, and the output variable takes on value `true`. Thus, we initialize valid to the opposite, i.e., `valid = false`. When there is a loop, we also initialize a loop count or condition variable. In the current example, the variable is the number of entry errors, `errorCount`, and we initialize it to zero.

Then, the diagram uses one action to get both user name and password and one action to check if the account exists in the data store. If yes, the process ends. If no, it will increment error count and then check if the login errors are more than 3. If it is more than 3, the process ends with a failure. Otherwise, it loops the control flow back to get a new entry.

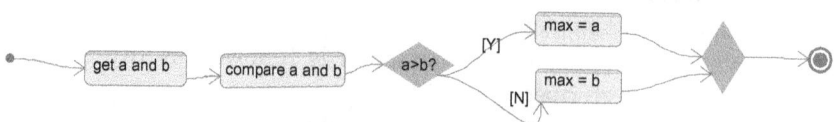

Figure 14.   Activity diagram for `FindMax`.

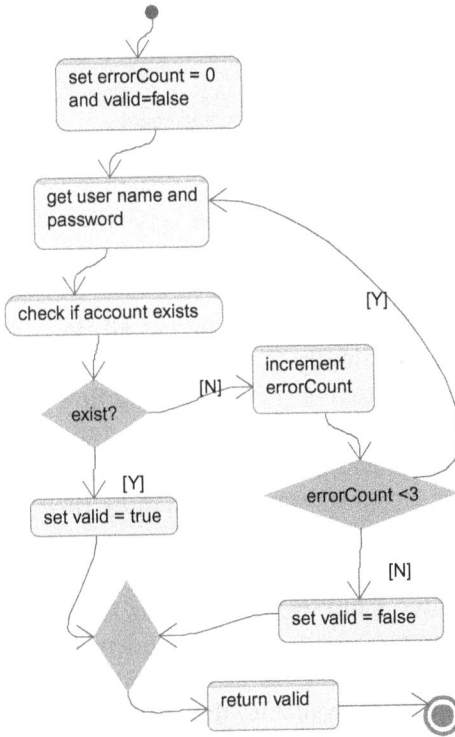

Figure 15. An activity diagram for validating account login.

Forks and *joins* represent concurrent actions. After a fork, one flow is broken up into two or more simultaneous flows, and the actions along all forked flows execute in parallel. At a join, all incoming actions must finish before the action can proceed past the join. Forks and joins look identical — they are both drawn with thick bars — but you can tell the difference because forks have multiple outgoing flows, whereas joins have multiple incoming flows.

Figure 16 shows the use of fork and join nodes to represent the parallel actions, Prepare Case and Prepare Motherboard, in computer assembly. Forks and joins may be drawn either horizontally or vertically. To change the orientation, we right click on the node and choose Flip Left or Flip Right in Rhapsody.

Note that both a decision and a fork have one inflow and multiple outflows, but they are different; one and only one of the multiple flows

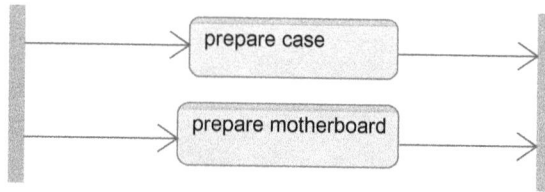

Figure 16.   Forks and joins.

out of a decision node is followed, whereas all flows out of a fork node must be followed concurrently. A *merge node* also appears like to a *join node* as they both have two or more incoming flows and a single outgoing flow. They are different as a merge is used to unite several possible incoming flows, only one of which presents a token to the merge; the join reunites concurrent flows where each incoming flow presents a token to the join.

Example 2 (Register Courses): To register a course for a student, as Figure 7 shows, the process needs both student ID and class ID. After having the inputs, the process can be split into three parallel flows, one checking if the course is available, the second assessing if the student has met the prerequisites of the course, and the third analyzing if the course has time conflicts with the other currently enrolled ones. Only if all checks are finished can we decide if the student can enroll into the course. Figure 17 uses fork and join nodes for parallel checks before a decision node.

There are times we need to model `for-loop` controls to represent the situation in which we perform some actions repeatedly a definite number of times. For example, to compute the sum of numbers in a list, we add each item in the list to the output variable one by one until all items are added. Rhapsody turns a decision node into a *for* loop by carrying out the following steps:

1. Double click on a decision node to open Features dialogue and select *FlowChartForLoop* as the stereotype (see Figure 18).
2. Enter the relevant loop initialization code in the Loop initialization field on the General tab of the Features dialogue.
3. Enter the relevant loop increment code in the Loop step field on the General tab of the Features dialogue.

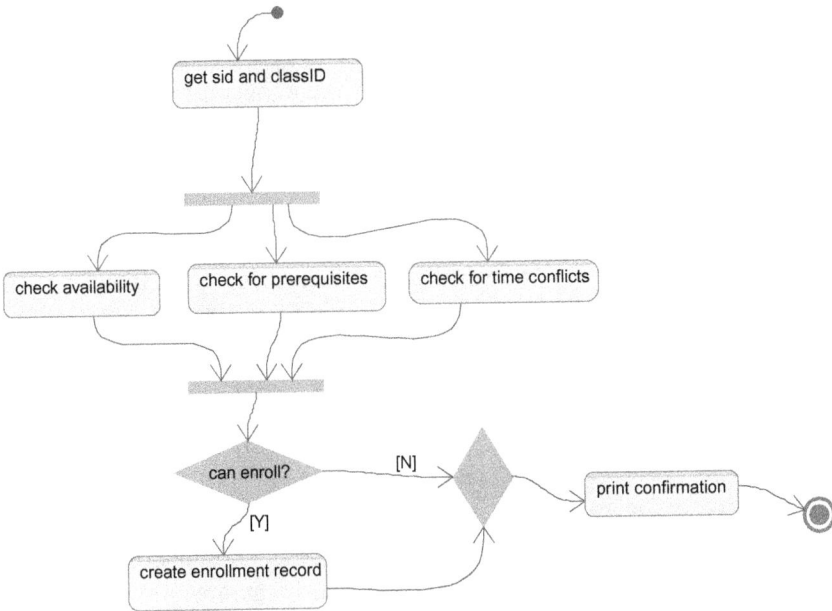

Figure 17.   The activity diagram for enrollment process.

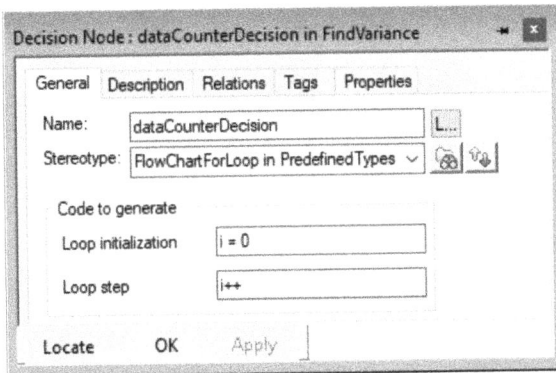

Figure 18.   Specifying for-loop in Rhapsody.

Example 3 (Compute Mean and Variance): To compute the mean or average value $\bar{x}$ of a list of numbers, $x_1, x_2, \ldots, x_n$, we need to compute the sum of the numbers and divide the sum by the item count:

$$\bar{x} = \frac{x_1 + x_2 + \cdots + x_n}{n}$$

As mentioned in Chapter 2, whenever we use a loop to compute a sum, we initialize the output variable to zero. When using a for-loop, there is no need to initialize the loop-step count or condition variable, which is set and initialized in the loop control node as in Figure 18. In this example, the input data is a data list, `dataList`. The total number of items in `dataList` is expressed as `dataList.Count`, and the `i`th item in the list is expressed as `dataList[i]`. The notations look like Java or C# codes and are very concise to use in diagrams. Figure 19 shows the activity diagram for computing the mean. Here, the for-loop node expresses the exit condition $i < dataList.Count$ like a decision node but hides the initialization and step increment behind. The hidden stuff can however be used for code generation.

Variance measures how wide data are spread. Given the data $x_1, x_2, \ldots, x_n$, assume the mean value is $\bar{x}$, and so the standard formula for computing the variance is:

$$var = \frac{(x_1 - \bar{x})^2 + (x_2 - \bar{x})^2 + \cdots + (x_n - \bar{x})^2}{n}$$

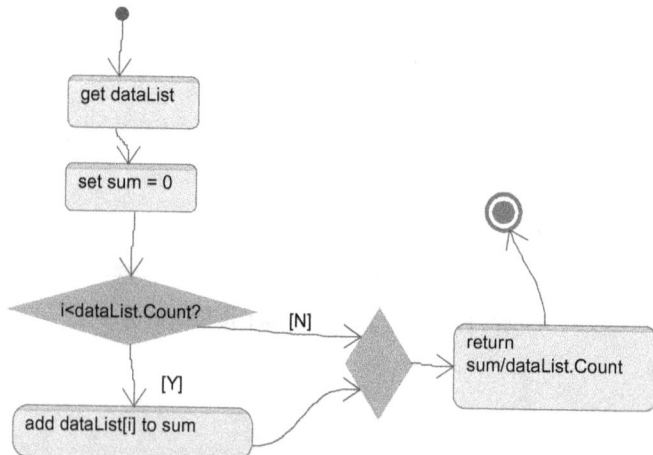

Figure 19.   Compute mean value.

According to the formula, we will need to first compute the mean value as in Figure 19, and then use another loop to compute the sum of squared differences between each item and the mean value. Thus, we would have to go through the same data list twice in order to compute the variance. Programmers are concerned with finding a better problem-solving strategy, or *algorithm*, that uses a smaller number of arithmetic operations to achieve the same goal. In the current problem, business users are statisticians, and they will probably tell a system analyst that there is yet another way to compute the variance, which is as follows:

$$var = \frac{(x_1 - \overline{x})^2 + (x_2 - \overline{x})^2 + \cdots + (x_n - \overline{x})^2}{n} = \frac{x_1^2 + x_2^2 + \cdots + x_n^2 - n\overline{x}^2}{n}$$

The analyst should take advantage of knowledge and design the algorithm via an activity diagram more efficiently. Figure 20 depicts such an activity diagram. It uses one loop to add each item to variable sum and adds the square of each item to variable sqSum. Thus, it initializes two outcome variables, sum = 0 and sqSum = 0.

The above examples involve one loop in the algorithm. More complex problems often entail the use of nested loops, or one loop inside the other loop. Examples include sorting items in a list and comparing two lists to see if they overlap or if one is a sublist of the other. The advanced reader may continue to the following examples on how to solve problems using nested loops.

Example 4 (Check Prerequisites and Check for Time Conflicts):
In Example 2, "check prerequisites" and "check for time conflicts" are two of three concurrent activities to ensure a student is eligible to take a course. How do we perform these activities in a more algorithmically detailed level? These activities essentially entail comparing two lists.

The "check prerequisites" activity is used to compare two lists — a list of required courses for taking a course and a list of completed courses in a student's transcripts — to see if the required courses are all in the finished courses. If not, the student does not meet prerequisites. Figure 21 shows an activity diagram. It first uses student ID and class ID, respectively, to retrieve the completed course list, compCourses, and the required course list, reqCourses, from the database. Then it checks if every item in reqCourses is in compCourses. It uses two

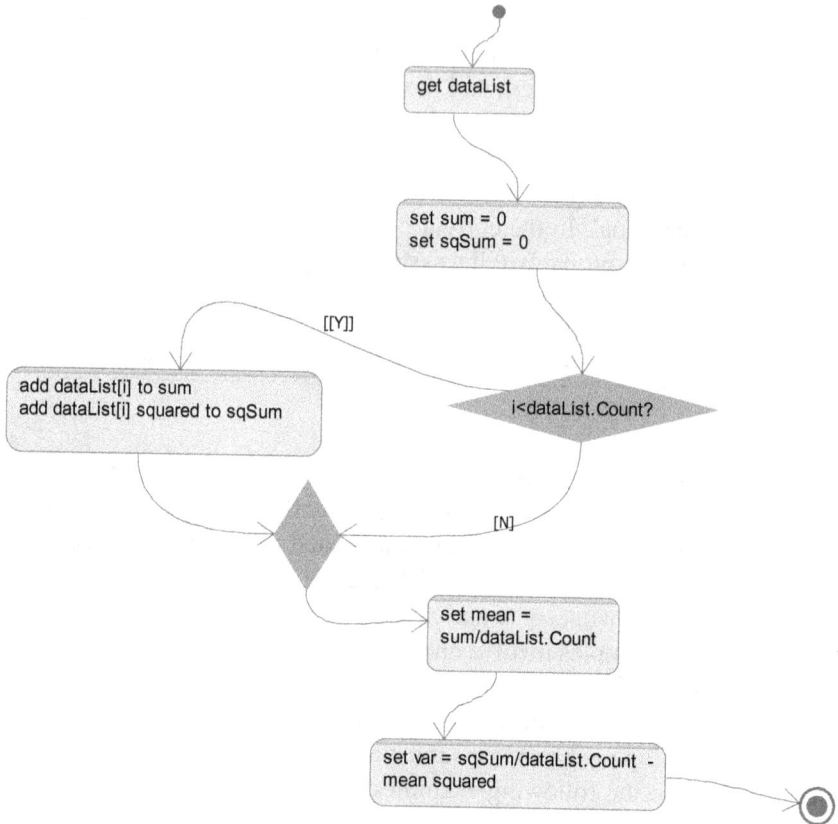

Figure 20.   Compute mean and variance.

nested for-loops to do so.[1] The outer loop selects one required course in each step, and the inner loop selects one completed course in each step. If there is a match, it means the required course selected in the outer loop is completed by the student, and so we will need to go back to the outer loop to select another required course to continue. If it is not a match, we will go back to the inner loop to select another completed

---

[1] A better solution strategy is to use hashsets instead of lists for reqCourses and comp- Courses since the courses in them don't have to be ordered. Using hashsets, it takes a constant amount of time to find if a required course is contained in compCourses regardless of the size of compCourses.

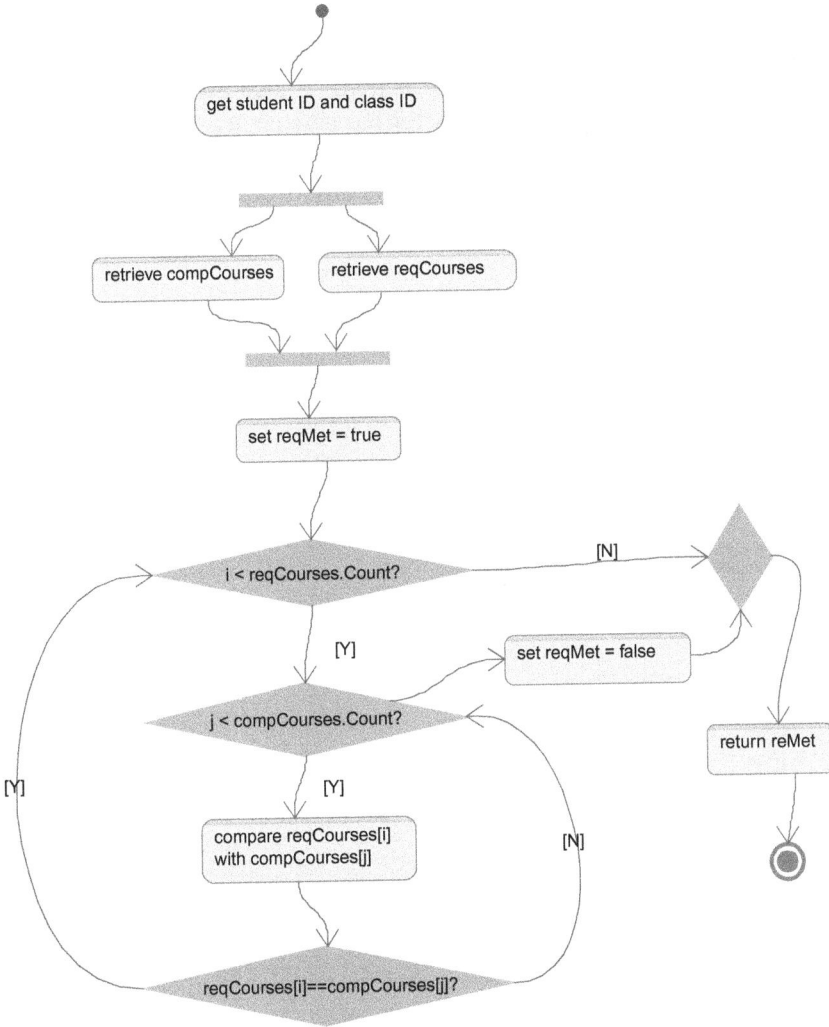

Figure 21.   Check prerequisites.

course to compare. If, after finishing the entire inner loop, there is still no match, it means the selected required course from the outer loop is not in compCourses list. In this case, because at least one required course is not completed, the whole activity must be aborted, and the loops can be broken. The outcome is false, meaning that prerequisites are not met. Thus, we initialize the outcome variable, `reqMet`, to its opposite, *true*.

The "check for time conflicts" activity is also used to compare two lists, a list of candidate time slots for the new course to be enrolled and another list of committed time slots for the existing courses that a student has already registered, to see if there is any overlap. If yes, the student will have time conflicts while taking the new course. Figure 22 shows

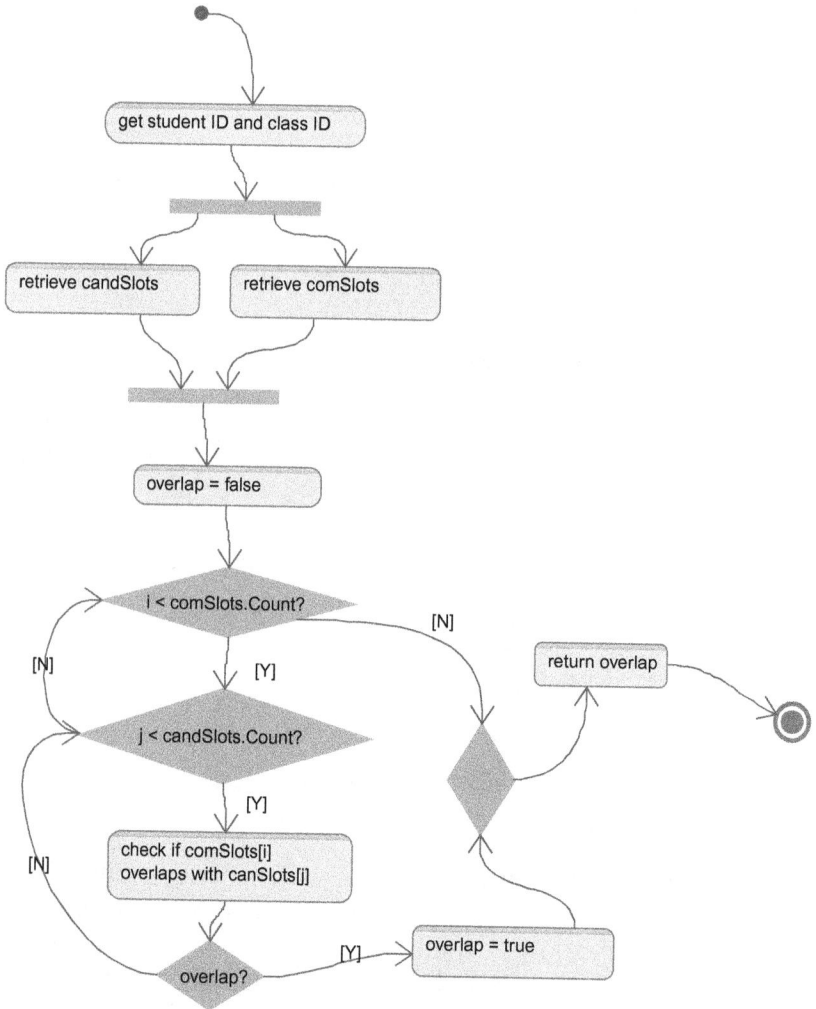

Figure 22.   Check for time conflicts.

an activity diagram. It first uses student ID and class ID, respectively, to retrieve the list of committed time slots from the existing courses, denoted by comSlots, and the list of candidate time slots from the new course to be enrolled, denoted by canSlots. Then it uses nested loops to check if each item in comSlots is overlapping with any item in canSlots. If there is one pair of time slots overlapping with another set, the activity can be terminated, and the outcome will be true, meaning there exists some time conflict if the student takes the new class. Thus, we initialize the outcome variable overlap to false. The outer loop picks one time slot from comSlots, and the inner picks one time slot from canSlots. We check if the two picked time slots overlap. If they do, we reset the outcome variable to true and bring the activity to the end. If the two do not overlap, then we go back to the inner loop to pick up another time slot from canSlots to continue. If, after going through the entire inner loop, we find no overlaps, meaning the time slot picked from the outer loop does not overlap with any of candidate time slots, we go back to the outer loop to pick out another committed time slot to continue. If, after going through the entire list of committed time slots, we find no overlaps, we bring the process to the end.

Example 5 (Bubble Sort): This example tackles a classic problem in an algorithm course in computer science. The problem is about how to sort a list of numbers in ascending order. There exist several problem-solving strategies or algorithms to tackle the problem. One of them is the so-called *bubble sort*. The strategy uses multiple passes to go through the list. In the first pass, it brings the largest number to the last place. In the second pass, it brings the second largest number to the second to last place. In general, in the ith pass, it brings the ith largest number to the ith to the last place. In each pass, it compares each number to its next neighbor and swaps them if it is larger. Assume the list is (1 5 4 2), and the following is a simple illustration of bubble sort. Note that the two numbers to be compared in each step are shown in bold, and the number of steps gets smaller as more passes are finished.

First pass (through the entire list):
Step 1: (**1 5** 4 2) → (**1 5** 4 2) by comparing 1 and 5, no swap is need because 1 < 5
Step 2: (1 **5 4** 2) → (1 **4 5** 2) by comparing 5 and 4, swap the numbers because 5 > 4

Step 3: (1 4 **5 2**) → (1 4 **2 5**) by comparing 5 and 2, swap the numbers because 5 > 2

Second pass (through the list without the last number):
Step 1: (**1 4** 2 5) → (**1 4** 2 5) by comparing 1 and 4, no swap is need because 1 < 4
Step 2: (1 **4 2** 5) → (1 **2 4** 5) by comparing 4 and 2, swap the numbers because 4 > 2

Third pass (through the list without the last two numbers):
Step 1: (**1 2** 4 5) → (**1 2** 4 5) by comparing 1 and 2, no swap is needed because 1 < 2

Three passes, and in total six comparisons and possible swaps, bring the four numbers in the list in ascending order. Bubble sort is one of the most efficient algorithms for sorting numbers. Figure 23 shows an activity

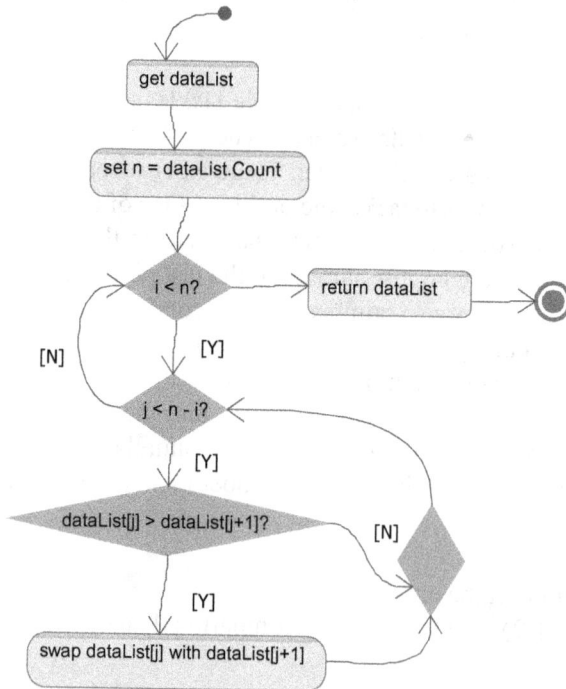

Figure 23.   Bubble sort.

diagram for the algorithm. It uses two nested loops: the outer loop steps through the passes, and the inner loop steps through the steps in each pass. If the list has n numbers, there will be n passes. Since the first pass will bring the largest number to the last place, the second pass just needs to go through the first $n - 1$ numbers. Similarly, the ith pass just needs to go through the first $n - i$ numbers. Therefore, in the ith pass, the inner loop will only go through steps $j = 0, j = 1, \ldots,$ and $j = n - i - 1$. In each step, we compare the jth number with the $(j + 1)$th number and swap them if the jth number is larger. Then we move to the next step and compare the $(j + 1)$th number with the $(j + 2)$th number, and the process continues to the last two numbers: the $(n - i - 1)$th number and the $(n - i)$th number. After the inner loop is finished, we move back to the outer loop to do the next pass. After all passes are finished, we bring the activity to the end.

Review Questions

1. What is a process, and how is it different from an activity?
2. Are activity and action the same thing?
3. What is the overall process, and what is special with major processes?
4. Among the three activities in process analysis, which will deliver the useful artifacts for the object-oriented method?
5. Why must each process have sufficient input?
6. Why must each process produce value-added output?
7. What are the tools for procedural modeling?
8. What is different between process representation and procedural modeling?
9. If a process entails only one command to be performed, will it be considered a complex or major process?
10. Identify whether each of the following is a process in the sense of performing data activities, and briefly answer why it is or is not. Note that if it is not entirely a process, can you re-frame or rephrase it to be one?
    a. Plow the field to grow rice
    b. Packing orders for shipping
    c. Receiving shipments
    d. Check inventory level
    e. Grade student projects
    f. Recruit employees

Exercises

1. Suppose you develop a point-of-sale system for a small retail store to manage its daily transactions and management. What are the major processes? If there is any process that is complex, decompose it and finally create a process hierarchy for the system.

2. An inventory system is responsible for generating orders if the actual stock falls below the minimum re-order level and for paying invoices to suppliers. It is responsible for updating inventory added if a new order is received and updating inventory used based on the inventory decrement data generated from the ordering system. It should also allow the manager to query inventory levels. Create a process hierarchy for the inventory system.

3. Create the process representations for the following business processes:
   a. Admit students.
   b. Check for order status.
   c. Handle customer returns.

4. Create activity diagrams for the following simple processes:
   a. Find the solution to a quadratic equation $ax^2 + bc + c = 0$.
   b. Check whether an integer is a prime number or not.
   c. Check whether today is your birthday.
   d. Find how many numbers of a certain character appear in a text.
   e. Find the max of three decimal numbers.
   f. Find the variance of three numbers.

5. Create an activity diagram for the following procedures:
   a. Check if two list of names are overlapping.
   b. Find the standard deviation of a list of decimal numbers.
   c. Sort a list of integers from smallest to biggest.
   d. Find the maximum of a list of decimal numbers.
   e. Check if a name is in a list of names.
   f. Find all the prime factors of any integer.

6. Create activity diagrams for the following business processes:
   a. Assume the existence of these database tables: students, classes, enrollment, instructors, and teaching assignments. Check if a student and a teacher are related.
   b. Assume the existence of these database tables: students, courses, prerequisites, classes, enrollment. Check if a student has a prerequisite for taking a new class.

   c. Assume the existence of these database tables: students, classes, enrollment. Check if a student has time conflict in taking a new class.

   d. Assume the existence of these database tables: borrowers, titles, copies, and rentals. Return a rental.

   e. Assume the existence of these database tables: sales, items, inventories. Handle customer returns.

7. Suppose there are missing values in an array of observations. Create an activity diagram to move the numbers to the beginning contiguously so that there is no missing block that interrupts the numbers.

8. Whenever a new patient is seen for the first time at Cyberdale Care Center, he or she has to finish a patient information form that asks name, address, phone number, insurance carrier, and yes/no answers to certain questions such as whether a patient is allergic to certain drugs, whether the patient has had any surgery in the last five years, etc. When a patient calls to schedule a new appointment or change an existing appointment, the receptionist checks the appointment schedule for an available time. Once a good time is found for the patient, the appointment is scheduled. If the patient is new, an incomplete entry is made in the patient file; the full information will be collected when the patient arrives for the appointment. Sometimes, appointments are made so far in advance that the receptionist will have to send a reminder postcard to each patient a week before the appointment. Develop an activity diagram to show the business process.

9. To enroll a student into a class, the registration system must check whether the student has all the prerequisites taken, whether the class is still open, and whether the total number of credit hours the student registers is not beyond the maximum allowed. After a student finishes her registration, she will need to pick up a printed confirmation that shows all the courses she has registered, the date/time, section number, credit hours, and instructor for each class. Also, the confirmation paper shows the student status, state of residence, the total number of credit hours, and the total amount to be paid to the college. The student will bring the confirmation to the business office and make a deposit, which is equivalent to 20% of the total amount to reserve her registration. If she fails to do so within 10 days, her registration will be canceled. The system also actively monitors the number of students signed up for each class. Three days before the class starts, if the number of registered students for a class is less than 15, the class will be

canceled. The registered students will be informed to find alternative classes. To better serve the students and departments, the system has functionality for students to make course requests for future terms. The requests will be summarized and sent to departments so that they can make informed decisions on what needs to be offered in the future. Create the following diagrams:

a. Create the process hierarchy to show the decomposition of the overall process.

b. Create the activity diagram to show the internal workings of the overall process.

10. The National Parks Association wants to track the attacks on visitors by animals in the parks. For each incident, the name and address of the person is recorded, along with the type of animal that attacked, the date of the attack, and the location of the attack. Answer the following sub-questions:

a. Identify the functions to be performed in the tracking system.

b. Create a data flow diagram for the function SearchForAccidents.

c. Create an activity diagram for the process SearchForAccidents.

11. Professor Johnson wants to set up an application to keep track of the attendance records of his classes. This is how he would like to check attendance. First, he will create a class meeting with specific beginning and ending times. Then he can print out a sheet that has all registered students so that he can check who is in and who is not.

a. Identify the functions to be performed in the tracking system.

b. Create a data flow diagram for the function CheckAttendance.

c. Create an activity diagram for the process CheckAttendance.

12. A warehouse receives supplies from various vendors and checks out the items to its customers, including individual employees and departments. The actual cost of each item is billed to the customers who use the supplies. Internally, as a convention of organizing inventories, supplies are organized into categories. For each supply, the maximum and minimum inventory levels are kept so that when the stock of a part is below the minimum, an order will be issued to get it refilled.

a. Identify the functions to be performed in the tracking system.

b. Create a data flow diagram for the function CheckOut.

c. Create an activity diagram for the process CheckOut.

13. The Board of Watson Town Memorial Hospital has recently decided to develop a new information system to manage their patient admissions and discharges. The hospital handles two types of patients: outpatient and resident patient. As typical, each time when a new patient comes, the data about his/her identification, address, phone, and issuance carriers are recorded. If a patient is resident, he/she will be assigned to a bed and an admission date is recorded. After the treatment, a nurse has to sign off the discharge card. For an outpatient, the nurse will set a check-back time after each treatment.

   a.   Identify the functions to be performed in the tracking system.
   b.   Create a data flow diagram for the function CheckIn.
   c.   Create an activity diagram for the process CheckIn.

Appendix: Algorithms

Besides following the order of DAMO (declaration–assignment–manipulation–output) and PEMDAS, we often need an algorithm, or problem-solving strategy, to write instructions to solve a problem. In fact, the most difficult task in computer programming is not memorizing, understanding, and/or applying programming principles, syntax rules, and programming language-specific constructs but rather how to design a problem-solving strategy. A computer programmer may be fluent in a programming language, but he or she may not know how to solve a domain problem. Systems analysts, with assistance from end users, play a critical role in systems development not only by capturing and modeling requirements but also designing algorithms. Then a programmer's job is to re-express the algorithms using code.

Algorithm design entails both domain knowledge and creativity; so no book can teach how to design algorithms. An algorithm course in computer science mostly teaches us how to evaluate the efficiency of an algorithm. This chapter teaches us how to represent an algorithm using activity diagrams. This Appendix uses examples and exercises to show that algorithm design is sometimes just about raw intelligence.

Example 1: Use a balance to single out the lightest egg among 100 eggs. Programmers may realize this problem is identical to finding the minimum of 100 numbers. A simple strategy would be: (1) compare 1st egg with the

2nd one, find the lighter one, called theLight; (2) compare theLight with the 3rd egg, and swap the 3rd one with theLight if the 3rd one is lighter; (3) repeat (2) to compare theLight with the 4th, 5th, ... , and 100th egg.

Of course, you may be able to come up with a more sophisticated strategy to solve the problem faster. Then your algorithm is more efficient than the simple one here. Unfortunately, for this problem, there is no better algorithm.

Example 2: Use a balance to sort 100 eggs from the lightest to the heaviest. We just learned an algorithm called bubble sort to sort 100 numbers in ascending order. Here is the strategy translated from the bubble sort: (1) compare 1st with 2nd and swap them if 1st is heavier than the 2nd; (2) compare the 2nd with the 3rd and swap them if the former is heavier; (3) repeat the same procedure for the 3rd with 4th, 4th with 5th, ... , 99th with 100th; (4) repeat the above steps 1–3 again and again until there is nothing to swap. Can you think of a better algorithm to solve this problem?

Programming exercises

1. You have 12 eggs of equal weight, with one of them being bad. How can you use the fewest number of balance uses to single out the bad egg?

2. You have a 4-gallon jug and a 3-gollon jug. No measuring tools are available. There is a pump that can be used to fill the jugs with water. How can you get exactly 2 gallons of water into the 4-gallons jug?

3. A jail has 20 locked cells numbered Cell 1, Cell 2, ..., Cell 20. There are much fewer prisoners, each occupying one cell, and so there are some empty cells. The jail manager wants to move the prisoners to consecutive cells, starting from Cell 1. After moving, the prisoners are still kept in the same order as they were before. What is the best approach to perform this operation?

4. A goat, a wolf, and a salad are on one side of a river and you need to get them to the other side using your boat. You can carry one item in your boat to the other side at any given time. However, when the goat and the wolf are left alone the wolf will eat the goat. If the goat and the salad are left alone the goat will eat the salad. As long as you are with them nothing will happen, i.e., the wolf won't eat the goat and the goat

won't eat the salad. How do you bring the goat, the wolf, and the salad across the river using the boat?

5. A salesman has a list of cities, each of which he must visit once. There are direct roads between each pair of cities on the list. Find the route the salesman should follow for the shortest possible round trip that both starts and finishes at any one of the cities.

	Boston	NY	Miami	Dallas
Boston		250	1450	1700
NY	250			
Miami	1450	1200		1600
Dallas	1700	1500	1600	

Chapter 4

Coding Functions and Procedures

Introduction

The primary role of a systems analyst is to bridge the gap between programmers and business users. Therefore, to be an effective communicator, a systems analyst needs to speak the language that a programmer speaks. In this chapter, we will elaborate on how a programmer speaks the language of business processes. As we will see, coding is nothing but an alternative, more rigorous representation of processes that we modeled in the last chapter. Another purpose of this chapter is to establish an intuition toward the understanding of concepts of objects or classes in later chapters.

Operations and Methods

An *operation* is the evolution of the external representation of a business process, whereas a *method* is the evolution of a procedural model. They are both evolving artifacts of the same business process in programming language.

From the programmer's perspective, a process is still a data processor, as was seen in Chapter 3. There are two differences however. The first is the terminologies to be used to name a process or its parts. The second is the method to be used to represent the functions, which is more rigid and cryptic.

In programming, a process may be called a *function, procedure,* or *module* that processes data inputs, which are called *parameters*, and generates outputs called *returns* or output parameters (see Figure 1).

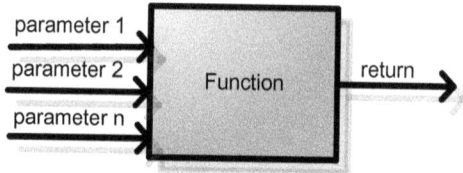

Figure 1.   A business process from a programmers' perspective.

Figure 2.   A function of adding two integers.

For example, if a simple process is to add two integers to obtain their sum, the two integer addends will be input parameters and the sum, which is also an integer, is the return (see Figure 2).

The actual program in Java or C# is as follows:

```
int Sum(int a, int b)
{
        return a + b;
}
```

Here, the code inside curly braces {} contains the actual instructions that do the job: it does the addition (a + b) and outputs the result using the `return` statement. The function is named `Sum` with two integer parameters `int a` and `int b`. The keyword `int` in front of the function indicates the return's data type as integer.

As we can see, a function is programmed as a code block consisting of two parts: head and body. The head includes the name of the function, a list of parameters of the function, and the return's data type. The body defines how the function works. In this book, we call the head of a function an *operation* and the body of the function a *method*. They are the terminologies to be used when we introduce the concepts of objects and classes in the later chapters.

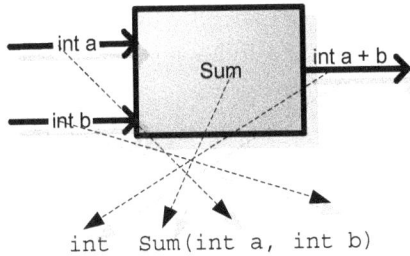

Figure 3. From data flow diagrams to code.

It is easy to see that an operation corresponds to the external representation of the process: inputs to parameters, outputs to returns, and process name to function name (see Figure 3).

A method corresponds to the internal representation of a business process, i.e., procedural model. A method can not only be expressed as a procedural model but also implemented in actual code. In fact, if a business analyst is fluent in a programming language, he or she could well use the language to express the internal logics of a business process in lieu of activity diagrams or structured English to bridge the gap between modeling and coding.

Code Functions

In this section, we will write code to implement a few process and procedure models to show how to convert business processes to operations and methods. Some of the example processes and procedures are depicted as diagrams in the last chapter.

Example 1 (Find Max): Suppose a process computes the maximum of two decimal numbers. The two inputs are any two decimal numbers, and the output is the decimal number that is the maximum of the original two numbers (see Figure 4(a) of Chapter 3). The operation can be written as:

```
decimal FindMax(decimal a, decimal b)
```

Here, the name of the function is FindMax. Parentheses surround two parameters *a* and *b*, both of which are decimal numbers. The decimal in front of the function name is the data type of the return values by the function. The function body or method is expressed as an activity diagram in Figure 14 of Chapter 3. The following is the Java or C# code of the procedure:

```
if (a >= b)
{
        return a;
}
else
{
        return b;
}
```

The method defines how the operation works through a sequence of instructions on how to find the result given two numbers a and b. The procedure is simple here; when a is bigger than b, the result is a. Otherwise it is b. Putting the operation and the method together, we obtain the whole function.

```
decimal FindMax(decimal a, decimal b)
{
        if (a >= b)
        {
                return a;
        }
        else
        {
                return b;
        }
}
```

Does every process require data inputs or generate outputs? Not necessarily. This may sound odd because a process is just a data processor, transforming inputs into outputs. Without data, what can it transform? Yet, we can think about it this way: a process still needs data inputs, but if it can get the data from an available source such as a computer system, it will not need them as explicit inputs. For example, to produce a greeting like "Good Morning," a process needs to know the current time, which is readily available from a computer system without an explicit input. In fact, when a function is located inside an object, all the data owned by the object do not need to be explicit inputs to the function. We call this feature *data flow reduction* in later chapters.

In a similar way, we can understand the scenarios in which a function does not produce outputs. A function still produces outputs, but the outputs are not in the form of data to be used by other processes. Examples of such outputs include printed documents, updated databases, email messages dispatched, or message boxes displayed. If a function is placed inside an object, its result, when merely changes the object data, will not be in the form of explicit outputs.

How do we code a process if it does not take inputs or does not return any outputs? If a process does not take any explicit inputs, we will attach a pair of empty parentheses () to the function name and leave out any parameters. If a process does not return any output to other processes, we will mark the return type as void.

Example 2 (Produce Greeting): Assume that this process takes the current time and produces a message like "Good Morning" to be displayed in a message dialogue. Because it gets the current time from the computer system, it does not need any input. Because it generates a result not in the form of data to be used by other processes, it does not have a return value. The operation or function head may be written as:

```
void ProduceGreeting()
```

and the method may be programmed by using DateTime object to get the current system time and hour.

```
{
        DateTime currentTime = DateTime.Now;
        int currentHour = currentTime.Hour;
        string myGreeting;
        if (currentHour > 6 && currentHour <= 12)
              myGreeting = "Good Morning";
        else if (currentHour >12 && curentHour <=18)
              myGreeting = "Good Afternoon";
        else
              myGreeting = "Good Evening";
        MessageBox.Show(myGreeting);
}
```

In the same token, data inputs from and outputs to data stores are usually considered as a system's internal data, and thus they will not be expressed as explicit inputs or outputs.

Example 3 (Check Prerequisites): The process of checking prerequisites takes class ID, student ID, course catalog, and enrollment records as inputs and generate Boolean value true or false as output (see Figure 6 in Chapter 3). Assume that both course catalog and enrollment records are tables in a relational database and assume class ID and student ID are both integers. Then the operation will be expressed as:

```
bool CheckPrerequisites(int classID, int studentID)
```

Here, data stores `Enrollment` and `CourseCatalog` are not expressed as parameters or returns in the operation. Also, `bool` means the return data type is a Boolean number, `true` or `false`. The method is expressed as an activity diagram in Figure 21 of Chapter 3, and its code implementation requires database instructions in SQL and code to connect to databases and run SQL commands.

```
{
        //get prerequisites from CourseCatalog using
          classID
        //get completed courses from Enrollment using
          studentID
        //check if prerequisites are completed
        //return true or false accordingly
}
```

Example 4 (Message Encryption): Encryption entails the use of a key to change a plain text message into a cryptic one so that no one can understand the message without the key. The process is diagramed in Figure 4.

Encryption, in general, is a complex process to code, but we will illustrate it using Caesar's cipher in what follows. We will use an integer as a key to increment the ASCII code for each character in the original text

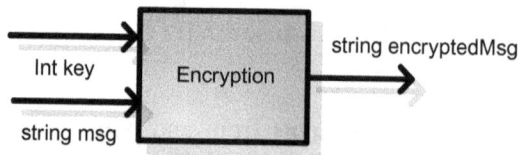

Figure 4.   Encryption process.

message. If the incremented code value is beyond the 0–127 range, we will add or subtract 128 to bring the code into the valid range. The input parameters are a key and a text message, and the output is an encrypted text message. In the function body, the function first converts a message into an array of characters, adds the key to the integer ASCII code of each character, and outputs the result as the encrypted message.

```
string Encryption(string message, int key)
{
    char[] a;
    a = message.ToCharArray();
    string encodedMessage = "";
    int length = a.Length;
    for (int i = 0; i < length; i++)
    {
        int j = (int) a[i]) + key;
        while(j > 128)
            j = j - 128;
        while(j < 0)
            j = j + 128;
        char c = (char) j;
        encodedMessage = encodedMessage + c.ToString();
    }
    return encodedMessage.ToString();
}
```

When a process takes multiple pieces of data of the same type as inputs, we can combine the inputs into either an array or a list as one input. Similarly, if a process outputs a list of values of the same type, we can use a list or an array as the return data type.

Example 5 (Factorization): Create a function that returns all the prime factors of an integer. Here, we use an array list (not array) to hold the list of possible prime factors. The reason to use ArrayList instead of Array is that ArrayList allows to dynamically extend the number of memory sub-blocks while Array has its size fixed at the time of declaration.

The strategy to solve the problem is to repeatedly divide factor 2 out of an integer until the quotient becomes an odd number. Then we use a for-loop to go through all the odd numbers starting from 3, 5, 7, etc., to extract out smaller prime factors one by one. Note that after all the smaller prime numbers are extracted, a larger odd number to be tested must be a prime number.

```
private ArrayList Factorization(int n)
{
    int quotient = n;
    ArrayList factors = new ArrayList();
    factors.Clear();
    while (quotient%2 == 0)
    {
        factors.Add(2);
        quotient = quotient / 2;
    }

    for (int i = 3; i <= quotient; i = i + 2)
    {
        while (quotient%i == 0)
        {
            factors.Add(i);
            quotient = quotient / i;
        }

    }

    return factors;
}
```

What if there is a process that returns data to be used by other processes while simultaneously executing actions such as prints documents and creates new database records? What if a process returns multiple values of different types? In these cases, we may use *out* parameters in a programming language like C# that supports the concept. The regular parameters we have used so far are *in* parameters; they carry data inputs to a process. We can also use parameters to carry the return values out. Such parameters are called *out* parameters. Since we can use multiple out parameters, we can use the device to return multiple different values. Also, there is no need to have the function return a value, we can make the operation as void and so the function can not only return values but also perform data actions that do not return any values.

For example, suppose a process is created to get what day of the week today is. It does not need any input since the system knows the current date and time. The return data is an integer from 1 to 7, with 1 for Sunday, 2 for Monday, etc. Thus, the operation can be written as:

```
int GetDayOfWeek()
```

The method may be expressed using the built-in `DateTime` class in C# as follows:

```
{
        DateTime today = DateTime.Now;
        return (int) today.DayOfWeek;
}
```

Note here that `(int)` is used to convert or cast the data type of `DayOfWeek` into an integer. In the following we describe another way to create the function using an out parameter:

```
void GetDayOfWeek(out int day)
{
        DateTime today = DateTime.Now;
        day =(int) today.DayOfWeek;
}
```

Here, we use the out-parameter `day` to return the day of week and change the whole function's return type to `void`.

Can we create two functions with the same name as above? The answer is no in structured programming. However, in objected-oriented programming, a special feature called *overloading* allows the reuse of a function name if the functions have different *signatures*. Two signatures are different if the functions have different names, different number of parameters, or different types of parameters at the same location. For example, `Sum(int a, int b)` can overload `Sum(decimal a, decimal b)`, `Sum(int a, int b, int c)`, `Sum(int a, decimal b)`, and `Sum()`. But it cannot overload `Sum(int x, int y)` because the two functions have the same number of parameters and have the same type of parameter at each place despite the differences in name.

Example 6 (Descriptive Statistics): In the last chapter, we modeled a procedure that can compute the mean and the variance using one pass through a list of data. We can do even better than that: we can compute all descriptive statistics, including max, min, average, sum, count, variance, standard deviation, etc., in one function by using out parameters. The data flow diagram of the process is shown in Figure 5, and its C# code is as follows:

```
public void DescriptiveStat(List<double> data, out
    int count, out double mean,
```

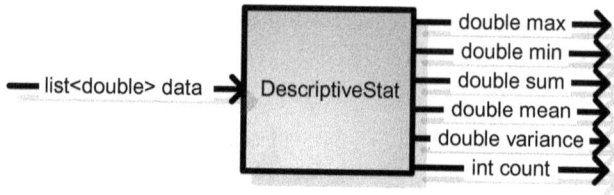

Figure 5.   Descriptive statistics.

```
                    out double variance, out double max,
                    out double min, out double stdev,
                         out double sum)
{
     sum = 0;
     count = data.Count;
     double sqSum = 0;
     max = data[0];
     min = data[0];
     foreach (double item in data)
     {
          sum = sum + item;
          sqSum = sqSum + Math.Pow(item, 2);
          if (max < item)
               max = item;

          if (min > item)
               min = item;
     }
     mean = sum / count;
     variance = (sqSum - count * Math.Pow(mean, 2)) /
          count;
     stdev = Math.Sqrt(variance);
}
```

Execute Functions

A function is created for reuse. A function encapsulates individual commands or instructions to perform a high-level task that needs to be executed repeatedly. The code that calls a function is called a *function caller*. Calling a function is simple. First, if a function requires input parameters, we will need to prepare for the inputs: declare the variables to hold input values and assign the parameter values to the variables. Next, depending on whether the function returns a value or not, the call can be handled in three different ways:

1. If a function returns one value, we will need to declare a variable so that it can hold the returned value and then run the function on the right-hand side of an assignment. For example, suppose we want to find the maximum of two decimal numbers e and π, we will call the function FindMax as follows:

```
decimal x, y;   //prepare for inputs
x = Math.E;     //assign Math.E to x
y = Math.Pi;    //assign Math.Pi to y
decimal result;  //prepare for the output
result = FindMax(x, y); //call the function
```

2. If a function does not return a value, calling the function reduces to the preparation of the inputs and the function is called as a command. For example, suppose we want to call ProduceGreeting() function, here is the code:

```
ProduceGreeting();
```

3. In case a function returns multiple values via out parameters, you will need to prepare values for input parameters and declare a variable to hold the value for each out parameter. Then call the function as a command, the output variables will bring the return values to be used by the caller. Here is the example:

```
int weekday;
GetDayOfWeek(out weekDay);
```

At this point, weekday will have the result for us to use in the code afterwards.

In structured programming, a function is the most basic unit of usable code, and a program is simply a juxtaposition of one or more functions. These functions are interdependent in the sense that some functions may call other functions in order to perform a more complex task. There is one main() function, which is the entry point of the program and will be called by the user. The main function will then start to call other functions, which can in turn call still other functions, toward the fulfillment of an entire task.

Example 7 (Find Maximum): Earlier we created a function that finds the maximum of two decimal numbers: decimal FindMax(decimal a, decimal b). Its procedure is very simple; just compare a and b and pick the greater. What if we need to find the maximum of more than two numbers? The reader may try to use the same algorithm to compare three numbers, four numbers, etc., but soon we will find out that it is not an easy job; we will need many nested if-else statements to compare three numbers, four, five, or more numbers.

A creative strategy is to reuse FindMax(decimal a, decimal b) function to compare the current max with additional number. Here is the implementation of the strategy to find the max of four numbers:

```
decimal FinxMax(decimal a, decimal b, decimal c,
    decimal d)
{
    decimal max;
    max = FindMax(a, b);
    max = FindMax(max, c);
    max = FindMax(max, d);
    return max;
}
```

Function calls demonstrate the advantage of functions. To use a function, there is never a need to know the internal code of the function; we only need to know the function head or operation. Another advantage is implicit. Modular functions are reusable, and there is never a need to copy and paste a same section of the code anymore. It allows an organization to create a library of functions and use them as pluggable modules for new development projects. Maintenance becomes easier since, if there is change, only one place needs to be changed as opposed to changing many places that duplicate the same code.

Review Questions

1. Compare functions, operations, and methods.
2. What is overloading?
3. Use an example to show how to convert a business process into the operation and method?

4. What are the key features of the structured systems analysis and design methodology?
5. What does a computer do when you declare a variable, assign a value to a variable, manipulate a variable, and output the value of a variable?
6. What should be the appropriate sequence of using the four types of instructions?

Exercises

1. For each process in Figure 3 of Chapter 3, create the external representation and then convert it into an operation.
2. Create the activity diagram for the process of message encryption as shown in the code of the text.
3. Use pseudo-code to express the method for the functions: CheckForAvailablity and CheckforTimeConflicts.
4. Use Activity Diagrams to express the method for the following functions:
   a. MakeAppointmentForPatient.
   b. ReceiveShipment.
5. Create a function that converts a numerical grade into a letter grade using the common scale used in your school. Then use Visual Studio to create a windows form with two text boxes so that when one box shows a numerical grade, the other box shows the letter grade.
6. Create a function that coverts a decimal integer into a binary integer.
7. Using Visual Studio to create a windows form with a large text box and then create two buttons to encrypt and decrypt the message in the text box. Make sure to use the function Encrypt in the book.
8. Code the following simple processes:
   a. Find the maximum of a list of decimal numbers.
   b. Check if a name is in a list of names.
   c. Find the solution to a quadratic equation $ax^2 + bc + c = 0$.
   d. Check if an integer is a prime number or not.
   e. Find all the prime factors of any integer.
9. Use activity diagrams to model the following procedures and code it in C#:
   a. Check if two list of names are overlapping.
   b. Find the standard deviation of a list of decimal numbers.
   c. Sort a list of integers from the smallest to the biggest.
   d. Given a list of names, find how many names are over 20 characters.
   e. Given a list of decimal numbers, find out how many of the numbers are zeros.

    f.   Given a list of dates, find a sub-list of dates that fall in 2018.

    g.   Given two lists of numbers, find a list of numbers that are common in both lists.

    h.   Encrypt a text message by adding a 32-bit binary number to each block of 10 characters in the message.

    i.   Given two time periods, determine if they are overlapping or not.

    j.   Given two time periods, determine the amount of overlapping time in days between them.

10. Code for the following business processes:

    a.   Assume the existence of these database tables: Students, Classes, Enrollment, Instructors, and TeachingAssignments. Check if a student and a teacher are related.

    b.   Assume the existence of these database tables: Students, Courses, Prerequisites, Classes, Enrollment. Check if a student has met the prerequisites for taking a new class.

    c.   Assume the existence of these database tables: Students, Classes, Enrollment. Check if a student has time conflict in taking a new class.

    d.   Assume the existence of these database tables: Borrowers, Titles, Copies, Rentals. Return a rental.

    e.   Assume the existence of these database tables: Sales, Items, Inventories: Handle customer returns.

Appendix: Text File Processing in C#

This chapter created a function to find descriptive statistics. A reasonable next step toward the use of the function is to get data from a text file. To read data from a text file using C#, we need to use the following three built-in classes, accessible by "using System.IO" directive:

- `File` — Use method `OpenText()` to open an existing file and `CreateText()` to create a new file.
- `StreamReader` — Use the method `ReadLine()` to read one line from the text file.
- `StreamWriter` — Use the method `Write()` and `WriteLine()` to write to a text file.

In addition, we may need `OpenFileDialog` object to open a text file to read or write. To use `OpenFileDialog` or any other dialog objects, there are three steps to follow: (1) create and show a dialog; (2) get dialog results using `DialogResult` object; and (3) process the dialog result. For example, the following is an example of using `FontDialog` to change the font for textbox `txtResult`:

```
FontDialog fontD = new FontDialog();
DialogResult myResult = fontD.ShowDialog();
if (myResult == DialogResult.OK)
  {
     Font f = fontD.Font;
     txtResult.Font = f;
  }
```

It is the same idea to use `OpenFileDialog`. The following example uses an `OpenFileDialog` object to browse a file to open and get the data into a list box and then call the function "Descriptive Statistics" to perform the calculation.

Example 1: Create a form with one list box that allows one to load all data from a data file to the box. Then, when the button "Compute" is pressed, use the textbox to show all the descriptive statistics.

The following uses Tab Control to design the user interface[1] to separate data from the result:

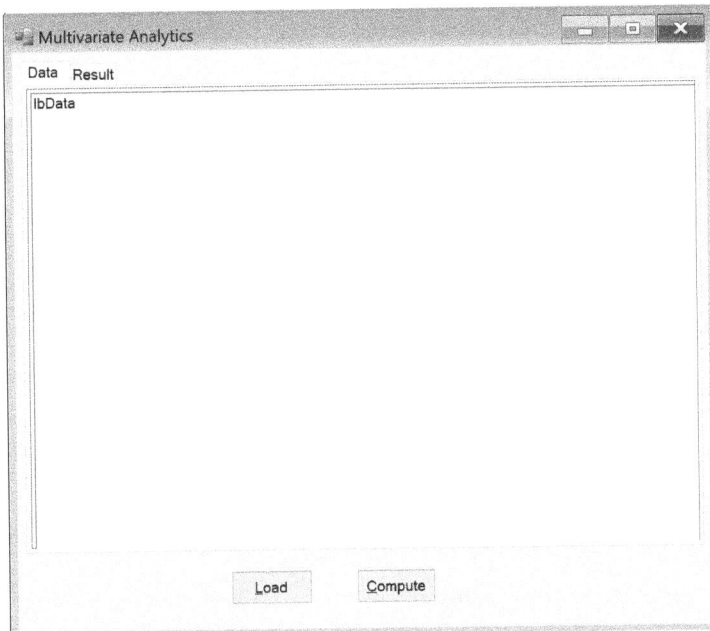

---

[1] The reader may refer to Chapter 10, Page 239, for user interface design techniques.

The Load button will open a file dialog to open a text file and load the data into the list box in Tab 1, and the Compute button will compute the statistics and display it in the result text box in Tab 2. Since both buttons will have to share a common list of data — the Load button that creates the list and the Compute button that uses the list — we will create an array: string[] strData to hold the data from a text file, outside the functions, to respond to the button clicks. Such a variable is called a *global variable*.

```
string[] strData;
private void btnLoad_Click(object sender, EventArgs e)
{
    OpenFileDialog mydialog = new OpenFileDialog();
    mydialog.InitialDirectory = @"c:\";
    mydialog.Filter = "Text Files (*.txt)|*.txt|All
        Files(*.*)|*.*";
    if(mydialog.ShowDialog()==DialogResult.OK)
    {
        strData = File.ReadAllLines(mydialog.FileName);
    }
    lbData.DataSource = strData;
}
```

Then, the Compute button will compute the result and show the result in Tab 2:

```
private void btnCompute_Click(object sender, EventArgs
    e)
{
    double[] dblData = new double[strData.Length];
    double sum = 0;
    for (int i=0;i<strData.Length;i++)
    {
        dblData[i] = Convert.ToDouble(strData[i]);
    }
    Universe myTool = new Universe();
    List lstData = dblData.ToList();
    //change array to list
    double variance = myTool.FindVar(lstData);
    txtResult.Text = "Variance:" + variance + "\r\n";
    tabControl1.SelectedTab = tabPage2; //show Tab 2
}
```

In the above code, we used ReadAllLines() function of File object to retrieve all the lines in a text file. We can also read the file line by line by using `ReadLine()` method of `StreamReader` object. In this case, a file is first opened, the file location pointer points to the first line, if there is any line, and each time when the `ReadLine()` function is called, the pointer automatically advances to the next line, if it exists. The end of a file is reached if `ReadLine()` retrieves a `null` value. The following is the code to load data into a list and then use the list as the data source of a list view.

```
//create a global list lstData to hold data
   List lstData;

//initialize lstData in Form_Load or form constructor
   lstData = new List();

//load data into lstData when form starts
private void cmdLoad_Click(object sender, System.
   EventArgs e)
{
     lstData.Items.Clear();
     OpenFileDialog myDialog = new OpenFileDialog();
     myDialog.InitialDirectory = @"C:\";
     myDialog.Filter = "text files (*.txt)|*.txt|All
         files (*.*)|*.*";

     if (myDialog.ShowDialog() == DialogResult.OK)
     {
           StreamReader myReader = File.OpenText
               (myDialog.FileName);
           string aLine;
           aLine = myReader.ReadLine();
           while (aLine != null)
           {
                 lstData.Add(Convert.toDouble(aLine.
                     Trim()));
                 aLine = myReader.ReadLine();
           }
           myReader.Close();

           //add lstData to a listbox lbData
           lbData.DataSource = lstData;

     }
}
```

Next, we can call `Universe` object to find variance.

```
private void cmdCompute_Click(object sender, System.
    EventArgs e)
{
    Universe myTool = new Universe();
    double variance = myTool.FindVar(lstData);
    txtResult.Text = "Variance:" + variance;
}
```

Finally, we can call `Universe` object to find descriptive statistics.

```
private void cmdCompute_Click(object sender, System.
    EventArgs e)
{
    Universe myTool = new Universe();
    double min, max, a, v, s;
    int c;
    myTool.FindStatistics(lstData, out c, out min,
        out max, out a, out v, out s);
    txtResult.Text += "Sample Size:\t\t" + c + "\r\n";
    txtResult.Text += "Min:\t\t" + min + "\r\n";
    txtResult.Text += "Max:\t\t" + max + "\r\n";
    txtResult.Text += "Mean:\t\t" + a + "\r\n";
    txtResult.Text += "Variance:\t\t" + v + "\r\n";
    txtResult.Text += "Std Deviation\t: " + s + "\r\n";
}
```

Example 2: Create a text file called users.txt that stores a list of user names and passwords. Then use Visual Studio to create a login form, called frmMain, with two text boxes, called txtUser and txtPassword, and a second blank windows form called frmSecond, to validate users against the file. Of course, you should have an exit button that allows the user to unload the application in case of failure to log in.

```
private void cmdLogin_Click(object sender, System.
    EventArgs e)
{
    StreamReader myStream = File.OpenText(@"c:\
        users.txt");
    string aLine;
    bool found = false;
    aLine = myStream.ReadLine();

    //aLine has two pieces of data to be split into uid and pwd
```

```
string[] userAccount;
  while (aLine != null && found == false)
  {
        userAccount = aLine.Split(',');
        if ((txtUser.Text == userAccount[0]) &&
           (txtPassword.Text == userAccount[1]))
              found = true;
        else
              aLine = myStream.ReadLine();
  }
  myStream.Close();
  if (found == true)
  {
        frmMain.ActiveForm.Hide();
        frmSecond myApp = new frmSecond();
        myApp.Show();
  }
  else
  {
        MessageBox.Show("You provided an invalid
           account. Try again");
  }
}
```

Let us once again use dialogues. Let us modify the exit button to allow the user to make a choice before exiting.

```
private void cmdExit_Click(object sender, System.
   EventArgs e
{
     string myQuestion = "Do you really want to
        exit?";
     string myTitle = "Exit or not?";
     MessageBoxButtons myButtons =
        MessageBoxButtons.YesNo;
     DialogResult myResult;
     myResult = MessageBox.Show(this, myQuestion,
        myTitle, myButtons);
     if (myResult == DialogResult.Yes)
     {
           Application.Exit();
     }
}
```

Programming exercises

1. Create a method to multiply two decimal numbers.
2. Create a method to test whether an integer is even or odd.
3. Create a method that changes weekday number 1–7 into weekday names. For example, 1 → Sunday, 2 → Monday, etc.
4. Create a method that returns the last character when given a string.
5. Create a method that finds whether today is your birthday.
6. Create a method that will return the mean and standard deviation of any three numbers.
7. Create a method that finds the number of a certain character in a text.
8. Create a method that computes the factorial of any integer.
9. Create a method SumToN(int n) such that it computes $1 + 2 + 3 + \ldots + n$.
10. Create the method ToCelsius that converts a temperature in Fahrenheit into one in Celsius. Then create a form with two boxes, respectively, for Celsius and Fahrenheit degrees and one button for the conversion.
11. Create a method to covert numerical grades (such as 94, 89, etc.) into letter grades.
12. Create a method that returns an income tax by giving an income. Assume income tax rate is 15% for income over $45,000, 18% for income over $65,000, and 25% for income over $100,000.
13. Create a method that returns a text in its reversed order (Hint: use x.subString(0,1) to get the first character of x; use x.length to get the number of characters in x).
14. Create a method to test if a number is prime or not; you would need to test if it divisible by any number greater than 1 and less than its square root. For example, to test if 34 is prime, you will test if it is divisible by 2, 3, or any number less than Math.Sqrt(34). If it is, then 34 is not prime. Create a method to test if an integer is prime.
15. Opening a data file to get a list of data and then call the function to get descriptive statistics into a text box.
16. Open a text file to load two-dimensional data where each row has multiple pieces of data separated by commas, and show the data using either ListView or Data Grid View.

Chapter 5

Objects and Classes

Introduction

In the last two chapters, we learned the concept of functions. The key difference between the traditional structured development and the modern object-oriented development lies in how to place functions. In structured development, functions are the basic unit of a program. The functions and the data, inputs to, and outputs from functions to be processed, are separated. In object-oriented development, functions are not the basic units of a program and are not separate from the data they process. Instead, data and functions are packed together into a higher level of program units called *objects*. A program is made of one or more *classes*. A class acts as a template to create objects of the same kind.

From the end user perspective, objects are nothing more than real-world objects such as a person, a car, an order, an account, an event, or a pen that have both attributes (data) and behavior (functions), and a class is simply a group of real-world objects of the same kind. The connection between real-world objects and programming objects justifies the method of object-oriented systems analysis: the very same artifacts, objects, and classes, evolve along the development process — end-users, analysts, and programmers all communicate using the same terminologies but refer to different underpinnings.

This chapter explains the basic concepts of objects and classes, including the end-user's conception, analysts' abstraction, and the programmer's creation. We will use the understanding of processes and functions to understand these concepts. We will also discuss a few principles on how to allocate functions into objects.

Programming Objects

From the programmer's viewpoint, an *object* is a memory block that stores both data and function code readily available to be executed. That is, an object encapsulates both data and functions. In this sense, an object is nothing but a variable. But unlike primitive variables that store an integer or Boolean number, an object may store many different pieces of data and the executable code of functions. Figure 1 is a schematic picture of a dog object and a person object. Most of the time, we cannot see objects because they live inside the computer memory. However, some special programming objects are observable. Examples include a running screen, a control like command buttons, or an instance of running system.

What data does an object possess and what functions can it execute? These are specified or defined by a program module called a **class**. A class is like a template that a programmer codes using a programming language such as C++, C#, or Java at the design time, and then it can be used for

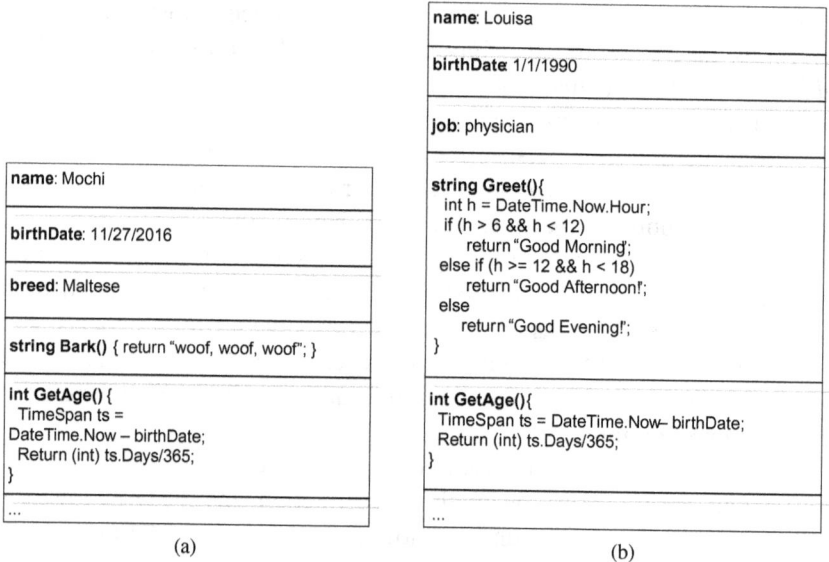

```
name: Mochi

birthDate: 11/27/2016

breed: Maltese

string Bark() { return "woof, woof, woof"; }

int GetAge() {
    TimeSpan ts =
    DateTime.Now – birthDate;
    Return (int) ts.Days/365;
}

...
```
(a)

```
name: Louisa

birthDate: 1/1/1990

job: physician

string Greet(){
    int h = DateTime.Now.Hour;
    if (h > 6 && h < 12)
        return "Good Morning";
    else if (h >= 12 && h < 18)
        return "Good Afternoon!";
    else
        return "Good Evening!";
}

int GetAge(){
    TimeSpan ts = DateTime.Now– birthDate;
    Return (int) ts.Days/365;
}

...
```
(b)

Figure 1.   Illustration of programming objects: (a) a dog and (b) a person.

creating objects that execute functions at the runtime. A class and its objects are analogous to a cookie cutter and cookies: how each cookie looks like is determined by the cookie cutter. How each object looks like and behaves is determined by the class.

A class is a basic reusable program unit in an object-oriented program, and a computer program is made of one or more classes. Each class is made of a list of data items (variables) and behavior items (functions). The following shows the `Dog` and `Person` classes used to create the sample objects in Figure 1. For example, the code defines that the `Dog` class has three variables: `name`, `birthDate`, and `breed`, and two functions: `Bark()` and `GetAge()`. A variables or function inside a class may also be called a member: a variable is a *data member*, and a function is a *behavior member*. Note that data members are also called instance or global variables, as opposed to those local variables declared inside a function.

```
public class Dog
{
        string name;
        DateTime birthDate;
        string breed;

        string Bark()
        {
                return "woof, woof, woof";
        }


        int GetAge()
        {
                TimeSpan ts = DateTime.Now - birthDate;
                return (int) ts.Days/365;
        }
}

public class Person
{
        string name;
        DateTime birthDate;
        string job;

        string Greet()
        {
```

```
              int h = DateTime.Now.Hour;
              if (h > 6 && h < 12)
              {
                      return "Good Morning";
              }
              else if (h >=12 && h < 18)
              {
                      return "Good Afternoon";
              }
              else
              {
                      return "Good Evening";
              }
      }

      int GetAge()
      {
              TimeSpan ts = DateTime.Now - birthDate;
              return (int) ts.Days/365;
      }

      Void ChangeJob(string newJobTitle)
      {
              job = newJobTitle;
      }
}
```

A class may contain a special kind of functions called **constructors**. Like any function, a constructor may have parameters and can be overloaded, i.e., multiple constructors can have the same name if they have different *signatures*. However, it has two differences from other functions. First, a constructor takes the same name as the class. Second, a constructor is used to create new objects rather than to perform data processing tasks, and so it does not return any values. Thus, there is no void, int, string, etc., in front of a constructor's signature. Here are two example constructors for the Dog class.

```
Dog()
{
      birthDate = DateTime.Now;
      name = "Mochi";
}

Dog(DateTime aDate, string aName)
```

```
    {
            birthDate = aDate;
            name = aName;
    }
```

The first constructor will create a default dog with a birth date as the current system time and "Mochi" as the name. To create a new dog object with this constructor, the code is as follows:

```
Dog aDog = new Dog();
```

Here, the left-hand side declares variable aDog as of Dog object, and the right-hand side really uses the constructor to create aDog object by calling the default constructor. After its creation, aDog will take "Mochi" as the name and the current system time as the birth date. The object can perform any of the two functions as the Dog class defined. For example, through the following code:

```
string greeting = aDog.Bark();
```

object aDog will greet a visitor, and through

```
int age = aDog.GetAge();
```

object aDog will tell his or her age.

The second constructor will allow one to create a dog object with a specific birth date and a specific name. For example, the following code will create bDog with "January 3, 2009" as the birth date and "Cloudy" as the name.

```
Dog bDog = new Dog(Convert.ToDateTime("January 3,
    2009"), "Cloudy");
```

After bDog is created, it will have two pieces of data — a name and a birth date — and, of course, it can execute any functions as the aDog object.

Data flow reduction

In a class, variables are ordinary variables, and functions are ordinary functions that we see in structured programs. A class simply packs them into a higher-level program module. There are, of course, differences, or there will be no object-oriented advantages. The most significant difference is that when a function is packed inside a class, it does not need a parameter when the class already has the data member or variable. For example, does `GetAge()` not need to know the birthday in order to be performed? Yes, it does. However, each dog knows his or her birth date: the class `Dog` has `birthDate` variable, and it will reserve a sub-memory slot to store a birth date value for each dog object created. Therefore, when a dog object executes `GetAge()` function, it does not need a birthday as an input. In contrast, to rename a job title for any `Person` object, we will need to tell what the new job title is because the object does not know the new title. Yet in this case, the function does not need to output any data because the function is to simply change the exiting data stored in the memory sub-slot for the `job` variable. Data flow reduction is one of the four important advantages of object-orientation: *A functions does not need a parameter if the parameter value is known by the object or provided by a variable in the class.* Other advantages are *encapsulations, inheritance,* and *polymorphism.*

Accessibility scope

Since variables and functions are located inside a class, they belong to the class. Privatization gives rise to ownership and so creates the issue of who can access what. Of course, each object can access its own data members and execute its own functions. What about other objects? Do they have the same privilege? As a matter of fact, this can be specified by adding a keyword — either public or private — called *accessibility scope,* in front each variable or function. Here, *public* means that all objects have access to a variable or a function whereas *private* means only the owner object has access.

Two additional scopes are *protected* and *package,* which we will use in later chapters. Here, *protected* means only the owner or its children (see Chapter 6 for the concept of children or subclasses) have access to the variable or function, and *package* means all the objects created by the classes inside the same package or project as the owner object have access.

How do we decide whether a variable or function should be declared as public or private? The encapsulation principle, one of the three pillars in object-oriented development, recommends *keeping all variables private or protected.* If there is a need to access an object's data, we should do so through functions. The rationale is to improve the manageability of code change: if a class adds or removes a variable, or changes a variable's data type or format, the change will create a ripple effect to all the places that access the variable. However, if we restrict its access to one class, then the code change will be restricted to the class locally. Ripple effects occur in structured development (see Chapter 17). The same problem will re-occur in object-oriented programs if data members are not private.

Real-World Objects

From the end-user's viewpoint, an *object* is any real-world entity, real or imaginary, tangible or intangible, that has both data and behavior. A *class* is a group of the objects of the same type. For example, in a university, individual students such as "Lisa Johnson" are objects while the group of all students constitutes the Student class. Similarly, courses, professors, events, classrooms, tools, meetings, accounts, transactions, products, tasks, assignments, projects, and enrollments are all business objects, and they can be grouped into respective classes.

There are times people may use different terminologies for classes and objects. For example, a class can be called an *entity set or a type*, and an object can be called an *entity or an instance.*

From the programmer's perspectives, objects are data holders and behavior executors, i.e., objects have both data members and behavioral members. In programming, data members are variables. From the end user's perspective, data members are attributes, properties, or characteristic of an object. Each object can be described by a set of characteristics. For example, we can describe a student by his or her student ID, last name, first name, major, address, etc. We call these properties *attributes* or *fields*. For another example, we can describe a course by a course number, a title, a credit hour, and a description.

From the end user's perspective, a behavioral member is a behavior that the object can perform. For example, a student object should be able to register for a course, an airplane should be able to fly, an account object should be able to credit or debit, and a printer object should be able to print. In this case, "Register", "Fly", "Credit", and "Print" are behavioral members.

Conceptual Objects

The goal of requirements modeling is to create artifacts that will lead to a computer program. Then what is the point of modeling real-world objects like a student sitting in the classroom, a meeting being held in the third floor, or an account in the book? The answer to the question is simple. Because real-world objects are data holders, a model of these objects will capture the data requirements for the system. Then what is the point of modeling functions such as Register, Fly, Credit, and Print? Correctly answering this question entails a mindset transformation for thinking in objects.

Let us first distinguish the following three concepts of objects: real-world object, conceptual object, and programming object. A real-world object is a business data holder and a behavior performer. The real-world object will stay in the real world. The conceptual object is a model of the real-world object that *captures the required data to be processed and func-tions that process the data.* It is the conceptual object that will eventually evolve into and be implemented as a programming object that stores both data and functions capable of processing the data. The following are the links of the related concepts:

Real-world objects → conceptual objects → programming objects

Real-world classes → conceptual classes → programming classes

Note that our conception of conceptual objects includes those of analysis objects and design objects, which some authors attempt to distinguish as the evolutionary differences of the same abstraction in two different stages. Analysis objects is the model of a real-world object with some technical details such as data types and function parameters left out. When evolving into the design stage, the model becomes more refined with all technical details included. Of course, at the design stage, additional objects such as user interfaces and controllers may emerge as design objects in addition to those refined conceptual objects.

As in any models, a conceptual object entails abstraction, i.e., take a perspective to view the real-world object and captures its relevant attributes and related functions or operations.

Capture attributes

As far as which attributes are to be captured, the judgment criterion is rel-evance. A real-world object has a lot of attributes, but *a conceptual object*

only captures a few relevant ones. For example, a real-world student has name, address, phone number, height, weight, blood type, color, genes, etc. Obviously, we do not need all these data for a student registration system. All but name, address, and phone number are possibly irrelevant. On the other hand, for a clinical system, the data on height, weight, and blood type become relevant.

The relevance criterion is easy to apply when we know that a certain real-world object contains a certain list of attributes. However, beginners tend to feel uncertain in the early stage of conceptualizing the objects. For example, should city and state be considered as objects or attributes? Is an account number an attribute of customers or accounts? Here are some of practical guidelines.

Singletons: *If a class has only one attribute, consider it an attribute of another class*. The scenario likely occurs when one takes an object attribute, e.g., city, temperature, weight, and creates its own class. Also related to this case is when one divides one object into two or more objects and ends up with several classes with attributes arbitrarily separated.

Derivatives: *If one attribute can be derived from other attributes, delete it unless there is a need to keep the intermediate computational result*. There is no need to store a value that can be calculated. Calculations, if needed, should be done by functions. Thus, derived values suggest the need for functions.

Transplants: Transplants are mislocated attributes. For example, an account object should not have attributes from its owners, such as customer ID or social security number. A course-offering object should not have attributes for the instructor such as office, phone, etc. Watch out for attributes like xNumber, xCode, or xId — they are often class X's attributes.

Capture functions

Which functions are to be captured? This is a big problem, and later chapters of this book on use case modeling will scientifically address this problem. For now, we just need to know one basic principle: *the functions to be captured must support business processes.*

The attributes in a conceptual object are a subset of the data items of the real-world counterparts. However, the functions of a conceptual object may not correspond to the behavior performed by the real-world counterparts. This is the very reason one school of thought proposes focusing on object data rather than functions when modeling business objects. Indeed, when data are captured, the functions to process data tend to be bundled together. We will follow this approach for modeling business objects.

A real-world student can go to class, study, and take tests. They can also eat, sleep, and listen to their iPod. If the system is to simulate students' learning experience, then going to class, studying, and taking tests will be the ones to be captured. However, they will be pointless if we develop management information systems like a student registration system in which functions are used to process object data. Functions to be captured must be the ones that support the mission of a system to be developed, i.e., they are the functions acquired through the process analysis. For example, a registration system needs the functions to check for prerequisites, check for availability, change student majors, update grades, etc. These functions — instead of Learn, Take Tests, or Sleep — should be captured.

Allocating operations to objects is the most difficult task for beginners. In fact, it is essentially the skill that differentiates a beginner from an expert. Our later chapters on collaboration models will deal with this problem extensively. For now, we will present one basic principle: *a function enables an object that processes its data.*

Enabling objects prescribes where each function will be allocated. In a student registration system, who changes student majors or updates grades? In the real world, a real person such as a registrar does it. Which object do we allocate this function to? It is tempting to pack it into the registrar object, but it is wrong to do so on several counts. First, which object holds data members such as a major? It is a student. A registrar object will have to communicate with the student object in order to get the major changed. Is it not simpler to ask the student object to perform the function without the registrar? Second, packing the functions like "change major" into the registrar object will eventually make one or two objects so powerful and perform all the functions, and the remaining objects do nothing but hold data. In the registration system, the powerful objects will be the ones like registrar, advisor, or department chair because they perform all the registration functions in the real world.

To enable an object, *a function that processes data shall be packed into the object that holds most of the data.* For example, since a student object has data on major and grade point averages (GPA), the functions to change a major and update GPA shall be packed into the Student class. Why? Because by doing so we archive the data flow reduction and avoid unnecessary data passing between objects. A student object can perform its functions independently without involving other objects. Therefore, objects become more autonomous, and classes become less coupled.

For another example, in the real world, a manager is responsible for annual salary increases for her employees. The function to update salary needs to be captured. Where do we pack the function: `Employee` or `Manager`? According to the same principle, it should be the `Employee` class because each employee object has the salary as an attribute.

Does this mean that each student can change her own GPA, and each employee can change his or her own salary? No. The objects that have the capabilities are conceptual ones, which will be implemented as a part of the computer system. They are not real-world objects.

A real-world object may be lifeless by itself. However, its conceptual counterpart must still be alive. For example, in the real world, an account does not have life. But the conceptual account object will have capabilities such as handling withdrawals and deposits. It can also answer account-related inquires. For another example, in the real world, an appointment is lifeless and maneuvered by a secretary. In the conceptual counterpart, it is alive, with the ability of creating new appointments and updating existing ones.

Enabling conceptual objects may be also used for discovering functions rather than merely allocating them. If an object has an attribute, what do we typically do with the attribute values? The primitive data activities are CRUD ones, i.e., creating new values, retrieving the values, updating the values, and deleting the values. For example, students have contact data and GPAs, and thus we should have functions to update contacts and GPA. Non-primitive operations may process several attributes or call other operations to fulfill a high-level task. For example, each order item object has quantity, price, and discount attributes, so why do we not create an operation to calculate subtotal?

Autonomous agent heuristics

As a corollary of the principle of enabling conceptual objects, we can stipulate the following important heuristics for object modeling: *Each conceptual object should be an intelligent autonomous agent with capabilities to process its own data.* We can apply the heuristics to analyze a few classic cases.

First, should a driver start a car, or should a car start itself? A car has full knowledge of itself, including its ignition system and engine, how to fire the ignition system, and how to connect ignition to engine. Thus, according to the autonomous agent heuristics, the function should be allocated to a car. In contrast, if we allocate the function into the driver object, the car object would have to pass all the knowledge to the driver object in order to program the function. This is too much to pass! Would it not be easier to define `Start()` in the car object and have the driver merely execute/call the `Start()` function by sending a message to the car object if it needs to drive the car?

A related case arises from a popular riddle among object-oriented developers. I expand and rephrase it like this. On an object-oriented farm lives an object-oriented farmer who raises object-oriented babies, along with many object-oriented cows that produce object-oriented milk for the babies. There are many functions these objects must perform. Let us consider two of them: Get Milk and Feed Milk. Should a cow be able to unmilk itself, should milk be able to uncow itself, or should a farmer squeeze milk from a cow? Should a child drink the milk or should a farmer feed the milk?

Since a cow has full knowledge of its milk, it is easy to program how to produce milk if the operation is inside the cow. Thus, a cow should have the unmilk operation to be autonomous. In contrast, a farmer must have full knowledge of a cow, including where the milk is stored, in order to program SqueezeMilk operation if we pack the SqueezeMilk function into the farmer object. Asking a milk object to unmilk is also too much: the milk object needs to know a lot about the cow object such as where the milk is stored and by which route the milk is supposed to leave the cow's body. Similarly, a baby knows his- or herself and can easily program how to drink milk. In contrast, for a farmer to feed milk, he or she must get hold of milk first and know the baby's internal organs very well, including where the mouth, the throat, and the lungs are located, and how to operate those organs. Thus, it is a baby's responsibility to drink milk, not a farmer's responsibility to feed the milk. Of course, a farmer knows how to raise a baby, and when it is time for the baby to eat, he or she can be lazy and send a message to the baby and ask the baby to drink milk.

Representing Conceptual Objects

We use a rectangular box with three compartments as the model of a class. In the first compartment, we use a single noun with the first letter in caps

Dog	Person	Account	Student
birthDate:DateTime breed:string name:string	birthDate:DateTime job:string name:string	acctNo:int balance:double openDate:DateTime	admitDate:DateTime sid:int firstName:string lastName:string gpa:double credits:double
Bark():string GetAge():int	GetAge():int Greet():string	Credit(amt:double):void Debit(amt:double):void GetBalace():double	GetGPA():double

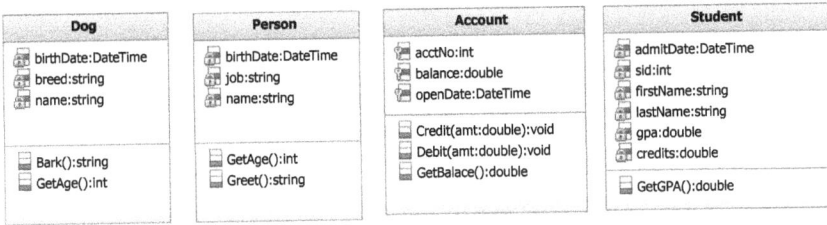

Figure 2.   Illustration of conceptual classes.

to label the respective class name. We list all the attributes in the second compartment. Each attribute has a name along with its data type such as int, string, decimal, etc. and accessibility scope as symbolized by – for private, + for public, # for protected, and ~ for package. Figure 2 shows four conceptual classes, Dog, Person, Account, and Student, along with a few sample attributes and operations, drawn using Rhapsody, which uses a lock icon for the private scope, a key icon for the protected scope, and window icon for the public scope.

To create classes in IBM Rhapsody, right click with the mouse on a package and choose the popup menu Add New → Diagrams → Class Diagram. Then two windows will be open (see Figure 3): the main window with a blank canvas for us to draw a class diagram and a side window to list the diagram tools. To create a class, drop the class icon in the toolbox to the canvas and give a class name. Then right click on the class and choose the popup menu Add New → Attributes or Add New → Operations to add attributes and operations. After an attribute or operation is added, double click on its name to change its name or data type. To change its accessibility scope, double click on the attribute or operation icon to open its features window. By default, all attributes and operations are set with public accessibility. This violates the encapsulation principle, and so we need to change the scope for each attribute by opening its features window. To remove an attribute or operation, double click on the box for a class to open the features window of the class, which has tabs for adding or deleting attributes and operations.

Attributes

Attributes are expressed in the attributeName:DataType format in class diagrams. Primitive data types such as int, double, char, bool, and

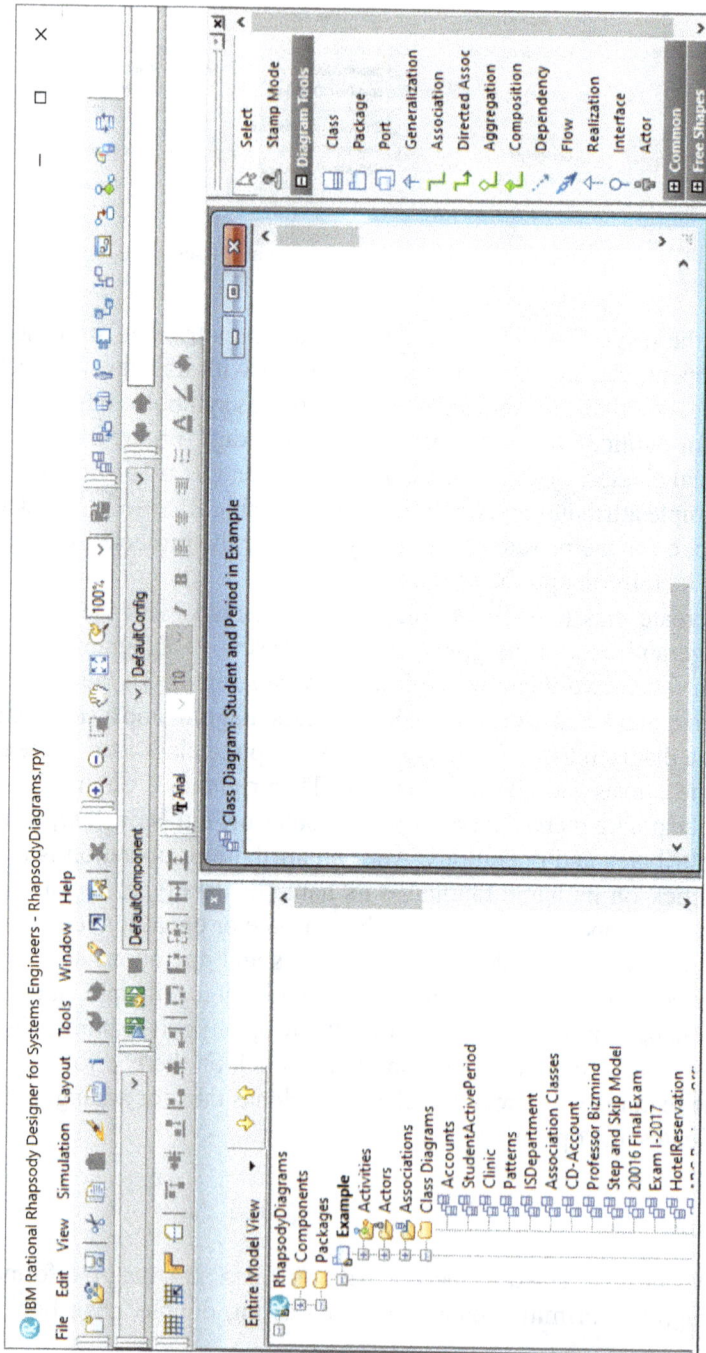

Figure 3.   Create classes in Rhapsody.

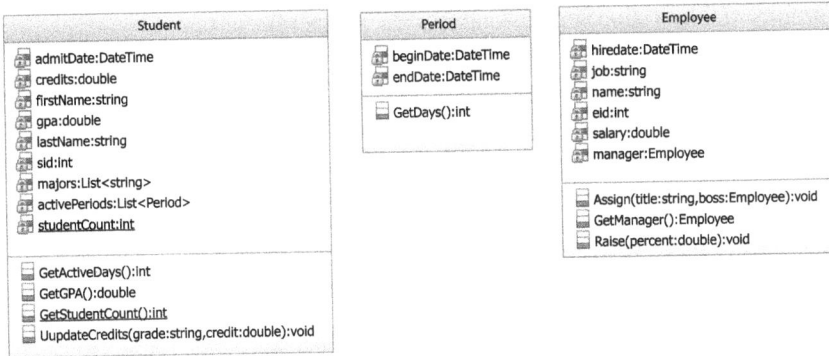

Figure 4.   Attributes with custom types.

string are available in most programming languages. Complex types are language dependent. For example, in C#, `DateTime` type is available while in Java, `Date` and `Time` type are used. For multi-valued attributes, we can use an array or a list as the data type. For example, a student object may have multiple majors. We can use `List<string>` as the data type for attribute `majors` (see Figure 4). If a type is not available or we want to be language independent, we may create custom types, which are classes themselves. For example, if we need to define an attribute called `activePeriods` for the `Student` class to specify the time periods a student is enrolled, you can create a class called `Period` with two attributes `beginDate` and `endDate`, and then use the type to define attribute `activePeriods` as of type `List<Period>` (see Figure 4).

Figure 4 shows the `Employee` class also with a custom type. In this example, employee objects have `eid`, `name`, `job`, `hireDate`, `salary`, and `manager` as attributes, and `Raise`, `GetManager`, and `Assign` as functions. Note that the data type for the manager attribute is `Employee`, the class in question. Operation `GetManager` is used to find the manager of an employee, and `Assign` is used to put an employee to a new job under a new boss.

Operations

Operations are expressed in the

```
FunctionName(Parameter 1: Data Type, Parameter 2:
    Data Type, …): Return Data Type
```

format in class diagrams. That is, each operation has a name along with a list of possible parameters and corresponding data types, and the data type of a return value such as void, int, decimal, string, etc. It also has a symbol for accessibility. Note the difference on how to express operations in class diagrams and how to write function heads. For example, the function "decimal ComputeMax (decimal a, decimal b)" will be written as ComputeMax(a : decimal, b : decimal) : decimal. The correspondence is shown in Figure 5.

As discussed in Chapter 4, a function has two parts: *operation* that defines what objects can do, and *method* that specifies a procedure on how the operation is carried logically and procedurally. The operations are listed in the third compartment, but the method will be documented separately using a procedural model such as an activity diagram.

Static attributes and operations

Typically, an attribute is about individual objects, and each object has a specific measure of the attribute value. For example, one student may have a name as John and another as Lisa. However, there are occasions we need to capture the attributes that are invariant across individual objects; they are attributes of the entire group of objects or class. For example, school name, average GPA, and the total number of students are not an individual student's attributes but of the Student class or group. We symbolize a class-level attribute, or *static attribute*, by underlining the attribute name as studentCount in Figure 4. The programmer will add modifier "static" to symbolize a class variable as static:

```
static private int studentCount;
```

By the same token, a function is usually the behavior of an object, i.e., only the object can execute the function. However, there are cases when

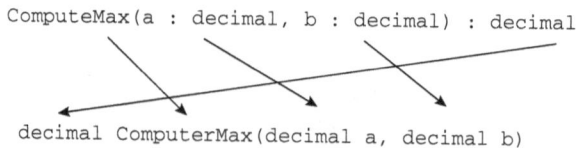

Figure 5.   Notation of operations in class diagrams.

a function is executed by the class. A programmer marks such a function as "*static*" as for a variable. In class diagrams, we underline class-level or static functions. For example, in the `Student` class, some examples of static functions include `GetStudentCount`, `UpdateAverageGPA`, etc. In Figure 4, we showed `GetStudentCount()` as an example static function. Note that a constructor, used for creating new objects, is by nature a static function even though it is not labeled so. In requirements modeling, however, there is seldom a need for documenting constructors. CASE tools such as IBM Rhapsody may be configured to create constructors automatically.

To specify class-level attributes and operations as static ones in Rhapsody, double click on the attribute or operation icon to open its features window.

Implementing Conceptual Objects

Real-world objects eventually evolve into programming objects. First, the real-world objects are abstracted into conceptual objects. We call this step *objects modeling*, in which relevant data are captured and functions obtained through business process analysis or use case descriptions are packed into the objects. The second step is to convert conceptual objects into programming ones. We call this step *implementation*. Implementation entails the choice of programming language and sometimes the platform on which the code will be deployed. In this section, we will use C# to code all the conceptual classes illustrated in the chapter.

The code for both `Dog` and `Person` classes has already been shown earlier in the chapter. The following is the code for the `Account` class in Figure 2. Note that each conceptual class maps into a programming class, each attribute is mapped into a variable declaration, and each operation is mapped into a function declaration.

```
public class Account
{
    private int acctNo;
    private double balance;
    private DateTime openDate;

    public void Credit(decimal amt)
    {
        //
    }
```

```
public void Debit(decimal amt)
{
    //
}

public double GetBalance()
{
    //
}
}
```

The body of each function is empty at this stage and will be filled in by implementing the respective procedure model as seen in last few chapters. Here, the GetBalance() function simply returns the balance that each account object has, and the Credit() and Debit() functions simply increment and decrement an amount from the balance. Thus, these functions can be implemented without any procedural model per the spirit of the agile development.

```
public class Account
{
    private int acctNo;
    private double balance;
    private DateTime openDate;

    public void Credit(double amt)
    {
        balance = balance + amt;
    }

    public void Debit(double amt)
    {
        balance = balance - amt;
    }

    public double GetBalance()
    {
        return balance;
    }
}
```

The Employee class in Figure 4 can be implemented similarly. The reader just notices that attribute manager is declared as a variable of

`Employee`; it is perfectly fine to use a class as a custom type to define attributes or operations inside the same class. Operation `Assign()` assigns a job title and a manager to an employee, which leads to a change of the values in the sub-memory slots corresponding to the `job` and `manager` variables of the employee object, and `GetManager()` returns whatever value is currently inside the sub-memory slot of the `manager` variable. Operation `Raise()` is used to raise an existing salary by a percent and to change the value in the sub-memory slot corresponding to `salary` variable of the employee object. These operations may be implemented as follows without procedural models:

```
public class Employee
{
    private int eid;
    private string name, job;
    private DateTime hireDate;
    private double salary;
    private Employee manager;

    public void Raise(double percent)
    {
        salary = salary + salary * percent;
    }

    public void Assign(string title, Employee boss)
    {
        job = title;
        manager = boss;
    }

    public Employee GetManager()
    {
        return manager;
    }
}
```

To implement the `Student` class in Figure 4, we need to code the `Period` class first. The operation `GetDays()` finds the number of days between `beginDate` and `endDate` of each period object. The following implementation sends a message to a built-in `TimeSpan` object, which finds the days as a decimal value.

```
public class Period
{
        private DateTime beginDate;
        private DateTime endDate;

        public int GetDays()
        {
            TimeSpan ts = endDate - beginDate;
            return (int) ts.Days;
        }
}
```

The following code is the complete implementation of the Student class in Figure 4. First, notice that we combined the declaration of two variables, lastName and firstName, because they are all the same type. Next, notice that studentCount is declared as a static variable and majors as a list. Finally, notice that, in order to compute the number of days a student is active in each active period, we delegate the job to a Period object because it is specialized in dealing with time periods, and then we use a for-loop to add the days of all the active periods.

```
public class Student
{
        private int sid;
        private string lastName, firstName;
        private List(<string> majors;
        private double credits;
        private double gpa;
        static private int studentCount;
        private List<Period> activePeriods;

        public static int GetStudentCount()
        {
                return stuCount;
        }

        public int GetActiveDays()
        {
                int days = 0;
                foreach (Period p in activePeriods) {
                        days = days + p.GetDays();
                }
                return days;
        }
}
```

```
    public void UpdateCredit(string grade, double
      credit)
    {
          //
    }
}
```

Operation `UpdateCredits()` is used to update total credits and GPA for a student when a new course is completed. It takes the credit hour and the grade of the finished course and changes the values of variables `gpa` and `credits`. This operation is not a simple one and deserves a procedural model delineating its logic. First, we need a solution strategy or algorithm. One such strategy is to compute the total number of points by multiplying existing GPA with existing total credits and adding the new points from the newly finished course. The new total points divided by the new total credits will be the new GPA. Second, we need a procedural model to implement the algorithm. An activity diagram or structured English, as we will use in describing use cases in later chapters, will help in this. The following is the updated code of the `Student` class, in which the boldfaced portion implements the above solution strategy to update GPA and credits:

```
public class Student
{
        private int sid;
        private string lastName, firstName;
        private List(<string> majors;
        private double credits;
        private double gpa;
        static private int studentCount;
        private List<Period> activePeriods;

        public static int GetStudentCount()
        {
        return stuCount;
        }

        public int GetActiveDays()
        {
                int days = 0;
                foreach (Period p in activePeriods) {
                        days = days + p.GetDays();
                }
                return days;
        }
```

```
public void UpdateCredit(string grade, double
    credit)
{
    double totalHours = credits + credit;
    double totalPoints = 0;
    double newPoints = 0;
    switch (grade)
    {
        case "A":
            newPoints = 4;
            break;
        case "B":
            newPoints = 3;
            break;
        case "C":
            newPoints = 2;
            break;
        case "D":
            newPoints = 1;
            break;
        case "F":
            newPoints = 0;
            break;
        default:
            throw new Exception("invalid
                grade");
    }
    totalPoints = gpa * credits + newPoints;
    credits = totalHours;
    gpa = totalPoints / totalHours;
}
}
```

Review Questions

1. What are the three types of objects and how are they related?
2. What are criteria to decide which functions are allocated into an object?
3. Give a few examples of real-world objects and list their data attributes and behaviors.
4. Where do you pack the function?
   a. A shelf object "shelfs" a book or a book "stands" on a shelf?
   b. A mower "cuts" grass or grass "shrinks"?
   c. A registrar "enrolls" a student, a student "registers" for a class, or a class "enlists" a student?

d. A manager "receives" a shipment or a shipment "updates" its status?

e. A pilot "lifts" an airplane or the airplane "rises"?

5. List three example conceptual objects for each of the following systems:
   a. Inventory system
   b. Online order system
   c. Airline reservation system
   d. Medical record system

6. Think of three examples of static attributes and static operations in any conceptual objects.

7. In the real world, a telephone object has attributes such as phone number and on or off status and functions such as dial, turn on, turn off, hang up, etc. Think of two situations or systems in which these data functions are captured.

8. What are the basic members in a class?

9. What is encapsulation? Should a data member (variable) be declared as public or private? Why?

10. True or False:
    a. An object has both data and behaviors. A class is a template that defines the data and behaviors.
    b. A class cannot have more than one constructor.
    c. All classes are used to create new objects.
    d. A property is a data member.

11. What are the three types of behaviors an object can have?

12. A method is declared as private in a class. Then who can invoke the method?

13. Who can invoke a public method?

14. What is the difference between protected and public?

15. Can a protected method or property be accessible to children, grand-children, and great-grandchildren classes?

16. What is overloading?

17. Can a property read and change the value of a private variable?

18. Does a property have to be tied to a data member?

Exercises

1. Use a CASE tool to create following classes and then use C# or Java to implement the classes:
   a. Order (data: oid, odate, oamount; behavior: changeAmount, computeTax).

b. Product (data: sku, desc, price, qty; behavior: changeQTY, computeValue).

c. Author (data: id, name, street_address, city, state, zip, country, phone, email).

d. Publisher (data: companyName, contact, address, phone, email).

e. Article (data: title, authors, page_ranage, journal, publicationDate).

f. Book (data: ISBN, authors, title, publisher, price, publicationDate).

g. Proceedings (data: ISBN, editors, title, publisher, price, publcationDate, ConferenceTitle, ConferenceDate, ConferenceLocation, articles).

h. Appointment (data: time, reason, location, participants).

2. Create the following classes using C# and make at least five attributes for each:

a. Flight

b. Course

c. Shipment

d. Customer

e. Apartment

3. Create Account class with appropriate attributes such as acctNo, balance, and status (indicating whether an account is flagged or not), then crate an operation to allow withdrawal of money from an account. Note that if the account is flagged or if the withdrawal is above the balance, the transaction should be denied. Use an activity diagram to describe the procedure used for implementing the withdrawal operation.

4. A classroom is a real-world object. It has data like room number, size, equipment, etc. It has functions such as Open, Close, Turn Light On, Dim Light, House Meetings, etc. Then create a conceptual object of a classroom for the student registration system. What data and functions should you capture?

5. A student club wants to have a database to manage its data on members. The club assigns members to its numerous committees. It is possible that one member can serve in more than one committee and will chair at most one committee. Each year, the club organizes many events. Each time, the club assigns one committee to oversee an event. List the real-world objects with the data and functions they perform. Then create the model of these objects.

6. Create User class with data members like username, password, status, and userRole and an operation Login(). Does the operation need parameters username and password? Why? Use a procedure model to describe the procedures necessary to perform an operation Login().

Basically, the user must enter a user ID and a password. The system will have to check whether the information matches one of the account records to allow the user to enter the system. The account is locked if the user tries over three times unsuccessfully.

7. Use Java or C# to create Inventory class as follows:
   a. Data members: productID, locationID, qty, minQty, maxQty.
   b. Behavioral members: checkout(int amount), checkin(int amount), isFull() — determine if the inventory is full, isTooLow() is to determine the inventory level is too low.

8. Use Java or C# to create Cat class as follows:
   a. Data members: name, breed, birthdate, weight.
   b. Behavioral members: eat (double amountOfFood), findAge(), exercise(int minutes) (assume a cat can reduce 0.1 grams of weight per minute of exercise but cannot exercise more than 60 minutes at a time without taking 12 hours to recover).

9. Use Java or C# to create Classroom class as follows:
   a. Data members: building, room, numberOfSeats, phone.
   b. Behavioral members: getName() — return something like "103C Simens Hall", getSeats().

10. Use Java or C# to create a balloon class as follows:
    a. Data members: x and y coordinates, size, color.
    b. Behavioral members: moveHorizontal(int delta), moveVertical(int delta), expand(int delta), show(Graphics paper), pop() — balloon will destroy itself.

11. Use Java or C# to create a rectangle class as follows:
    a. Data members: coordinates for the top left corner, width, length.
    b. Behavioral members: getPerimeters(), getArea(), move(int xChange, int yChange), resize(int widthChange, int lengthChange).

12. Use Java or C# to create a triangle class as follows:
    a. Data members: vertexA, vertexB, vertexC (use Point structure).
    b. Behavior members: getPerimeters (the total length of three size), isIsosceles — check if the triangle is isosceles, isTriangle() to check if the sum of any two sides is greater than the other side.

13. Use Java or C# to create Order class as follows:
    a. Data members: oid, orderDate, promisedDate, shipDate, deliveryDate, amount, status.
    b. Behavioral members: isOnTime() — check if the delvieryDate is ahead of promisedDate, timeToDelivery() — get number of days from order to delivery, getLateFees() — company pays 1% fee to the customer for every day of delay in delivery according to the contract.

Chapter 6

Class Diagrams

Introduction

In the previous chapter, we learned how to model individual objects, including their data and behavior, using conceptual objects and then implement them into programming objects. Real-world objects do not live in isolation, they are related. For example, a building object contains classroom objects, a student object registers for courses, and instructor objects are a kind of employee objects, which in turn are a kind of person objects. Real-world objects relate to each other through two general types of relationships: *association and inheritance.* Inheritance is a simpler relationship merely symbolizing that one object is a special kind of another object. Inheritance enables a child object to behave like its parent object. Association has a broad scope: it includes contacts, ownerships, assignments, containments, etc. Associations enable the objects to communicate with one other.

A class diagram is a graphical model representing conceptual objects, as groups or classes, and their relationships. Therefore, in class diagramming, we must abstract not only individual objects but also their relationships, which link objects together. In this chapter, we learn how to model these relationships and introduce the concepts of *cardinality (or multiplicity)* and *navigability.* Then we will discuss how the relationships may be implemented into computer programs.

Associations

Real-world objects may participate in relationships with each other through contacts, ownerships, communications, correspondences, containments, assignments, etc., which can be all modeled as *associations*. For examples, a customer places an order, a pilot flies an airplane, a document is printed by a printer, a student takes a course, a customer has an account, a building has a room, a student lives in a dormitory, a professor is assigned to teach a course, a professor advises a student, a transaction is done under an account, etc. All these are represented as associations in class diagrams.

Since we group conceptual objects into classes, we use lines connecting classes to show the associations between individual objects in the classes. For example, to show that a customer takes an order, we use a line to connect `Customer` and `Order` classes. The result is a class diagram (see Figure 1).

Note that the line in the class diagram represents all the associations between individual customers and orders. In the real world, each order is made by one customer, but a customer may take many orders. One simple line will not be able to show these details, but with the concept of cardinality, we can capture a portion of this nature.

Cardinality

To capture the multiplicity of associations, we express the maximum and the minimum numbers in an association line. The maximum cardinality expresses the maximum number of associations that one object at each end can possibly participate in. Typically, we do not need to express the exact

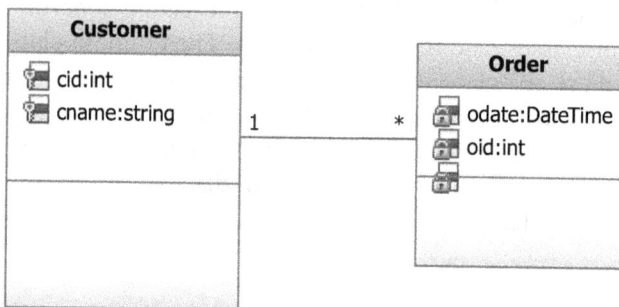

Figure 1.   Association between customer and order.

magnitude of the maximum cardinality. Instead, we just need to express whether the maximum number is one or multiple using symbols 1 or *, respectively, because in computer programming, the maximum cardinalities of two or more will be treated equally using a list or array.

The minimum cardinality expresses the minimum number of associations that one object at each end can possibly participate in. Again, we typically do not specify the exact magnitude. Instead, we want to know whether the minimum cardinality is 0 or 1; in computer programming, the minimum cardinality 0 or 1 means whether a *foreign data member* can be optional or not, i.e., has missing values or not.

We use a symbol consisting of the minimum and maximum cardinalities separated by two dots to show the range of cardinality. For example, if a customer may take zero or more orders, then the cardinality shown at the `Order` side of the relationship will be 0..*, which may be shortened as *. On the other hand, if each order is made by one and only one customer, we express it as 1..1, which may be shortened as 1 (see Figure 1). In general, the range of cardinalities is shown as *m..n*, with *m* being the minimum and *n* the maximum cardinality, but not all CASE tools support the general range notation.

Navigability

The second characteristic of an association is navigability. The notion of cardinalities is the same as that in data modeling, but the notion of navigability is unique to class diagramming. Both data modeling and class diagramming are concerned with navigability, but they have different concerns. In data modeling, navigability is concerned with whether, given the knowledge of one record, users can query the related other records. The navigability is always bidirectional; if two tables have an association, users can search for the associated records in both directions. For example, given an order, users can find what items are in the order. On the other hand, given an item, the users need to find what orders contain the item.

In class diagraming, navigability is concerned with whether one object can delegate a task to the other one or send messages to the other one for help to perform a task. The navigability may be unidirectional sometimes. For example, for an order object to compute the total amount of the order, it needs to ask an item object for help to obtain the item price. However, we could not contemplate any need in the opposite direction, i.e., an item object needs help from an order object. Therefore, in class diagrams, we need to explicitly mark whether the navigability of an association is unidirectional or bidirectional.

Besides the need for asking help, navigability may be also reframed to be concerned with whether one object needs to have knowledge of the other one. Here, knowledge means data; by saying that a customer object has knowledge of customer ID, name, and address, we mean the customer object has those data members. When modeling individual objects like orders and customers, the reader very likely makes customer as a data member of Order and orders as a data member for Customer. Why does the reader think we need those foreign data members? It is because of our common sense that an order object needs to know who makes the order and a customer needs to know the orders he or she has made. In a class diagram, *associations replace the foreign data members*. For example, a student may have one or more active periods on campus, and thus we created a custom type Period and captured activePeriods as a foreign data member of the Student class in the last chapter. Alternatively, we can use an association to replace activePeriods data member as in Figure 2.

The need for knowledge may be also unidirectional. In the above example, a student object has one or more active periods, but a period object does not have knowledge of any student object. Therefore, the association is drawn as unidirectional, pointing from Student to Period.

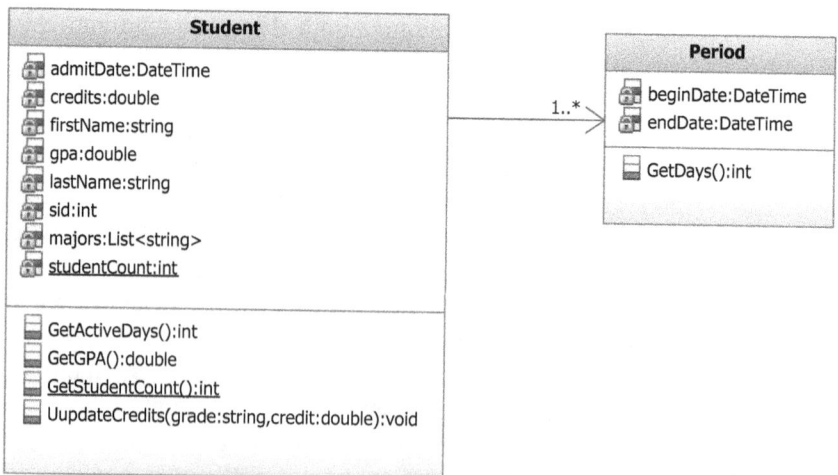

Figure 2.   Unidirectional association to replace data member activePeriods.

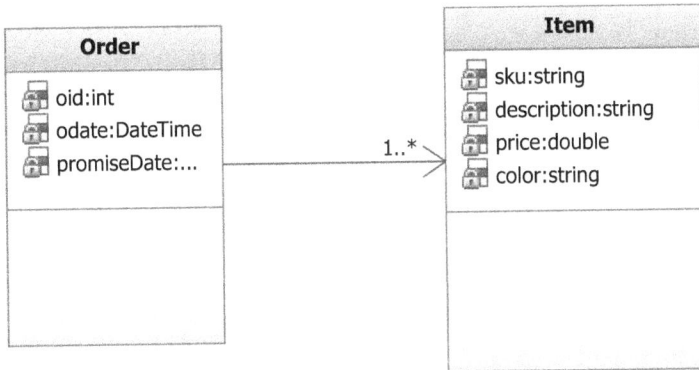

Figure 3. Unidirectional association between `Order` and `Item`.

Similarly, an order object needs to have the knowledge of the items included in the order; when we print an order confirmation, for example, we need to list item descriptions, prices, etc. In contrast, there is hardly a need for an item object to know the orders that contain the item. To print an item catalog, for example, does each item need to refer to the list of orders that contain the item? Therefore, based on the need for knowledge, the association between `Order` and `Item` should be also considered to be unidirectional (see Figure 3).

Needs for knowledge and needs for help are, in fact, equivalent. We can understand the equivalence as follows. In a class diagram, each object is designed to handle one area of concerns: it has a set of native data members, and all functions are designed to be specialized in processing the native data members. If these data members are needed somewhere else, the tasks of processing these data members must be delegated to the host object who owns the native data. For example, an item object is specialized in handling item information such as stock keeping unit (SKU), description, price, color, quantity on hand, etc. If an order needs to acquire or process item information such as price or quantity on hand, it needs to ask an item object for help. Similarly, when a student object needs to compute the total number of active days on campus, it delegates the job to a period object.

In the real world, only when two people know each other can they help each other. This same protocol applies to the programming world: two objects can communicate or send messages to each other only if they know each other. A class diagram models the need for knowledge and need for help between two conceptual objects using associations; two conceptual objects are associated if and only if their programing counterparts can command each other to provide services. When and only when `Customer` and `Order` are associated in a class diagram can a customer object in a computer program call an order object to perform its functions like `GetOrders()`, `ComputeOrderTotal()`, `TrackOrderStatus()`, etc. Similarly, an order object can ask the customer object to perform functions like `UpdateCustomerInfo()`, `NotifyCustomer()`, etc.

By default, an association between two classes is two-way navigable, meaning the objects both know and serve each other. In other words, the navigability is bidirectional when an association line has no direction, for example, as the association between `Customer` and `Order` in Figure 1. In the case in which one object needs to know or use the other but not vice versa, we use an arrow head explicitly at the end of an association to represent unidirectional *navigability*: If object A has an association pointing to object B, it means that A has knowledge about B, or A can send messages to B asking for help, but B does not need to know A or B cannot send messages to A. Figure 2 illustrates the unidirectional association between `Order` and `Item` objects. For another example, a pilot needs to use an airplane and thus needs to ask the airplane to provide service, to take off, for example, but the airplane does not need to know the pilot and does not ask the pilot to provide services (see Figure 4).

To create associations in Rhapsody, select Association in the diagram toolbox (see Figure 3 of Chapter 5) and click on each class box to be connected by the association line. Then double click on the association line to open its features window, which has two tabs, respectively, for the two

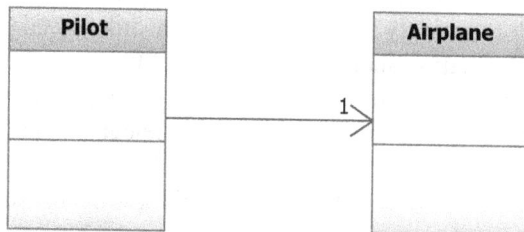

Figure 4.   Unidirectional association between `Pilot` and `Airplane`.

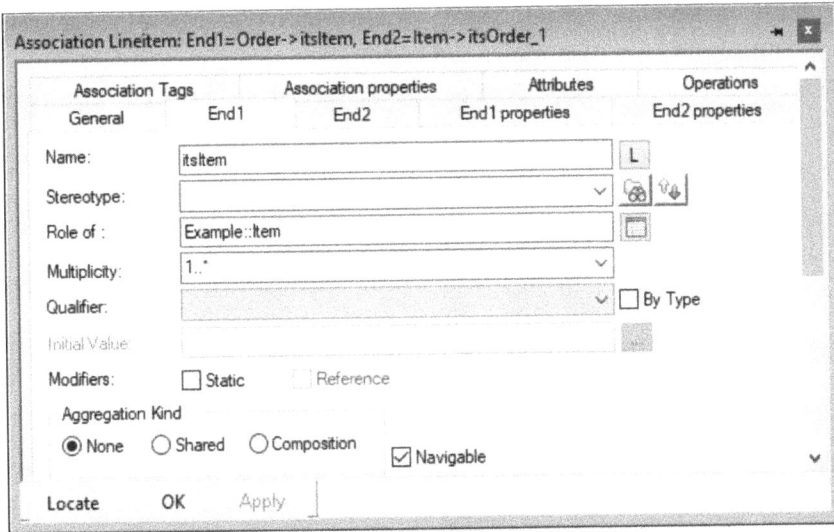

Figure 5. Feature window of associations.

ends to specify the multiplicity and navigability of each end (see Figure 5). If an association is one-way navigable, for example, from Order to Item, we will check the Navigable box at the Item end and uncheck Navigable box at the Order end. Note that we *check the Navigable box at the end we want the direction to point to*. Also, for unidirectional associations, *we only need to specify the cardinality at the end the arrow is pointing to*. It might be more convenient to use Directed Association in the toolbox to draw unidirectional associations, whose features window will have only one end, the one arrow points to, to specify cardinalities.

Inheritance

The second type of relationships among objects is *inheritance*, which may be equivalently called generalization/specialization or supertype/subtype. It is used to represent "is-a-kind" relationships between objects. For examples, a dog is a kind of animal, a car is a kind of vehicle, a square is a kind of shape, a desk is a kind of furniture, and full-time employees and part-time employees are both kinds of employees. In class diagrams, we use

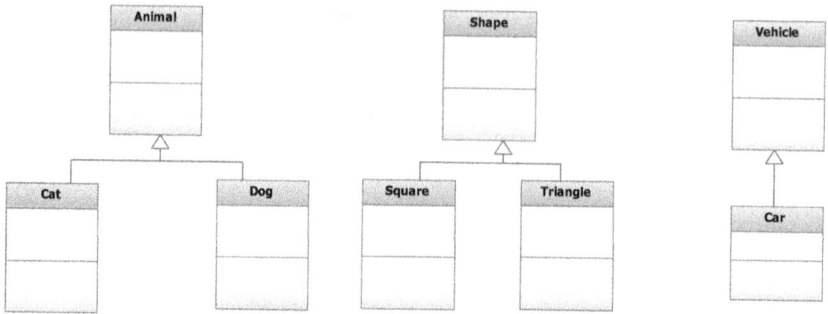

Figure 6.   Example inheritance relationships.

a line with a hollow triangle arrow to model the inheritance relationship. The triangle arrow is pointing from a child (or specialized or subtype) class to the parent (or generalized or supertype) classes, indicating that the child object is a kind of a parent object (see Figure 6 for examples).

Inheritance may be equally considered for a containment relationship between classes, where a superclass contains subclasses. As such, it allows one to model a large group of objects that are identical overall but with minor differences. For example, all employees have attributes like employee ID, name, job, hire date, etc. Yet full timers have salary, but part timers have hourly rates and work hours. To emphasize these minor differences while not duplicating many identical data and behavioral members, we could factor the common attributes and operations out of all employees into a generalized class called Employee and then put the specialized attributes and operations that are only relevant to full timers and part timers into two specialized classes: FullTimer and PartTimer (see Figure 7).

Inheritance is one of four features of object-orientation: it makes objects "rich." Is inheritance not making a lot of people rich? Just imagine if someone has designed and programmed a powerful class that can create objects capable of flying, swimming, shooting enemies, or launching rockets, etc. By merely inheriting from the powerful one, a new class will possess the same capabilities; implicitly, *all child objects inherit all the attributes and operations of a parent object along with the association relationships that the parent object may have with other objects.*

To take advantage of inheritance, there are two enhancements that need to be added to a class diagram. First, inheritance cannot override the accessibility scope; if an attribute or operation is declared as private in the parent class, it will not be accessible by its child — needless to say, it will be inherited. Therefore, to enable inheritance, data and operation to be

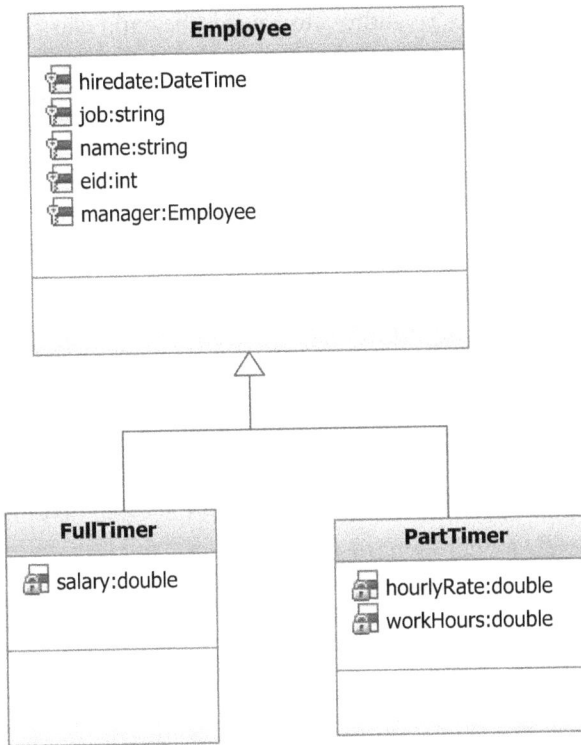

Figure 7.   Employee and its child classes.

inherited cannot be private. The public accessibility scope for data will be too dangerous. In the middle is the *protected* scope. Here, *protected* means only the owner or its child objects have access to the data or operation. Therefore, *all the attributes and private operations to be inherited will be declared as protected,* symbolized by #.

Second, a child object can inherit an operation from its parent object and at the same time may have a choice to modify the method of executing the operation. Note that, from the programming perspective, functions process object data. The child object inherits its parent's data members and, in addition, owns its own special data. Thus, inherited functions may need to process additional data owned by the child object. Thus, a function programmed for the parent class may be insufficient or irrelevant to the child class. For example, Employee class may define function ComputePay(), and PartTimer and FullTimer will inherit these functions. However, the child classes will perform the operation differently; a full timer uses salary, but a part timer uses wage and work hours to obtain the pay amount. To address this need, the child class may choose to inherit an operation (function head) but not how the operation is performed (or method/function body). The child objects can override an inherited function by modifying its inner logic or method. A function that is anticipated to be modified by a child must be declared as `virtual,` and then the child class must re-declare the operation as `override`. The class diagram may not show *virtual* or *override*. Instead, when an operation is shown in both parent and child classes, the qualification of `virtual` in the parent and `override` in the child classes are default.

Extending the second enhancement further, we can even declare a function in the parent class without the function body or method, i.e., the function has only the function head or operation. Such an operation is called abstract. If a class has one or more abstract operations, the class itself becomes abstract. If all operations of a class are *abstract*, and in addition there are no data members in the class, the class is called an *interface*. Here is an example interface:

```
interface IFlyable {
        void Fly();
        void TakeOff();
        void Land();
}
```

An interface essentially defines a set of operations or property signatures as a common protocol for child classes to implement. There are two reasons for abstract operations: (1) an interface or abstract class does not have data for the methods to be defined, and (2) a lead programmer may just want to specify a common operation for all child classes to follow as the common standard for implementation.

In programming, we use the modifier "abstract" in front of an abstract operation or class. The following code declares an abstract function `PrintCheck` without a function body:

```
protected abstract void PrintCheck(DateTime ckDate);
```

In class diagrams, the names of abstract operations and classes are shown in italics. For example, in Figure 8, `Employee` class is abstract, and the class has one abstract operation `PrintCheck()`.

In this example, an employee object cannot define the function body or the method; generic employee objects do not have data on how much

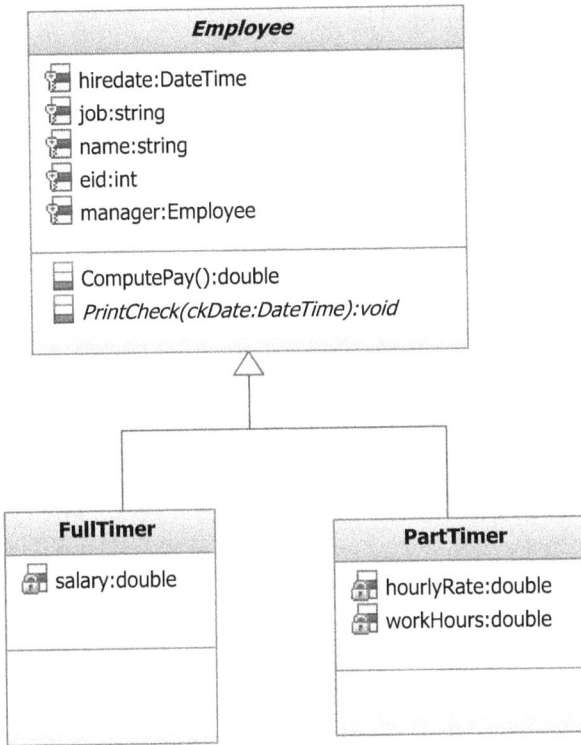

Figure 8. Abstract classes and operations.

needs to be paid. The full timer will have a regular salary to be printed on check along with other benefits, while a part timer has a wage rate and gets paid depending on the number of hours of work. So, we declare `PrintCheck()` as an abstract operation for all employees, and then implement the methods separately for full timers and part timers. When implementing the function for full and part timers, we will declare the operations as *override*, indicating the function to be defined is going to modify the one from generic employee objects.

```
protected override void PrintCheck(DateTime ckDate);
{
        //code for printing checks
}
```

What is good about declaring an abstract operation? The question is related to another basic feature of object-orientation that programmers all love to use: *polymorphism*. The word means that one thing has the ability to morph into many forms in different contexts. In computer programming, by declaring an abstract operation, one programmer sets a standard for all others to follow when implementing the operation. In other words, the operation may have different ways of implementation. When it is the time to use the operation, there is no need to know the differences of how the operation is implemented. A uniform message like `employee.PrintCheck(DateTime.Now)` can be sent to all employee objects, regardless of whether each one is a part timer or a full timer. The compiler will decide which implementation is to be used when printing each check; if the employee is a full timer, the operation implemented in the `FullTimer` class will be used or else the method in the `PartTimer` class is used. Therefore, abstract operations are polymorphic at both the time they are implemented and when they are used.

Implementation

In the previous chapter, we demonstrated how to implement individual conceptual objects, attributes, and operations. Essentially, a conceptual class becomes a programming class, an attribute becomes an instance or global variable, and an operation becomes a function head. Now let us learn how to implement association and inheritance relationships.

Using C# language, inheritance relationships are implemented by appending the parent class name to the child using a colon (:). For example, suppose A is the parent class of B (B ⇢ A), the implementation is:

```
public class A
{
        //
}

public class B : A
{
        //
}
```

Associations are implemented as *foreign instance variables*, but details depend on maximum cardinality and navigability. First, if navigability is bidirectional such as A↔B, then we will implement the association using two foreign instance variables: one is of B in A, and the other is of A in B. However, if the association is unidirectional such as A → B, then we implement it by creating a foreign instance variable of B in A but no variable of A in B. Foreign instance variables may be declared as a single variable or a collection depending on the maximum cardinality. For example, each order is taken by one customer. The foreign instance variable of `Customer` in the `Order` class will be a single variable.

```
public class Order
{
        Private Customer aCustomer;
}
```

On the other hand, each customer can take many orders. The foreign instance variable of `Order` in the `Customer` class will be an array or list.

```
public class Customer
{
        Private List<Order> orders;
}
```

Note that array is one of collection types in C#. It may not be the best collection type because the number of objects contained in the array variable cannot be dynamically changed. Also, it has limitations in terms of the method to retrieve or search for a list item. To avoid these limitations, programmers generally prefer the following collections, such as `List` or `ArrayList`, and `Dictionary` or `HashTable`, in C#:

1. Use `List` if there is a sequence order among the list items or if we want to use an integer sequence number to retrieve a list item;
2. Use `Dictionary` if we want to retrieve or search for an item using an arbitrary key.

Figure 9 is a list of implementation rules based on unidirectional navigability; if it is bidirectional, we will just need to follow the same rule to create an instance variable of A in B.

Note that, in implementing a class diagram, we should *implement all relationships before implementing operations*. Otherwise, we would face difficulty in implementing some operations because of a lack of some

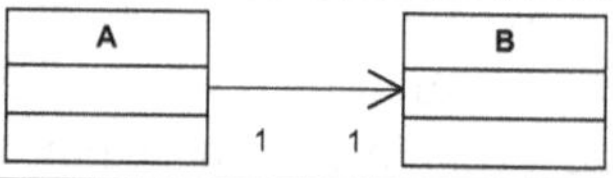	public class A
	{
	Private B b;
	}

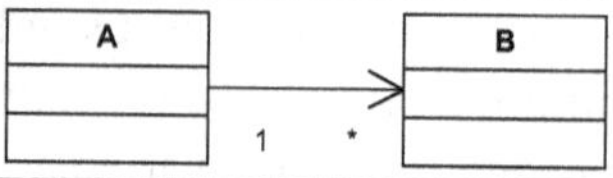	public class A
	{
	Private B[] b;
	}

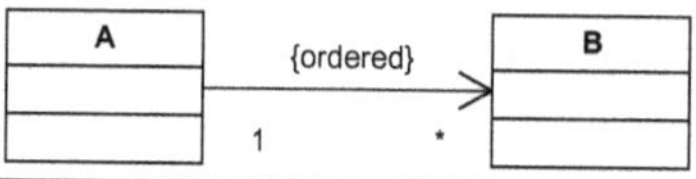	public class A
	{
	Private List<B> b;
	}

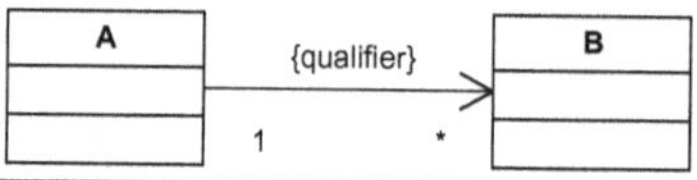	public class A
	{
	Private Dictionary<string, B> b;
	}

Figure 9. List of implementation rules based on unidirectional navigability.

foreign instance variables. The general order of implementing a class diagram is as follows:

1. Implement individual classes and their native attributes. Implement simple operations, especially those that do not require foreign instance variables.
2. Implement association and inheritance relationships. There is no preferential order of implementation between inheritance and association relationships.
3. Implement operations. This step usually involves heavy coding, and the reader may be required to have prior programming experience to perform this step.

Example 1: Figure 10 shows a slightly modified class diagram for employees and their child classes by adding a fringe benefit as an attribute and `Salesman` as a child class of `FullTimer`. Since salary is going to be inherited by `Salesman`, we make it protected.

First, let us implement individual classes. `Employee` has four data members: eid, name, hire date, and fringe benefit, one virtual method to compute payment, and one abstract method to print check. Note that we use keyword "abstract" to indicate that the class is abstract and the operation `PrintCheck` is abstract. Also note that the abstract operation does not have a function body.

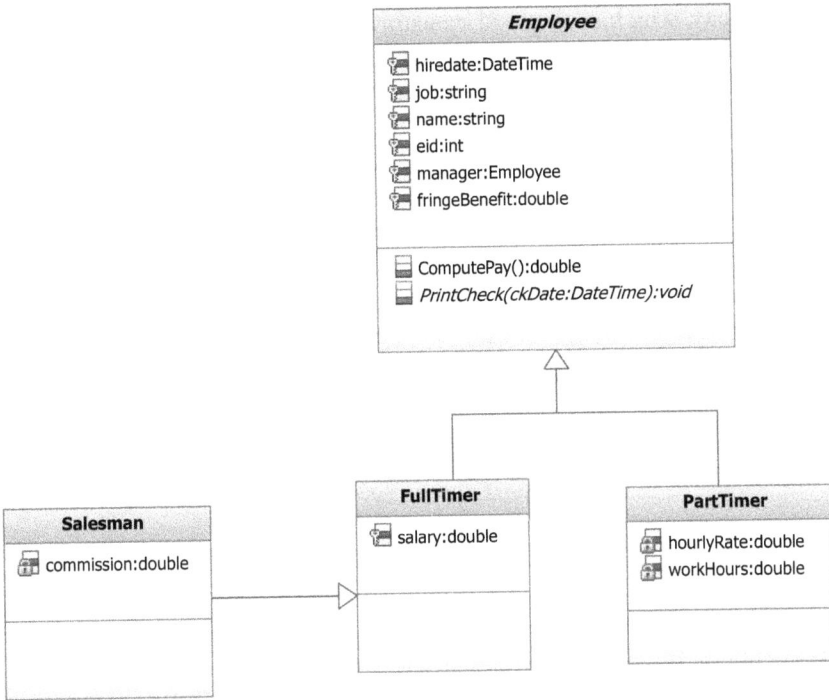

Figure 10. Class diagram for employees.

```
abstract class Employee
{
            protected string eid;
            protected string name;
            protected DateTime hireDate;
            protected double fringeBenefit;

            public virtual double ComputePay()
            {
                    return fringeBenefit;
            }

            public abstract void PrintCheck(DateTime
                chDate);
}
```

The following implementation creates the child class `FullTimer`. The child class overrides `computePay` method by adding salary to the base pay, which is obtained by using `base.ComputePay()`. Note that if a function is created in the parent class, the child class can call a function in the parent by referring the parent as `base`. To invoke function `ComputePay()`, for example, use `base.ComputePay()`. Since `PrintCheck()` is an abstract operation, the child class must provide implementation, or we will get `NoImplementation` exception. The following code uses a message box to show words about what is the total pay and what is the fringe benefit. Of course, we could create a more elaborate display looking like a real check, but we shall not be overly concerned with it for now; it is better to be deferred to the time when we learn how to design and code graphic user interfaces in later chapters.

```
class FullTimer : Employee
{
        protected double salary;

        public override double ComputePay()
        {
                return base.ComputePay() + salary;
        }

        public override void PrintCheck()
        {
                MessageBox.Show("The total salary
                    is " + this.ComputePay() + ", which
                    include a fringe benefit of " + base.
                    ComputePay());
        }
}
```

The following code defines child class `PartTimer`, which has hourly rate and work hours as special data members. It also overrides `ComputePay` method by adding the product of hourly rate and work hours to the fringe benefit as calculated by the parent class.

```
class PartTimer : Employee
{
        private decimal workHours;
        private decimal hourlyRate;

        public override double ComputePay()
        {
```

```
                return base.ComputePay () + workHours *
                    hourlyRate;
        }

        public override void PrintCheck()
        {
                MessageBox.Show("Your fringe benefit
                    is " + base.ComputePay() + " and your
                    total pay is " + this.ComputePay());

        }

}
```

Finally, we implement `Salesman` as a child class of `FullTimer`. To modify the method for `ComputePay`, we add commission to whatever payment is computed by the parent, which in this case is `FullTimer`.

```
class Salesman : FullTimer
{
        private decimal commissions = 0;

        Public override double ComputePay()
        {
                return base.ComputePay() + commission;
        }

}
```

Example 2: Figure 11 extends the `Account` class introduced in the last chapter by including account owners, and three special types of accounts: checking accounts, savings accounts, and certificate deposit (CD) accounts. The association between `Owner` and `Account` is easy to understand; an owner may have one or more accounts, but each account has only one owner. The association between `CheckingAccount` and `SavingsAccount` represents the assumption that each checking account has one or more associated savings accounts.

Assume that each checking account has a quota for the number of checks to be cleared, a CD has a definite yearly interest rate and maturity date, and a savings account has varying interest rates over time, with each rate having a valid period.

There are a few ways to model the interest rates. For example, we could represent interest rates using two multi-valued attributes: one for a list of interest rates and one for a list of valid periods. But the problem

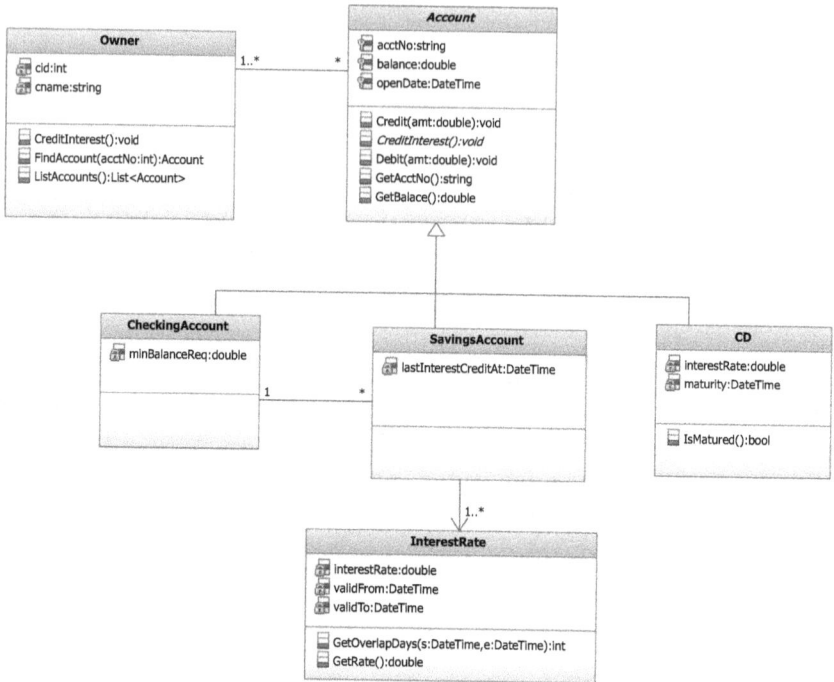

Figure 11.   The class diagram for bank accounts.

with this method is that the correspondence between the two lists must be always maintained in the same order, or interest rates may not correspond to their respective valid periods. Instead of this method, here we create a custom type, called `InterestRate`, that ties each interest rate with a valid beginning date and a valid ending date. Then, we represent the multiple interest rates of a savings account by a unidirectional association from `SavingsAccount` to `InterestRate`, meaning each savings account may have one or more `InterestRate` objects as foreign data members. To compute the interest for a period, we may have to use multiple interest rates because during the period there may be multiple valid rates. Thus, we need to get the number of days for each rate to be applicable and create two operations in `InterestRate` class for that purpose.

A related point to mention is that different types of accounts have different ways to credit interests: a checking account does not accrue interests, a CD account credits interests only at the maturity date, and a savings account credits interests periodically. Thus, we create an abstract operation,

CreditInterest(), in Account class to take advantage of polymorphism; every day, we can send one uniform message to all accounts to credit their interests, but it is up to each account to decide how they will respond to the message according to account type. CreditInterest() can take many forms: a checking account will simply add zero interest to the balance, a CD will only add the total interest to the balance if the day falls on the maturity, and a savings account will add interest if the day is 30 days after the last credit date.

Now let us implement the class diagram. First, we create skeleton classes with their native instance variables and operations. For the operations that can be easily coded, we may just write the code as we see in the following code for Credit, Debit, GetAcctNo, GetBalance, IsMatured, GetRate, etc. For operations that cannot be coded, leave the function body empty for now. Note that the Debit function will deduct the withdrawal amount from the account balance if there is sufficient funds. Otherwise, the function will throw an exception.

```
public class Owner
{
    private string cid;
    private string cname;

    public List<Account> ListAccounts()
    {
        //
    }

    public Account FindAccount(string no)
    {
        //
    }

    public void CreditInterest()
    {
        //
    }
}

public abstract class Account
{
    protected string acctNo;
    protected double balance;
    protected DateTime openDate;
```

```csharp
    public string GetAccountNo()
    {
        return acctNo;
    }

    public double GetBalance()
    {
        return balance;
    }
    public virtual void Debit(double amt)
    {
        if (balance > amt)
            balance = balance - amt;
        else
            throw new Exception("insufficient fund");
    }

    public virtual void Credit(double amt)
    {
        balance = balance + amt;
    }

    public abstract void CreditInterest();
}

public class CD
{
    private double interestRate;
    private DateTime maturity;

    public bool isMatured()
    {
        if (DateTime.Now >= maturity)
            return true;
        else
            return false;
    }
}

public class CheckingAccount
{
    private double minBalanceReq;
}

public class SavingsAccount
```

```
{
    private DateTime lastInterestCreditAt;

}

public class InterestRate
{
    private double interestRate;
    private DateTime validFrom;
    private DateTime validTo;

    public double GetRate()
    {
        return interestRate;
    }

    public int GetOverlapDays(DateTime s, DateTime e)
    {
        //

    }
}
```

Next, let us implement relationships. For the bidirectional association between Owner and Account, we create foreign instance variables (see the two bold lines as follows).

```
public class Owner
{
    private string cid;
    private string cname;
    private List<Account> accounts;

    public List<Account> ListAccounts()
    {
        //

    }

    public Account FindAccount(string no)
    {
        //

    }

    public void CreditInterest()
    {
        //
```

```
        }
    }

    public abstract class Account
    {
        protected string acctNo;
        protected double balance;
        protected DateTime openDate;
        protected List<Owner> owners;

        public string GetAccountNo()
        {
            return acctNo;
        }

        public double GetBalance()
        {
            return balance;
        }
        public virtual void Debit(double amt)
        {
            if (balance > amt)
                balance = balance - amt;
            else
                throw new Exception("insufficient fund");
        }

        public virtual void Credit(double amt)
        {
            balance = balance + amt;
        }

        public abstract void CreditInterest();
    }
```

For the three inheritance relationships, we will code them along with the required implementation of abstract operations (see the boldfaced code).

```
    public class CD : Account
    {
        private double interestRate;
        private DateTime maturity;

        public bool isMatured()
```

```
    {
      if (DateTime.Now >= maturity)
          return true;
      else
          return false;
    }

    public override void CreditInterest()
    {
        //
    }
}

public class CheckingAccount : Account
{
    private double minBalanceReq;
    public override void CreditInterest()
    {
        //
    }
}

public class SavingsAccount : Account
{
    private DateTime lastInterestCreditAt;

    public override void CreditInterest()
    {
        //
    }
}
```

The bidirectional association between CheckingAccount and SavingsAccount is implemented by creating foreign data members (see the boldfaced code).

```
public class CheckingAccount : Account
{
    private double minBalanceReq;
    private List<SavingsAccount> associatedSavings;

    public override void CreditInterest()
    {
        //
    }
```

```
        }

    public class SavingsAccount : Account
    {
        private DateTime lastInterestCreditAt;

        private CheckingAccount associatedChecking;

        public override void CreditInterest()
        {
                //
        }
    }
```

The unidirectional association between `SavingsAccount` and `InterestRate` is implemented by adding a list of `InterestRate` objects as a foreign data member inside `SavingsAccount` (see the boldfaced code).

```
    public class SavingsAccount : Account
    {
        private DateTime lastInterestCreditAt;

        private CheckingAccount associatedChecking;
        private List<InterestRate> rates;

        public override void CreditInterest()
        {
            //
        }
    }
```

Now some operations may become easy to code. All operations in `Owner` class can be coded because of foreign data member `accounts` (see the boldfaced code). Here, `CreditInterest()` simply asks each account in the `accounts` list to perform its `CreditInterest()` operation, and `FindAccount()` will go through each account to find one that has a matching account number.

```
    public class Owner
    {
        private string cid;
        private string cname;
```

```
private List<Account> accounts;

public List<Account> ListAccounts()
{
    return accounts;
}

public Account FindAccount(string no)
{
    foreach (Account a in accounts)
    {
        if (a.GetAccountNo() == no)
            return a;
    }
    throw new Exception("no account found");
}

public void CreditInterest()
{
    foreach (Account a in accounts)
        a.CreditInterest();
}
}
```

After implementing inheritance relationships, we can modify any virtual function coded in the parent class if we so desire. For example, we modified the code for Debit() function inside the SavingsAccount class, assuming a savings account must maintain $1,000 minimum balance (see the boldfaced code).

```
public class SavingsAccount : Account
{
    private DateTime lastInterestCreditAt;

    private CheckingAccount associatedChecking;
    private List<InterestRate> rates;

    public override void Debit(double amt)
    {
        if (balance > 1000 + amt)
            base.Debit(amt);
        else
            throw new Exception("minimum balance
                error");
    }
```

```
public override void CreditInterest()
{
    //
}
}
```

Finally, let us code `CreditInterest()` function for each type of account. A checking account does not accrue interest, and so nothing needs to be done. A CD account will credit interest on the maturity date based on the number of years the CD has been opened using the standard compound formula as follows:

$$balance = balance \times (1 + rate)^{years}$$

The following is the implementation of `CreditInterest()` function in both `CheckingAccount` and CD classes (see the boldfaced code).

```
public class CD : Account
{
    private double interestRate;
    private DateTime maturity;

    public bool isMatured()
    {
        if (DateTime.Now >= maturity)
            return true;
        else
            return false;
    }

    public override void CreditInterest()
    {
        if (DateTime.Today == maturity.Date)
        {
            TimeSpan ts = maturity - openDate;
            double years = ts.Days / 365.0;
            balance = balance * Math.Pow(1 +
            interestRate, years);
        }
    }
}

public class CheckingAccount : Account
{
```

```
    private double minBalanceReq;
    private List<SavingsAccount> associatedSavings;

    public override void CreditInterest()
    {
        //
    }
}
```

It is more complex to code the CreditInterest() function for a savings account because different rates may be applicable for different segments of the interest period. First, for any interest period between times s and e, let us ask InterestRate object to find the number of overlapping days between the interest period and the interest rate applicable period. The code would involve a lot of comparisons among four different dates, but the following code uses a smart solution strategy: find the maximum start date between s and validFrom and the minimum end date between e and validTo, and then find the number of days between the maximum start date and the minimum end date (see the boldfaced code).

```
public class InterestRate
{
    private double interestRate;
    private DateTime validFrom;
    private DateTime validTo;

    public double GetRate()
    {
        return interestRate;
    }

    public int GetOverlapDays(DateTime s, DateTime e)
    {
        if (s > validTo || e < validFrom)
            return 0;
```

```
        else
        {
            DateTime minEnd, maxStart;
            if (s < validFrom)
                maxStart = validFrom;
            else
                maxStart = s;

            if (e < validTo)
                minEnd = e;
            else
                minEnd = validTo;

            TimeSpan ts = minEnd - maxStart;
            return ts.Days;
        }
    }
}
```

With the `GetOverlapDays()` function, we can now go through each interest rate in a savings account to find its applicable days that overlap with the interest period and use the days to find the interest to be added. After finding the total interest during the interest period, update the balance and the last interest credit date (see the boldfaced code).

```
public class SavingsAccount : Account
{
    private DateTime lastInterestCreditAt;

    private CheckingAccount associatedChecking;
    private List<InterestRate> rates;

    public override void Debit(double amt)
    {
        if (balance > 1000 + amt)
            base.Debit(amt);
        else
            throw new Exception("minimum balance
                error");
    }

    public override void CreditInterest()
    {
```

```
double sum = 0;
TimeSpan ts = DateTime.Now -
    lastInterestCreditAt;
if (ts.Days >= 30)
{
    foreach (InterestRate ir in rates)
    {
        double rate = ir.GetRate();
        double years = ir.GetOverlapDays(lastIn
            terestCreditAt, DateTime.Now) /
            365.0;
        sum = sum + balance * Math.Pow(1 +
            rate, years);
    }
    balance = balance + sum;
    lastInterestCreditAt = DateTime.Now;
}
}
}
```

Review Questions

1. What is the cardinality of the relationship between Product and Order?
2. Provide a justification that Customer and Account must have bidirectional association.
3. What is the relationship between Order and Shipment?
4. If the knowledge of Product is important to Order but not vice versa, what kind of navigability exists between Product and Order?
5. Make up three real-world objects that can be modeled as inheritance relationships.
6. What is the difference between overload and override?
7. When do you use virtual to modify a method?
8. What is an interface? How is it different from an abstract class?
9. Give example of objects or classes that use the following concepts:
   a. Abstract operation.
   b. Static attribute.
   c. Interface.
   d. Private operation.
   e. Recursive relationships.
   f. Generalization/specialization relationships.

Exercises

1. Complete the list of implementation rules for the case when A and B have bidirectional associations.
2. Create the following classes and subclasses and draw the class diagrams, and then create the classes in Visual Studio using C#.
   a.   Student (as an abstract class that has name, birth date, address, campus fees, and an abstract method of computing the tuitions and fees).
   b.   PartTimeStudent is a special kind of student whose tuition is calculated based on the number of credit hours taken and tuition per credit.
   c.   FullTimeStudent is a special kind of student who pays the whole amount regardless of the number of credits taken.
   d.   Course is a catalog entry with information like cno, title, description, and credit.
   e.   Section is offered under a catalog Course entry to be offered for students to enroll. It has attributes like Section Number, Date and Time, Capacity, etc.
   f.   Section needs to know information about Course such as credit hours; however, a Course catalog entry does not need to know anything about Sections offered under the course.
3. Build a class diagram for a factory that represents the following requirements and business rules. Make up your own attributes for each object involved. Then create all the classes using C#.
   a.   A component can be used to make any of several other components.
   b.   A component can be made of several other components.
   c.   A component can be constructed from several raw materials.
   d.   A raw material is used in several components.
   e.   There are two kinds of raw materials, some are perishable and have definite expiration date and others are nonperishable.
   f.   Each component is produced by a single worker.
   g.   A worker can produce many kinds of components.
4. A student club wants to have a database to manage its data on members. The club assigns members to its numerous committees. It is possible that one member can serve in more than one committee and will chair at most one committee. Each year, the club organizes many events. Each time, the club assigns one committee to oversee an event.

Build a class diagram to model the club's business objects and then convert the diagram into skeleton code using C# or Java.

5. A professor wants to keep data for all his books, publishers, and authors so that whenever he has comments he can communicate with the authors using US Mail, phone, or email. He also likes to have data about the publishers because he may need to go to the company's website or call the company to look for more information. Create a class diagram to model the business objects and their relationships and then convert the diagram into skeleton code using C# or Java.

6. A tiny library contains thousands of publications of different kinds such as books, journals, conference proceedings, and videos. For each publication, the library tracks its title, publication date, and publication type, the publisher, and the authors. In addition, different publications will have additional data required. For example, each book has an ISBN and page count; a video has an ISBN and length; a conference proceeding has ISSN, ISBN, page-count, and the total number of articles included; a journal has an ISSN. As a convention, the library catalog also contains extensive contact data about publishers and authors such as their mailing addresses and phone numbers. Draw a class diagram to model the domain objects and business rules for the library.

7. The board of Watson Town Memorial Hospital has recently decided to develop a new database to manage their patient admissions and discharges. The hospital handles two types of patients: outpatient and resident patient. As typical, each time a new patient comes, the data about his/her identification, address, phone, and issuance carrier are recorded. If a patient is a resident, he/she will be assigned to a bed and an admission date is recorded. After the treatment, a nurse must sign off the discharge card. For an outpatient, the nurse will set a check-back time after each treatment. Develop a class diagram to model the objects requirements for the hospital.

8. Implement the following class diagrams using C# or Java:

   a. Service transactions

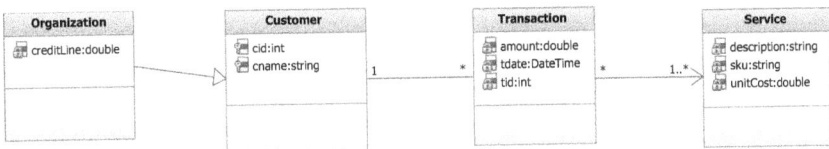

b.   Insurance

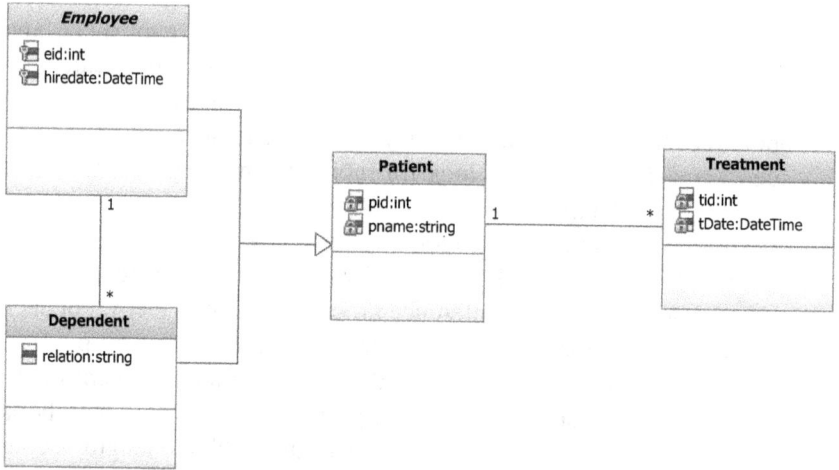

c.   Medical records

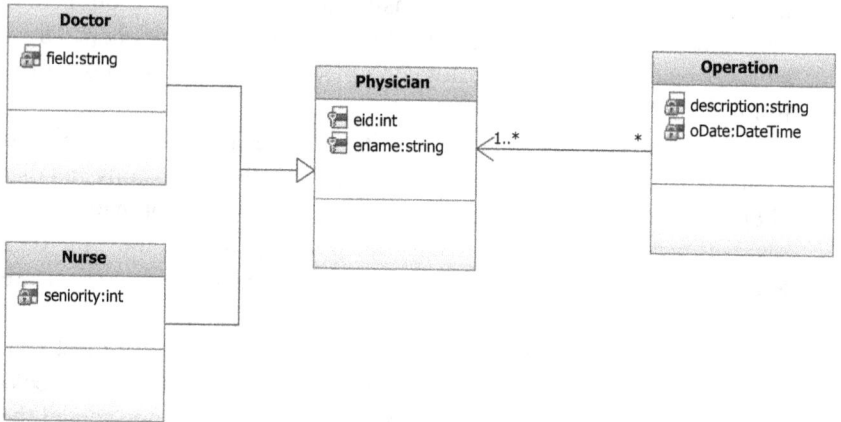

9. Implement the following class diagram. Note that the length of the time is measured in minutes.

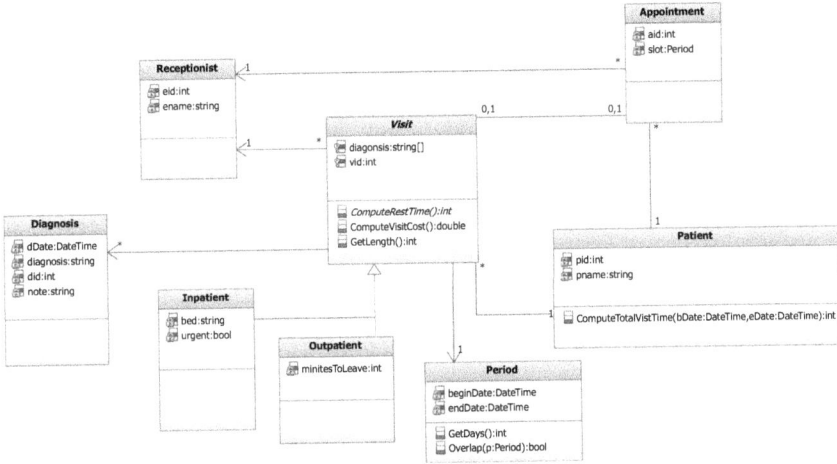

10. Write C# code to implement the following class diagram. Note that to code the function ComputeOrderAmount(), we assume that the quantity of each ordered product is one unit.

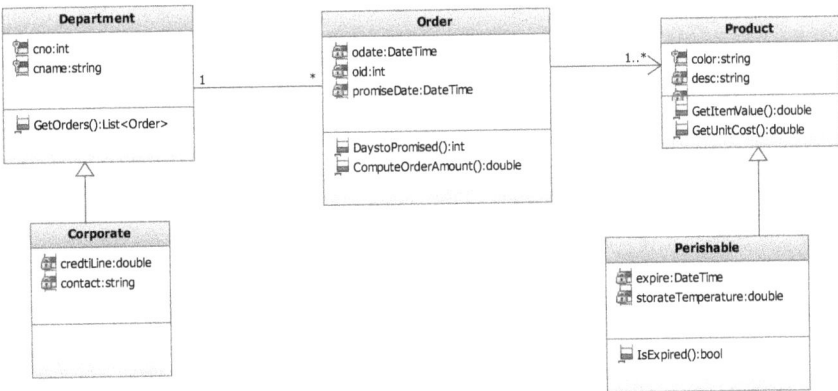

Chapter 7

Advanced Associations

Introduction

In this chapter we will learn how to create class diagrams using advanced associations, including composition and aggregation, association objects, reflexive association, multiway associations, and exclusive and dependent associations. These advanced concepts are often optional; not all CASE tools support their modeling, and not all programming languages support their implementation, but they tend to add richer semantics to generic associations and can improve the efficacy of the implementation using an advanced object-oriented language such as C++.

Composition and Aggregation

Composition and aggregation are two special kinds of associations — containment or whole-part associations. They are often used to model the associations between big objects and small objects in a way that a big object contains or has a small object as a part, or a small object belongs to or is a part of the big object. To be precise, let us call the big object *containing object* and the small object *contained object*.

Although both composition and aggregation model containment, composition is stronger than aggregation. A containment is composition if the contained objects cannot exist without the containing object; if the containing object is destroyed, all its contained objects will be also destroyed. Otherwise, a containment is aggregation. There is another way to tell the difference: a containment is composition is the containment is exclusive; each contained object can only be contained by one containing object, i.e., the maximum cardinality is one on the containing object side. The following are a few examples of both types of containment relationships:

Composition: A car contains wheels, a folder contains other folders and files, a house contains rooms, a watermelon contains seeds, a pen contains barrels, an order contains line items.

Aggregation: A room contains walls, a department contains employees, a country contains cities, a bank association contains banks.

In the above examples, a wall may belong to several rooms, an employee may work for several departments, a city may belong to several countries (like Jerusalem; not meant to take a political side), and a bank may belong to several associations. Thus, all these containment relationships are aggregations.

Composition and aggregation are represented by attaching a diamond to the containing end of the association. A solid diamond is for composition, and a hallow one for aggregation. In Rhapsody, we can refine a generic association by selecting either shared or composition as the aggregation kind at the end of contained objects (see Figure 5 of Chapter 6). We may also more conveniently pick Aggregation or Composition tool in the toolbox in Rhapsody (see Figure 3 of Chapter 5) to draw a containment relationship directly. Figure 1 shows four examples of containment relationships.

Containment relationships can be nested in several levels. For example, a car contains an engine, and the engine contains cylinders. Thus, a cylinder is a composite part of a car and an engine. In this case, their relationship should be modeled in a gradual or transitive containment order as the one shown in Figure 2(a) rather than 2(b).

Sometimes, it may not be easy to tell whether an association is of containment or not and whether a containment should be modeled as composition or aggregation. For example, in the case that a person owns a pen, should we model it as containment or not? For another example, in the case that a ballpoint pen has a top, should we model the relationship as composition or aggregation? Here is a general guideline: use a more general relationship against a special one. There is not much loss in using aggregation instead of composition, and there is no harm in using association instead of containment. For example, Figure 3 models the ownership relationship as a usual association, and a ballpoint aggregates tops, assuming the tops for a ballpoint may be used for other pens.

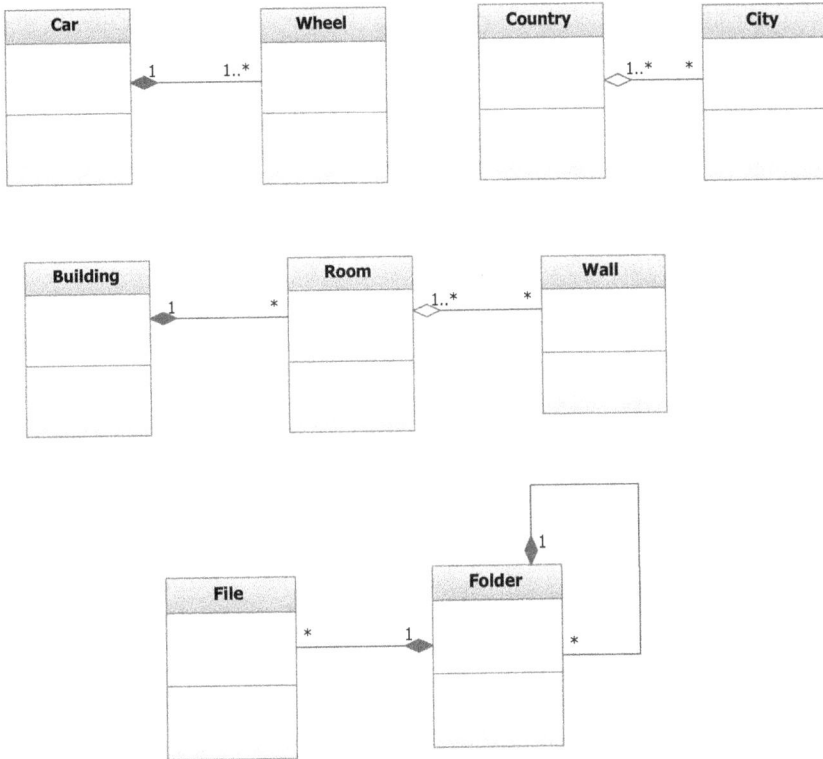

Figure 1.   Examples of composition and aggregation.

The reader should not confuse containment with inheritance; inheritance may be understood as a containment relationship between classes, or groups of objects, whereas composition or aggregation is a relationship between individual objects. For example, the Animal class contains the Dog class, but it does not make sense to say an animal contains a dog. Similarly, a car contains an engine, but it will make no sense to say the Car class contains the Engine class because an engine is not a kind of car.

Composition and aggregation are special kinds of associations, and so they are implemented in the same way as associations in Java or C#. When using an advanced language such as C++ that supports the notion of pointers, the subtle difference lies in how to control the life of contained objects in memory management. For aggregations and generic associations, an instance variable may be of a pointer to a contained object so that when the containing object is destroyed, the pointed contained one can be still alive.

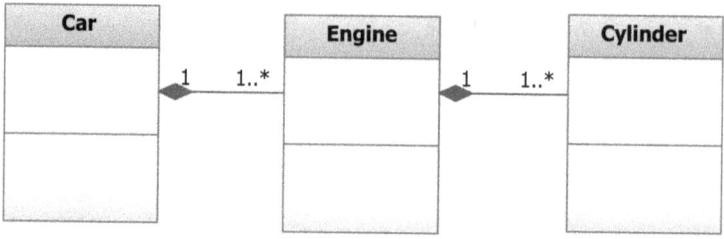

(a)

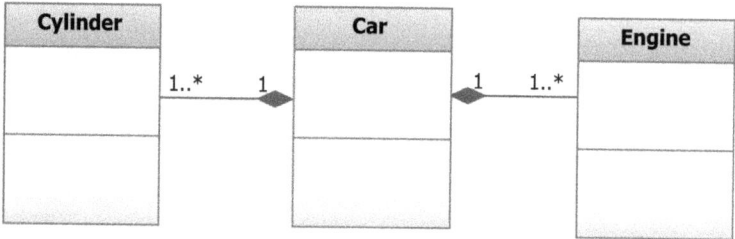

(b)

Figure 2.   Nested containment relationships.

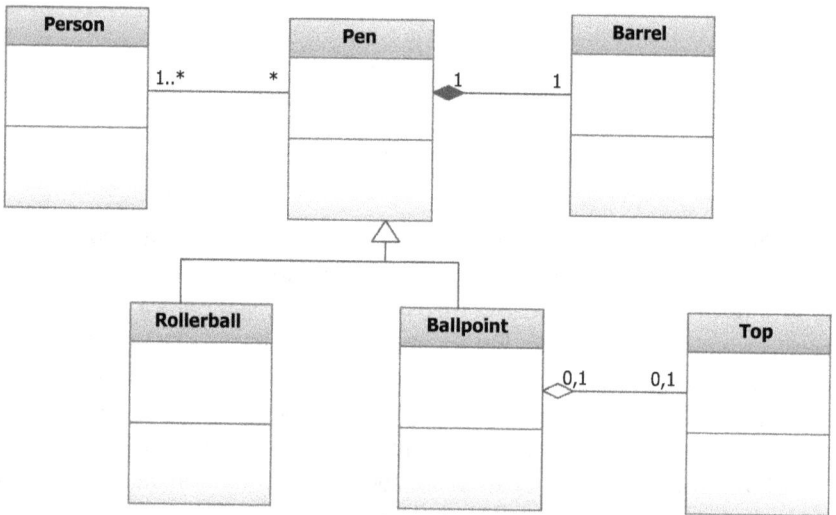

Figure 3.   Example with composition, aggregation, and inheritance.

However, for compositions, the instance variable is the actual contained object so that when the containing object is destroyed, the contained object is also destroyed.

Multivalued attributes

A multivalued attribute is the one that may have multiple values for one object. For example, if diploma is an attribute for employees, then we will have situations wherein some employees have more than one diploma, say BA, MS, MBA, etc. while others may have none. For another example, in the health insurance applications, "dependents" is a multivalued attribute for policy holders (see Figure 4).

In data modeling, attributes are required to be single valued to conform to the first normal form, and multivalued ones are converted into weak entities. In class diagramming, there is no such restriction. In fact, we may have a collection type like `array` and `list` for multivalued attributes, as seen in earlier chapters.

Another approach to handle a multivalued attribute in class diagramming is to take the attribute out of the hosting object and make it a contained object. In this way, each value becomes an object contained by the hosing object. For example, dependents become contained objects, contained by a policyholder object. In this approach, there is no limit on how many values the multivalued attribute can have. In addition, in this approach, we may have more attributes, rather than just dependent names, to describe the dependents. To represent the fact that each dependent is insured under one policyholder exclusively, we should use a composition relationship between the containing objects and the contained objects (see Figure 4).

Using this approach, composition essentially corresponds to the concept of weak entities in data models. Weak entities cannot exist independently and must depend on a strong entity to exist; if the record of the strong entity is destroyed, all the records on its associated weak entities will be also destroyed. Also, the cardinality between the strong and the weak entities must be 1 to many. In a sense, the strong entity is a containing

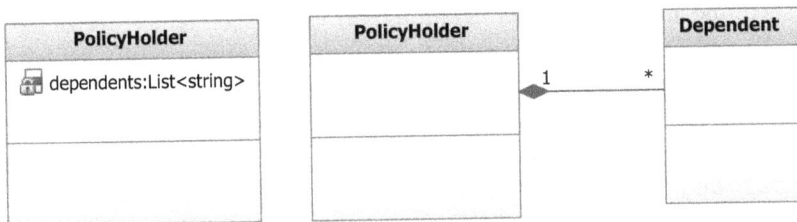

Figure 4.   Two ways to model multivalued attributes.

object, and the weak entities are contained objects. Their relationship is of composition.

Association Class

Unlike regular objects such as students, courses, classrooms, etc., an *association object* does not correspond to real-world objects. Instead it is an association relationship that is reframed as an object. There are two scenarios in which we may need association objects in lieu of associations. First, an association object is needed when we have attributes to describe the characteristics of an association or want the association to be able to perform some operations on the attributes. For example, in a university registration system, there are associations between students and course sections (or offerings). Where can we keep data on students' grades for each section they finish? Which objects perform operations such as searching for finished courses of a student or updating the enrollment status? Note that a grade is neither an attribute of a student nor an attribute of a course offering alone; it ties to both a student and a course offering. In this sense, we may think the grade is an attribute for the association between a student and an offering. Thus, we turn the association into an association object. For another example, a containment relationship between an order and an item should be also transformed into an association object if we need to capture the data on the quantity of the item in the order.

The second use case of association objects is to model dynamic associations. For example, a facility may be assigned to departments for meetings. The assignment is changing over time, and we need to keep track of the changes for the planning purpose; we need to dynamically create, modify, or remove the associations. Modeling these associations as association objects is probably the best approach in handling such dynamic associations.

In Rhapsody, associative classes are drawn as a regular class with a dashed line attached to an association line. The dashed line is drawn using the Anchor tool in the diagramming toolbox (see Figure 3 of Chapter 5). Note that, after an associative class is attached to an association by an anchor, the association will take the name of the association class. Figure 5 illustrates two association classes, `OrderLine` for aggregations between `Order` and `Product`, and `Enrollment` for associations between `Student` and `CourseOffering`. Here, a course offering means a section or class offered under a course.

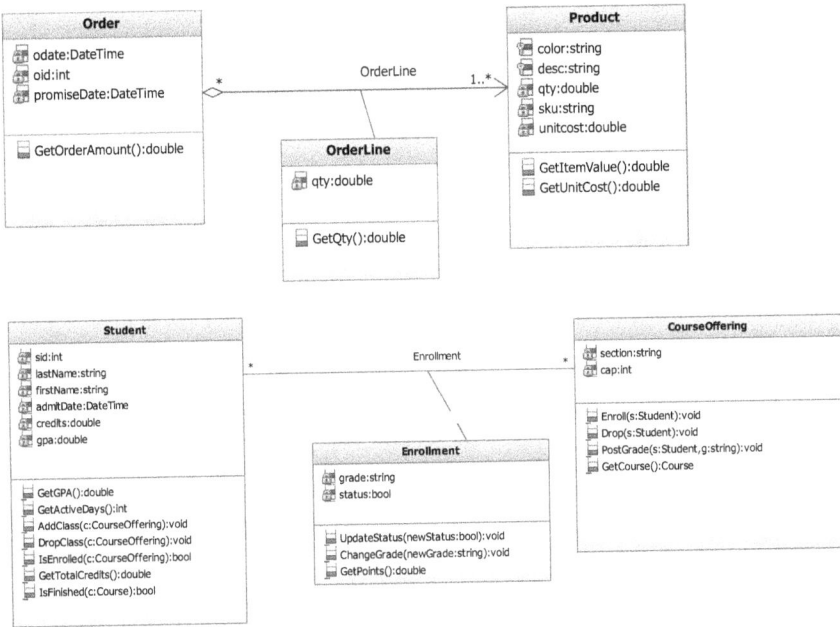

Figure 5.   Association objects illustrated.

Association objects are very similar to associative entities or gerunds in data models. Yet the reader shall note a significant difference between them: in data models, associative entities are used only for many-to-many relationships, as in the case between Student and CourseOffering. In contrast, association objects may be used for relationships of any cardinality. For example, Figure 6 uses association objects to represent marriage relationship between a man and a woman.

Implementation

To code a class diagram with association classes, we may imagine each association class as a regular one that cuts into the middle of the two classes that the original association ties, and the association class is then removed from the original association (see Figure 7).

The reader may notice that the equivalent class diagram in Figure 7 has two redundant relationships between Student and CourseOffering:

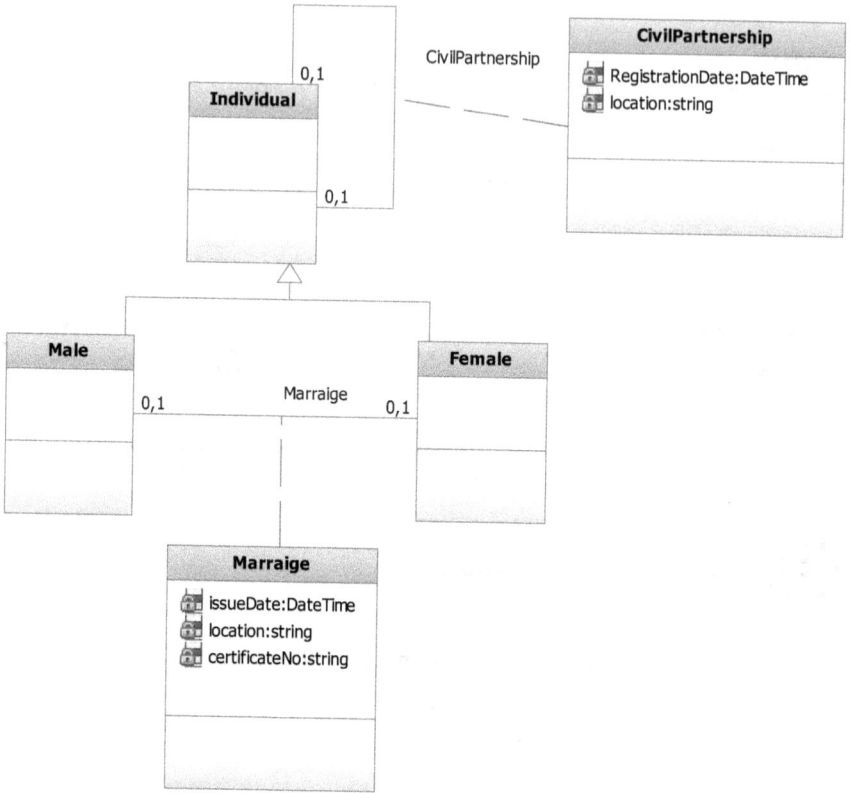

Figure 6.   Association objects for 1:1 relationships.

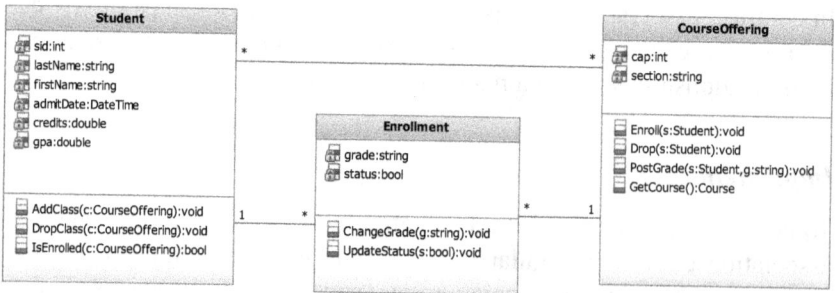

Figure 7.   Equivalents to association classes.

the original direct association and the indirect one connected by the Enrollment class. Do we still need to keep the original direct association when the association is converted into an association class? In data modeling, the answer is negative. However, in class diagramming, there appears to be an advantage of keeping the original direct association. For example, it allows two related objects to send messages directly without going through a middleman.

The reader shall also note the cardinality and the navigability of the two new associations to Enrollment in Figure 7. Cardinalities can be obtained by analyzing the nature of association objects: each Enrollment object connects one Student object and one CourseOffering object, and so the cardinalities at the Student end and at the CourseOffering end are both 1. On the other hand, each Student object is associated with multiple CourseOffering objects via multiple associations, and thus the cardinality of its relationship with Enrollment at the Enrollment end is zero or more. Similarly, the cardinality at the Enrollment end of the relationship between CourseOffering and Enrollment is zero or more. The cardinalities may be also directly derived from those of the original direct associations by cross-copying, as shown in Figure 8.

The navigability can be obtained as follows. First, each association object needs to have knowledge of who is participating in the association. Thus, it is always navigable from an association class to the classes at both ends. Second, if one object has knowledge of the other one, it should also have the knowledge of its association with the other one, and vice versa. Therefore, the two new associations with the Enrollment class are bidirectional. In the case of unidirectional associations, the object

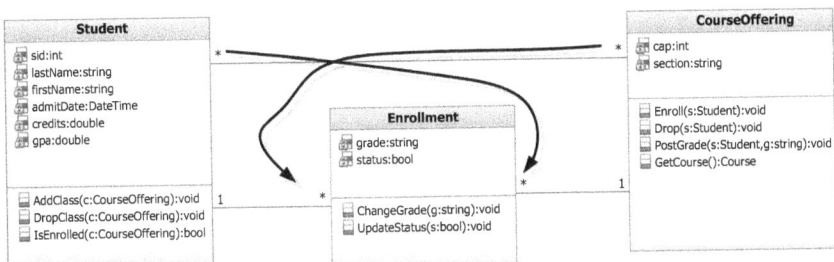

Figure 8. Deriving cardinalities for bidirectional association classes.

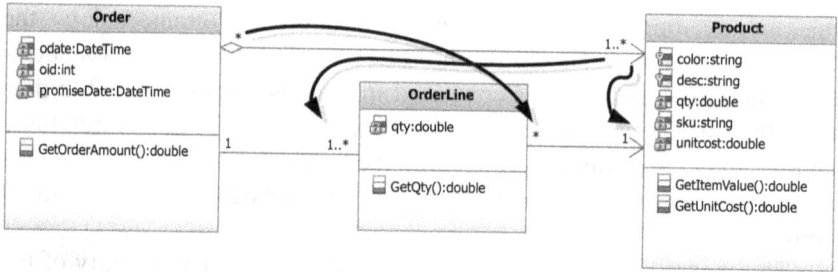

Figure 9.   Deriving cardinality and navigability for unidirectional association classes.

that does not have the knowledge of the other one also does not have the knowledge of its associations with the other one. For example, Figure 9 shows an equivalent class diagram for the association class `OrderLine` in Figure 5. A simple rule on the navigability may be summarized as follows: the navigable arrowhead, if any, is copied to the same end from the original association to a new association with the association class.

With the equivalent representation, an association class can be implemented as usual. Now let us implement the association class, `Enrollment`, based on its equivalent representation in Figure 7. First, we create a skeleton code for each class, including native attributes, functions that can be easily coded, and the heads of the functions that cannot be coded.

```
public class CourseOffering
{
    private int section;
    private int cap;

    public void Enroll(Student stu)
    {
        //
    }

    public void Drop(Student stu)
    {
        //
    }

    public void PostGrade(Student stu, string g)
    {
        //
    }
}
```

```
public class Enrollment
{
    private int status;
    private string grade;

    public void ChangeGrade(string g)
    {
        grade = g;
    }

    public void UdpateStatus(bool s)
    {
        status = s;
    }
}

public class Student
{
    private DateTime admitDate;
    private int sid;
    private string lastName, firstName;
    private double credits, gpa;

    public bool IsEnrolled(Section sec)
    {
        //
    }
    public void AddClass(Section sec)
    {
        //
    }

    public void DropClass(Section sec)
    {
        //
    }
}
```

Next let us implement associations as instance variables per the same rules in Chapter 6. The implementation of the `Enrollment` class is straightforward; each `Enrollment` object has one instance of `Student` and one instance of `CourseOffering` because the maximum cardinalities of its relationship with both `Student` and `CourseOffering` are 1 (see the boldfaced code).

```
public class Enrollment
{
    private int status;
    private string grade;
    private Student student;
    private CourseOffering section;

    public void ChangeGrade(string g)
    {
        grade = g;
    }

    public void UdpateStatus(bool s)
    {
        status = s;
    }
}
```

In the real world, each course offering has multiple students, and accordingly multiple association links with the students. Thus, each CourseOffering object corresponds to multiple instances of Student and multiple instances of Enrollment. We must use a collection type, such as an array, a list, or a dictionary, to define the instance variables. Here, we use a list for the Student objects and a dictionary for the Enrollment objects (see the boldfaced code).

```
public class CourseOffering
{
    private int section;
    private int cap;
    private List<Student> enrollees;
    private Dictionary<int, Enrollment> roster;

    public void Enroll(Student stu)
    {
        //
    }

    public void Drop(Student stu)
    {
        //
    }

    public void PostGrade(Student stu, string g)
```

```
    {
        //
    }
}
```

Note that in C#, a `List` variable contains a list of values, but a `Dictionary` variable contains a list of key-value pairs. The following code shows their difference in handling a list of text values. A dictionary uses a hashed key value as the address for each item and is thus more efficient to search for an item. To find an item in a plain list, the computer must exhaustively go through every item in order to find an item we need. For example, in the following code, to remove "Ohio" from `lstStates`, the computer will need to check each value in the list. In contrast, to remove the value from `dicStates`, it just needs to use the hashed value of 1 as the index to find the item and remove the item.

```
List<string> lstStates = new List<string>();
lstStates.Add("Ohio");
lstStates.Add("Illinois");
lstStates.Remove("Ohio");

Dictionary<int, string> dicStates = new Dictionary
    <int, string>();
dicStates.Add(1, "Ohio");
dicStates.Add(2, "Illinois");
dicStates.Remove(1);
```

The code

```
private Dictionary<int, Enrollment> roster;
```

creates instance variable `roster` as a dictionary to hold a list of student ID — `Enrollment` object pairs so that we will be able to use a student ID to refer to and search for an `Enrollment` object. For example, we can refer the `Enrollment` object for the student with ID 1234567 as `roster[1234567]`. Of course, to access the private `sid` variable of the `Student` class, we need to create either a data accessor or property in the `Student` class (see the boldfaced code).

```
public class Student
{
    private DateTime admitDate;
```

```
private int sid;
private string lastName, firstName;
private double credits, gpa;

public string SID
  {
     get {return sid;}
     set {sid = value;}
  }

public bool IsEnrolled(Section sec)
  {
      //
  }
public void AddClass(Section sec)
  {
      //
  }

public void DropClass(Section sec)
  {
      //
  }
}
```

With the declared instance variables for the two associations to Student and Enrollment, we can now implement the DropStudent and PostGrade functions in the CourseOffering class (see the boldfaced code).

```
public class CourseOffering
{
   private int section;
   private int cap;
   private List<Student> enrollees;
   private Dictionary<int, Enrollment> roster;

   public void Enroll(Student stu)
   {
       //
   }

   public void Drop(Student stu)
   {
       enrollees.Remove(stu);
       roster.Remove(stu.SID);
```

```
    }

    public void PostGrade(Student stu, string g)
    {
        roster[stu.SID].ChangeGrade(g);
    }
}
```

Note how convenient it is to remove an `Enrollment` object when dropping a student and to change grade for a specific student. If the instance variable roster were declared as a plain list, the code to implement the functions would involve a loop to search for the `Enrollment` object as follows.

```
public class CourseOffering
{
    private int section;
    private int cap;

    private List<Student> enrollees;
    private List<Enrollment> roster;

    public void Enroll(Student stu)
    {
        //
    }

    public void Drop(Student stu)
    {
        enrollees.Remove(stu);
        foreach (Enrollment e in roster)
        {
            if (e.GetStudent() == stu)
                roster.Remove(e);
        }
    }

    public void PostGrade(Student stu, string g)
    {
        foreach (Enrollment e in roster)
        {
            if (e.GetStudent() == stu)
                e.ChangeGrade(g);
        }
    }
}
```

To add a student, we will need to create a new `Enrollment` object for the student and add it to `roster`. To this end, we will create two constructors, one default and one non-default, for the `Enrollment` class as shown by the boldfaced code.

```
public class Enrollment
{
    private bool status;
    private string grade;
    private Student student;
    private CourseOffering section;

    public Enrollment()
    {
        //
    }

    public Enrollment(Student stu, CourseOffering sec)
    {
        student = stu;
        section = sec;
    }

    public Student GetStudent()
    {
        return student;
    }

    public void ChangeGrade(string g)
    {
        grade = g;
    }

    public void UdpateStatus(bool s)
    {
        status = s;
    }
}
```

Then to add a student, we just need to add the student to the `enrollees` list and add a new `Enrollment` object to the `roster` dictionary (see the boldfaced code).

```
public class CourseOffering
{
```

```
private int section;
private bool status;

private List<Student> enrollees;
private Dictionary<int, Enrollment> roster;

public void Enroll(Student stu)
{
    enrollees.Add(stu);
    Enrollment e = new Enrollment(stu, this);
    roster.Add(stu.SID, e);
}

public void Drop(Student stu)
{
    enrollees.Remove(stu);
    roster.Remove(stu.SID);
}

public void PostGrade(Student stu, string g)
{
    roster[stu.SID].ChangeGrade(g);
}
}
```

The Student class is implemented similarly. Since each Student object corresponds to multiple CourseOffering and multiple Enrollment objects, we use a list type for the instance variable of CourseOffering and a dictionary type for the instance variable of Enrollment (see the boldfaced code). Note that the dictionary transcript uses CourseOffering objects as keys and Enrollment objects as values. To see if the current student object is enrolled into a specific section, we simply check if enrolledSections contains the section. To enroll a student into a course offering, the code is simply to add the CourseOffering object to the enrolledSections list and to create and add a new Enrollment object to the transcript dictionary for the Student object. To drop a course offering, the code just does the opposite.

```
public class Student
{
    private DateTime admitDate;
    private int sid;
```

```
private string lastName, firstName;
private double credits, gpa;

private List<CourseOffering> enrolledSections;
private Dictionary<CourseOffering, Enrollment>
    transcript;

public string SID
{
    get {return sid;}
    set {sid = value;}
}

public bool IsEnrolled(CourseOffering sec)
{
    return enrolledSections.Contains(sec);
}

public void AddClass(Section sec)
{
    enrolledSections.Add(sec);
    Enrollment e = new Enrollment(this, sec);
    Transcript.Add(sec, e);
}

public void DropClass(Section sec)
{
    enrolledSections.Remove(sec);
    transcript.Remove(sec);
}
}
```

Of course, when a student enrolls into a course offering, the course offering should also add the student to its roster. Similarly, when the student drops a course offering, the section should also remove the student from its roster. Thus, we will need to update the instance variables enrollees and roster in the CourseOffering class. To do so, we may create a synchronization function in the CourseOffering class as follows.

```
public void Sync(string type, Student s, Enrollment
    e = null)
{
    if (type == "add")
```

```
    {
        enrollees.Add(s);
        roster.Add(s.SID, e);
    }
    else if (type == "drop")
    {
        enrollees.Remove(s);
        roster.Remove(s.SID);
    }
}
```

Then, when adding or dropping a course offering, the student object can ask the CourseOffering object to perform a synchronization (see the boldfaced code). Note that the above Sync() function has an *optional argument* e since synchronizing a drop does not need the input of an Enrollment object.

```
public class Student
{
    private DateTime admitDate;
    private int sid;
    private string lastName, firstName;
    private double credits, gpa;

    private List<CourseOffering> enrolledSections;
    private Dictionary<CourseOffering, Enrollment>
        transcript;

    public string SID
    {
        get {return sid;}
        set {sid = value;}
    }

    public bool IsEnrolled(CourseOffering sec)
    {
        return enrolledSections.Contains(sec);
    }

    public void AddClass(Section sec)
    {
        enrolledSections.Add(sec);
        Enrollment e = new Enrollment(this, sec);
        Transcript.Add(sec, e);
```

```
        Sec.Sync("add", this, e);
    }

    public void DropClass(Section sec)
    {
        enrolledSections.Remove(sec);
        transcript.Remove(sec);
        sec.Sync("drop", this);
    }
}
```

Synchronization must be also performed when a `CourseOffering` object adds or drops a student. Thus, we also need to create a synchronization function in the `Student` class as follows.

```
public void Sync(string type, CourseOffering sec,
    Enrollment e = null)
{
    if (type == "add")
    {
        enrolledSections.Add(sec);
        transcript.Add(sec, e);
    }
    else if (type == "drop")
    {
        enrolledSections.Remove(sec);
        transcript.Remove(sec);
    }
}
```

And then call a `CourseOffering` object to perform the function when the object adds or drops a student (see the boldfaced code).

```
public class CourseOffering
{
    private int section;
    private bool status;

    private List<Student> enrollees;
    private Dictionary<int, Enrollment> roster;

    public void Enroll(Student stu)
    {
        enrollees.Add(stu);
```

```
        Enrollment e = new Enrollment(stu, this);
        roster.Add(stu.SID, e);
        stu.Sync("add", this, e);
    }

    public void Drop(Student stu)
    {
        enrollees.Remove(stu);
        roster.Remove(stu.SID);
        stu.Sync("drop", this);
    }

    public void PostGrade(Student stu, string g)
    {
        roster[stu.SID].ChangeGrade(g);
    }
}
```

Before we conclude this section, let us revisit whether we should keep the two redundant relationships when implementing association classes. We should note that in the above implementation, the two lists in the Student class — enrolledSections and transcript — and two lists for the CourseOffering class — enrollees and roster — are redundant. In fact, transcript.Keys will be a list of enrolled sections for a student and roster.Keys will be a list of enrollees for a course offering. The question is whether transcript.Keys is identical to enrolledSections and whether roster.Keys is identical to enrollees. If they are, the instance variables enrolledSections and enrollees become unnecessary. In the above code for adding and dropping a student or a section, both lists for the association class are always synchronized, and so only transcript and roster are needed. As an exercise, the reader may update the above implementation accordingly.

Recursive Associations

Associations may relate objects from the same class or group. For example, one course may be a prerequisite for other courses, an employee may be a supervisor of other employees, a person may be married to another, and a part is a component to make another part. To model these associations, we will use an association relationship to link a class to itself in class diagrams. We call this type of association unary, reflexive, or recursive. Figure 10 shows two examples of recursive associations, respectively, for the relationship between courses and their prerequisites and between employees and their supervisors. A prerequisite is also a course, and so the association is a recursive one between courses. Note that a course must have prerequisites, but a prerequisite does not need to know which courses use it as a prerequisite. Thus, the association is a unidirectional recursive one. Similarly, supervisors are employees, and so the relationship between supervisors and employees are recursive.

Recursive associations are implemented in the same manner as usual associations as in the previous chapter. For example, to implement the `Course` class in Figure 10, we first create a blank class as follows:

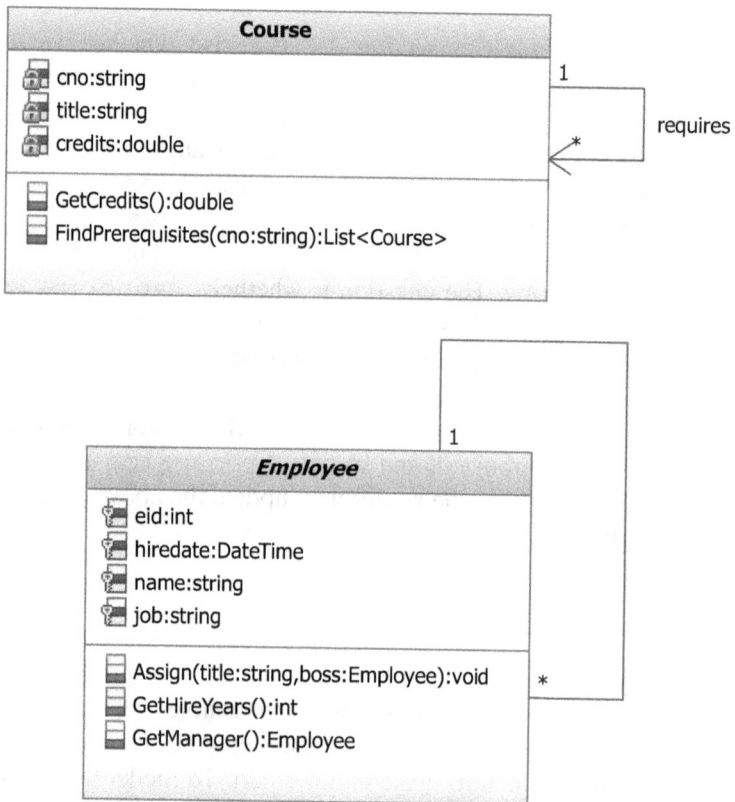

Figure 10.   Recursive associations.

```
class Course
{
    private string cno;
    private string title;
    private double credits;

    public void GetCredits()
    {
        return credits;
    }

    public List<Course> FindPrerequisites(string cno)
    {
        //;
    }
}
```

To implement the recursive association, we just imagine two separate copies of Course class connected by the association relationship. Since the navigability is unidirectional, we just need to create objects of one end as the instance variable, named prerequisites, in the second end. The maximum cardinality is *, and so we use a list type for the instance variable. Thus, we have an instance variable in the class as follows:

```
private List<Course> prerequisites;
```

With this new foreign data member, the code for GetPrerequisites is as simple as telling the value of a data member. The following is the complete implementation:

```
class Course
{
    private string cno;
    private string title;
    private double credits;

    private List<Course> prerequisites;

    public void GetCredits()
    {
        return credits;
    }

    public List<Course> GetPrerequisites()
```

```
        {
            return prerequisites;
        }
}
```

Note that due to unidirectional navigability, we create one instance variable. If the recursive association is bidirectional, we will have to create two instance variables. For example, to implement the Employee class in Figure 10, we will create two foreign instance variables, manager for an employee and subordinates for a manager (see the boldfaced code).

```
public abstract class Employee
{
    private List<string> addresses;
    protected string eid;
    protected string ename;
    protected DateTime hiredate;
    protected string job;

    protected Employee manager;
    protected List<Employee> subordinates;

    public int GetHireYears()
    {
        TimeSpan ts = DateTime.Now - hiredate;
        return ts.Days / 365;
    }

    public Employee GetManager()
    {
        return manager;
    }

    public void Assign(string t, Employee boss)
    {
        manager = boss;
        job = t;
    }
}
```

Like any other associations, a recursive association may also have its own data or functions and so is modeled as an association class. For example, between each pair of a supervisor and a supervisee, we may need to

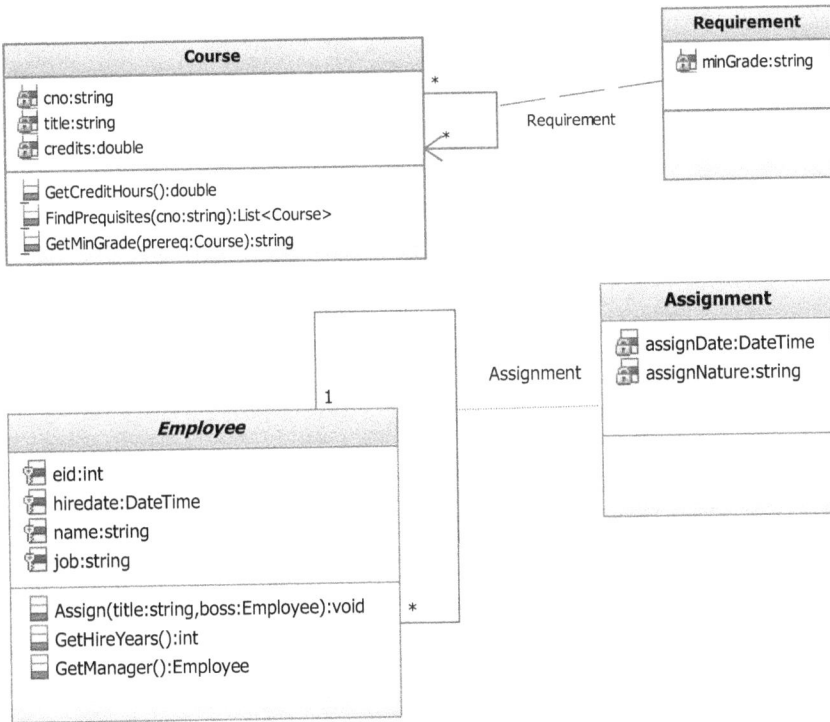

Figure 11. Recursive association classes.

have data on the assignment date and nature. Similarly, a course may not only require another course as a prerequisite but also students must pass the prerequisite with a minimum level of proficiency or grade. Figure 11 shows the use of association classes to capture the need for data in recursive associations.

Multiway Associations

Each binary association connects two objects at a time. Sometimes, three or more objects may simultaneously join into a single high-order or *multiway association*. Figure 12 shows two of such examples, where a diamond is a symbol for a multiway association and the dashed line connects it to an association class. Note that IBM Rhapsody does not support the modeling

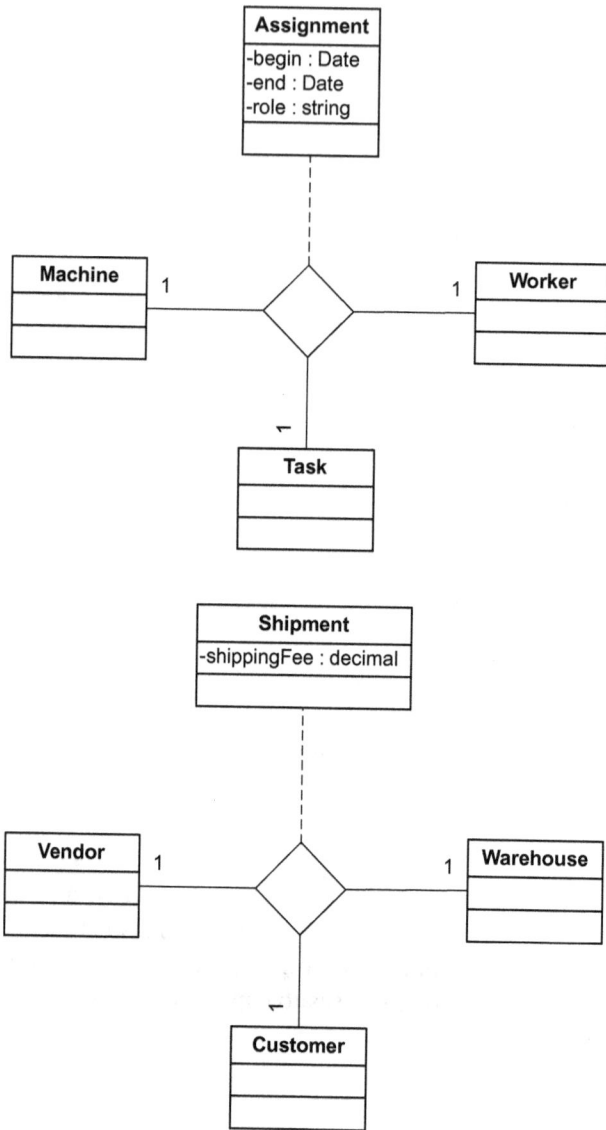

Figure 12.   Multiway association classes.

of multiway relationships. So, a multiway association may be drawn as an association class, and all the participating objects are linked to the association. The alternative representation, along with the *constraint* that requires all the associations to the association class hold simultaneously, will be equivalent to the original multiway association.

The essence of multiway relationships, as opposed to multiple binary relationships, is that all the objects must participate in each relationship simultaneously. If one quits, the relationship does not exist. Workplace assignments may be modeled as a 3-way association, where each assignment must be for one worker, on one task, and at one machine. Without any of them, the assignment is incomplete, and the association does not exist. If multiple shifts exist as in a manufacturing setting, assignments may be 4-way associations among workers, tasks, machines, and shifts. Attributes such as performance and quality must be declared in the 4-way association class.

Shipments may also be modeled as 3-way relationships among vendors, warehouses, and customers. A possible attribute for each shipment is shipping fee, which is assessed based on the vendor from which the order is made, the warehouse from which the order is shipped, and the customer to whom the shipment is to be delivered. In other words, it is determined simultaneously by all three objects: `Customer`, `Warehouse`, and `Vendor`. Therefore, a shipping fee is an attribute for the 3-way association class. It cannot be an attribute for a vendor, a warehouse, or a customer alone. It cannot be an attribute of a binary association between any two of them either.

Constrained Associations

When using an association class to replace a multiway association, we realize that, sometimes, we need to add constraints to associations. The results are *constrained associations*. In this section, we introduce a few optional techniques to represent constraints among associations, including exclusive relationships, conjoint associations, dependent relationships, and order and changeability constraints. All these techniques are parts of the UML standard. However, they are not supported by all CASE tools.

Exclusive and conjoint associations

Exclusive associations are those relationships that cannot hold simultaneously in the sense that if an object participates in one of them, it cannot participate in the others. For example, in a college, each office can be assigned to either one full-time teacher or several part-time ones. However, it cannot be assigned to both. In other words, the two association relationships between an office and a full-time teacher and between the office and

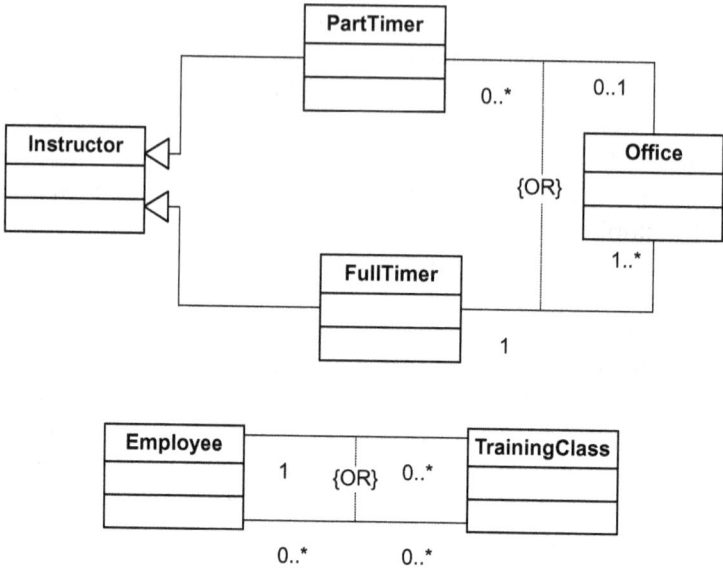

Figure 13.   Exclusive associations.

a part-time teacher are exclusive. For another example, in corporate trainings, some employees teach a training class while other employees take the training class. However, one employee cannot take and teach a class at the same time. Thus, the two associations are exclusive. To represent exclusive associations, we can use predefined OR constraint to link the associations (see Figure 13).

Conjoint associations are just the opposite of exclusive associations. Two or more associations are conjoint if they must hold or exist simultaneously. For example, to represent a multiway association using an association class, we will connect each participating object to the association class with an association relationship, and these relationships must be conjoint. To model conjoint associations, we use predefined AND constraint to link the conjoint associations.

Dependent associations

Dependent associations involve two relationships wherein one is dependent on the other in the sense that if one changes, the other will be affected. Like dependencies between classes, use cases, and packages, we use a

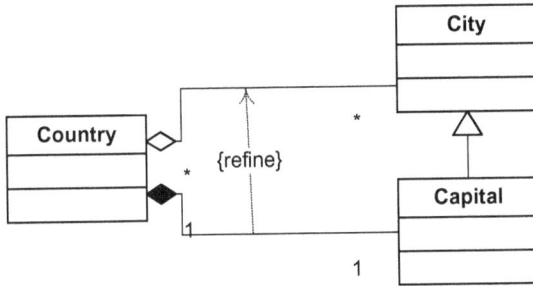

Figure 14.   Dependent associations.

dashed arrow pointing from the depending association to the dependent association to model the dependency. In addition, one shall attach a text to indicate the nature of the dependency. For example, in Figure 14, we show an example of dependent relationships, where the association between `Country` and `Capital` refines the association between `Country` and `City`.

Order and changeability constraints

Order and changeability constraints apply to attributes and associations as well. Since associations will be eventually implemented as foreign data members, these two constraints essentially govern whether the data members should be ordered or not if a data member takes on a list of values and whether the values of a data member are changeable, frozen, removed only, or added only.

For example, suppose an employee has zero or more diplomas. The attribute diploma will be of an array type. It may make sense to require that, if there are more than two diplomas, we order the values from highest degree to lowest one. Thus, we shall make diploma attribute ordered. Similarly, each employee's dependents, if any, shall be ordered too. For another example, a flight involves two or more airports, i.e., departure, arrival, and stopover locations, which shall be ordered. Thus, we can apply order constraint to the airport end of the association (see Figure 15).

Changeability allows one to specify whether an attribute value or an association end should be changeable, frozen, add only, or remove only. For an employee object, the id attribute should be set as frozen and diploma attribute should be set "add only" because, once obtained, a diploma cannot

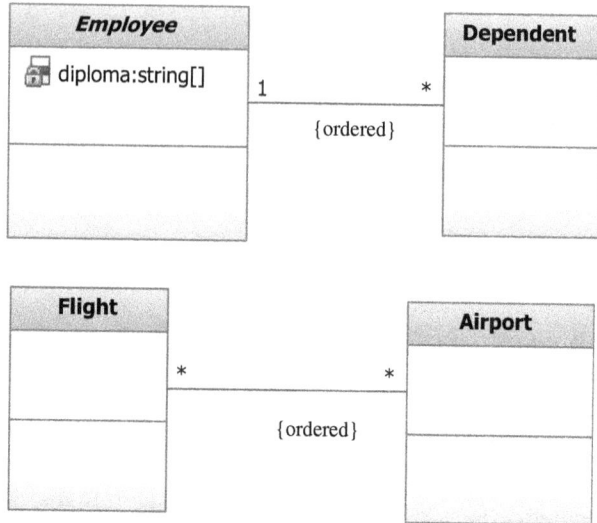

Figure 15. Order constraints.

be modified or removed. Similarly, if an employee's dependents can only be added but not removed or modified, then the Dependent end of the association in Figure 15 shall be set as "add only." For another example, if a flight is set to fly a fixed route from a departure airport, through zero or more stopovers, and arrive at the destination airport, then the airport end of the association in Figure 15 shall be set frozen.

Exercises

1. Give two examples for each of the following relationships and draw the corresponding class diagrams:
   a. Composite relationships.
   b. Aggregate relationships.
   c. Generation/specialization relationships.
   d. Recursive relationships.
2. Give one example of the following relationships and draw the corresponding class diagrams:
   a. Exclusive relationships.
   b. 3-way relationships.
   c. Dependent relationships.
   d. {frozen} relationships.
   e. {ordered} relationships.

3. Use class diagrams to model the following business needs:

   a. An order contains multiple items and sometimes may contain multiple quantities of the same item.

   b. A medical treatment often involves multiple procedures, with each one having a unique id and cost. Sometimes a treatment may apply the same procedure multiple times; clinics need to know the quantity in order to bill customers correctly.

   c. In the assembly industry, raw materials and ready-to-use parts may be used to build other parts. How would you handle the quantity of raw materials and parts used in the assembly? (Of course, raw materials cannot be used to make raw materials.)

   d. Test results, including x-ray files and various blood test indicators along with expert reading comments, are kept for each test performed on patients.

   e. A national merchandise club sells products at different prices at different locations and to different members. Where do we keep data about the prices?

4. A college has a few meeting facilities that can be reserved for individuals and departments, called customers. Here are the ongoing tasks for the facility manager: make and change reservations, print out the list of customers who will be using each facility for each day, and if a customer calls in, query the facility the customer reserved. In addition, after the facility is used as scheduled, its reservation status will be updated, and an invoice will be sent to the customer for payments. Create a class diagram with appropriate operations in each class and then program the classes using C# or Java.

5. Use Java or C# to implement the following class diagram:

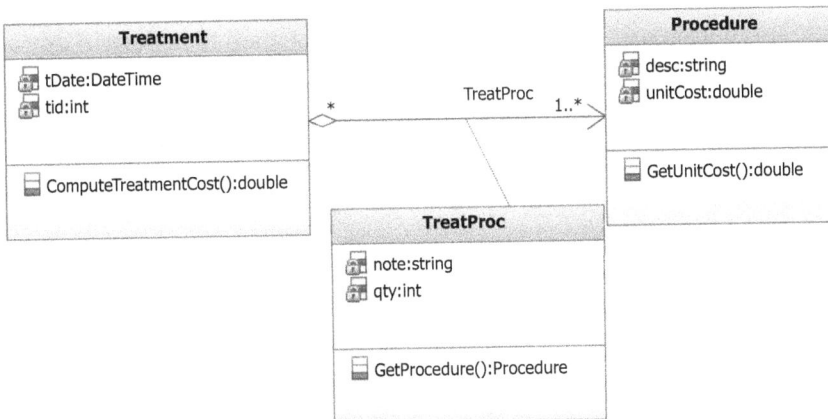

6. An academic department wants to put its annual report online. It has information about individual faculty: interest areas (AI, Software Engineering, Networking, etc.), recent publications, information about degrees (year granted, institution), and personal information (address, phone number, email, photograph, webpage, etc.). It has information about publications too: authors (may be by more than one faculty member and/or student), title of publication, date published, where published, and interest area(s). There is information about research groups: faculty and students in the group (student info includes interest areas, name, phone number, email, photograph, webpage, and faculty advisor), name of group, interest area(s). There are overview pages with: all faculty interested in different interest areas (e.g., all software engineering faculty), all faculty publications, arranged by year, pictures, and names of all faculty. Develop a class diagram to capture the domain knowledge in the text.

7. A gymnastics team wants to keep records of its gymnasts' scores in competition in addition to personal information about them (name, age, height, weight, sex, etc.). Female gymnasts compete in four exercises: balance beam, uneven bars, vault, and floor. Male gymnasts compete in six exercises: horse, parallel bars, high bar, rings, vault, and floor. Each competition has the gymnasts compete in all their exercises once, except for vault, which is done twice, and the best score is kept. The team wants to track the coaches, too, to be able to evaluate if gymnasts do better with certain coaches. Develop a class diagram to capture the domain knowledge in the text.

8. Whenever a new patient is seen for the first time at Cybercare Center, he or she must finish a patient information form that asks name, address, phone number, insurance carrier, and yes/no answers to certain questions such as whether a patient is allergic to certain drugs, whether the patient has any surgery in the last five years, etc. The patient can provide data on one insurance carrier so that the clinics can file claims on her behalf. The claim must have information about the visit, such as the date, purpose, a list of procedures performed, and the cost. Develop a class diagram to model the objects requirements for the clinic.

9. Use Java or C# to implement the following class diagram involving association classes:

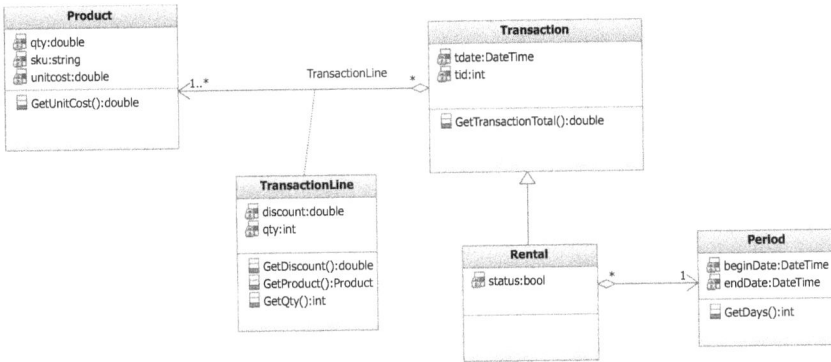

10. Whenever a new patient is seen for the first time at Cyberdale Care Center, she has to finish a patient information form that asks name, address, phone number, insurance carrier, and yes/no answers to certain questions such as whether a patient is allergic to certain drugs, whether the patient has any surgery in the last five years, etc. When a patient calls to schedule a new appointment or change an existing appointment, the receptionist checks the appointment schedule for an available time. Once a good time is found for the patient, the appointment is scheduled. If the patient is new, an incomplete entry is made in the patient file; the full information will be collected when the patient arrives for the appointment. Sometimes, appointments are made so far in advance that the receptionist will have to send a reminder postcard to each patient a week before her appointment. Develop a class diagram to model the objects requirements for the clinic.

11. Professor Bizmind does a lot of consulting in his life. He used to use FastBook to manage his bills and payments. Now he feels that the software cannot be customized to fit his needs. In particular, he would like his clients to be able to make job requests using the Internet. The client can get feedback immediately if the requested time conflicts with his existing schedule. The request can then be modified for another time, canceled, or sent regardless. Bizmind then looks at all the requests every day. If the requested time can be honored, he will update his schedule and send a confirmation to the client. Otherwise, he will talk to the client using email or phone to set up another time. Then he updates the schedule on the agreed date/time and sends an automatic confirmation. In terms of request details, the professor has

itemized a list of standard activities such as Data Analysis, Systems Administration, IT planning, etc. A client can just simply select one or more activities when she makes a job request. The professor also has standard unit fee associated with each activity. He may give discounts based on the quantity (e.g., number of hours) performed on an activity. A bill will be sent after each job is finished and at the beginning of each month, if a client has outstanding balance. A minimum payment and a due date will be specified on the bill. Late fee may be levied if a payment is overdue. Develop a class diagram to capture the domain objects required by the professor.

12. When members join OMCA health club, they pay a fee for a certain length of time. Most memberships are for 1 year, but memberships for short periods are available. Due to various promotions throughout the year, it is common for members to pay different amounts for the same length of membership. The club wants to mail out reminder letters to members to ask them to renew their memberships one month before their memberships expire. Some members have been angry when asked to renew at a much higher rate than their original membership contract. So, the club needs to keep track of the price paid so that the managers can override the regular prices with special prices when members are asked to renew. The system must keep track of these new prices so that renewals can be processed accurately. One of the problems in the health club industry is the high turnover rate of members. Although some members remain active for many years, about half of the members do not renew their memberships. This is a major problem because the club spends a lot in advertising to attract each new member. The manager wants to track each time a member comes into the club. The system will identify heavy users and generate a report so the manger can ask them to renew their memberships early at a reduced rate. Likewise, the system should identify those who do not come to the club often so that the manager can call them and attempt to attract them in the club. Create a class diagram for the problem.

13. To enroll a student into a class, the registration system must check whether the student has all the prerequisites taken, whether the class is still open, and whether the total number of credit hours the student registers is not beyond the maximum allowed. After a student finishes her registration, she will need to pick up a printed confirmation that shows all the courses she has registered, the data/time, section number, credit hours, ecourse.org access code, and instructor for each

class. Also, the confirmation paper shows the student status, state of residence, the total number of credit hours, and the total amount to be paid to the college. The student will bring the confirmation to the business office and make a deposit, which is equivalent to 20% of the total amount, to reserve her registration. If she fails to do so within 10 days, her registration will be canceled. The system also actively monitors the number of students signed up for each class. Three days before the class starts, if the number of registered students for a class is less than 15, the class will be canceled. The registered students will be informed to find alternative classes. To better serve the students and departments, the system has functionality for students to make course requests for future terms. The requests will be summarized and sent to departments so that they can make informed decisions on what to be offered in the future. Create a class diagram for the problem.

Chapter 8

Practical Class Diagramming

Introduction

We have learned the basic mechanics of class diagramming and some advanced features. To apply the mechanics and features requires practice, experience, and domain knowledge. However, there are certain valuable heuristics, design patterns, and practical methods that can help overcome some of the difficulties. In this chapter, we will introduce a few design patterns to improve the efficiency of class diagramming. Note that there are specialized books on those topics, but here we introduce a few simple aspects. Then we will study the phraseology, a linguistic analysis approach to object modeling, to model large practical problems.

Design Patterns

From time to time, people often find that some class diagrams are strikingly similar despite the differences in their problem domains. The similarity is not due to coincidence. Rather, it is because of the existence of common patterns. In this section, we study a few such common patterns.

The Transaction Pattern: This pattern comes from the model that shows a customer making 0 or more orders, each of which in turn contains one or more items, and an item may be ordered by 0 or more orders. The class diagram is shown in Figure 1.

Transaction pattern applies to all transaction or request situations such as rentals, returns, shipments, bids, requests, applications, reservations,

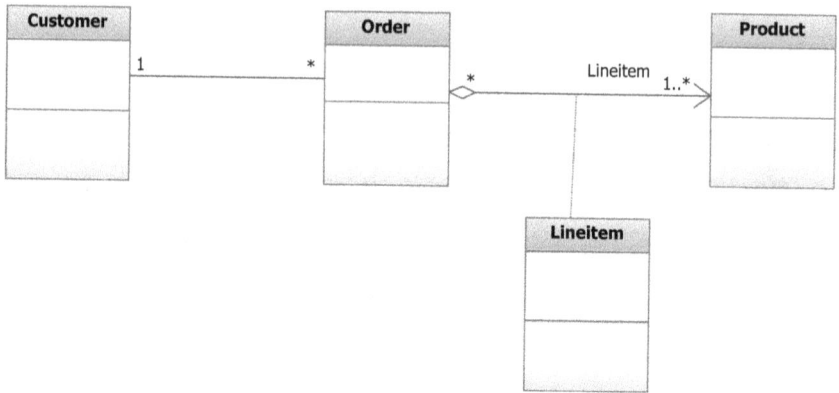

Figure 1.   The transaction pattern.

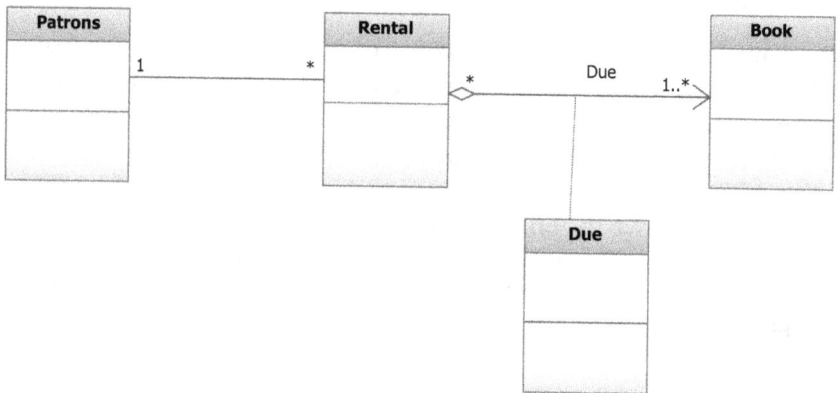

Figure 2.   Library rentals.

etc., although some may have nothing to do with orders. The following are some examples modeled using this pattern:

Example 1: Patrons make rentals (or returns), each of which contains one or more books (Figure 2). Because different rentals and items may have different due times, an association object is used to keep the data on dues.

Example 2: In procurement, employees or departments propose purchase requests to the purchasing department (Figure 3). Each proposal requests one or more products to buy.

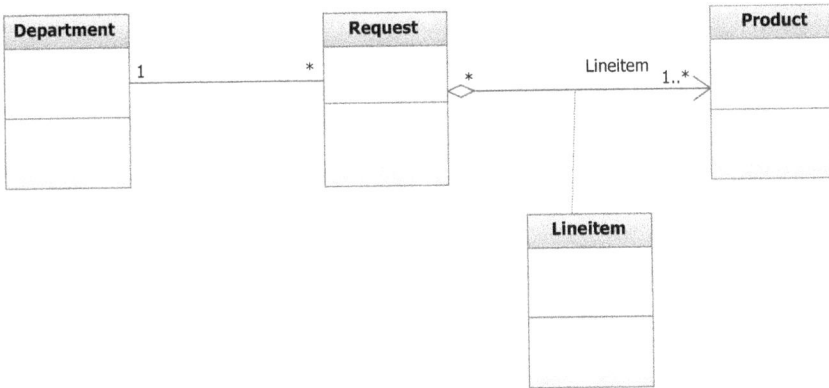

Figure 3.   Purchase requests.

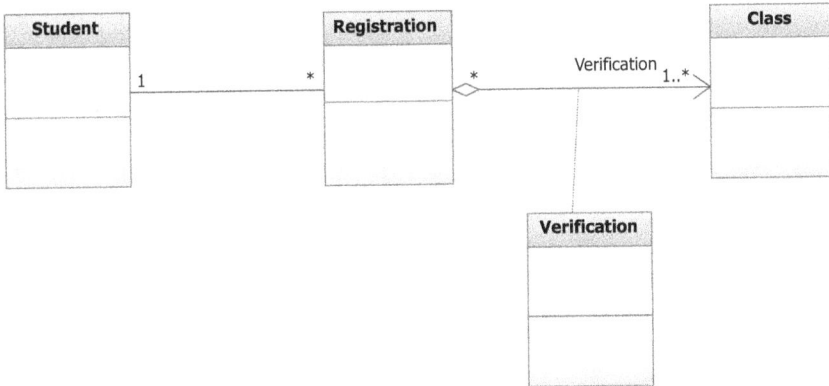

Figure 4.   Student registrations.

Example 3: During March of every year, students sign up for course offerings to take for the following academic year (Figure 4). For each pair of registration and class, the system needs to keep track of its various verification or approval status on whether it has special permissions, whether it must be approved, etc.

Example 4: Each company has 0 or more job positions. Each position gets 0 or more applicants, and each position requires one or more skills with specific proficiency requirements (Figure 5).

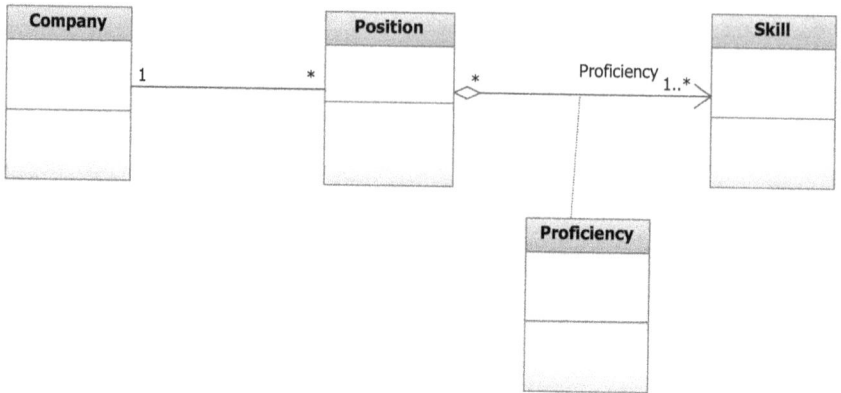

Figure 5.   Job advertisements.

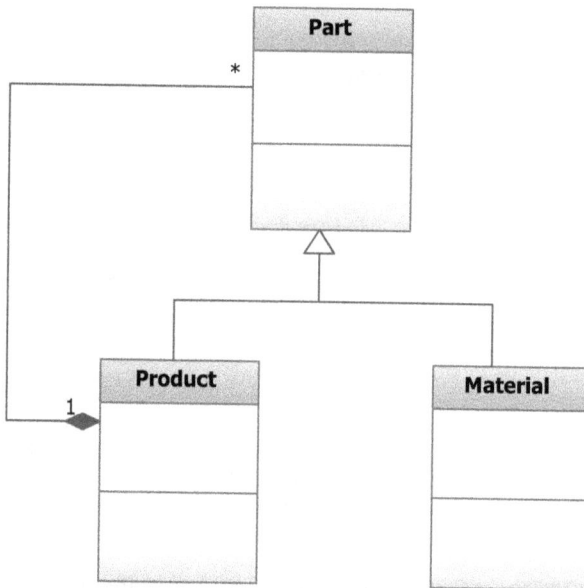

Figure 6.   The assembly pattern.

The Assembly Pattern: This pattern, which some authors called the composite pattern (Gamma *et al.*, 1995), comes from a manufacturer setting, where some parts and/or materials are used to assemble other parts or products. Figure 6 is the class diagram for the pattern.

The assembly pattern is useful for situations with hierarchical compositions from an individual element up to more and more complex objects. The following are a few examples of this pattern:

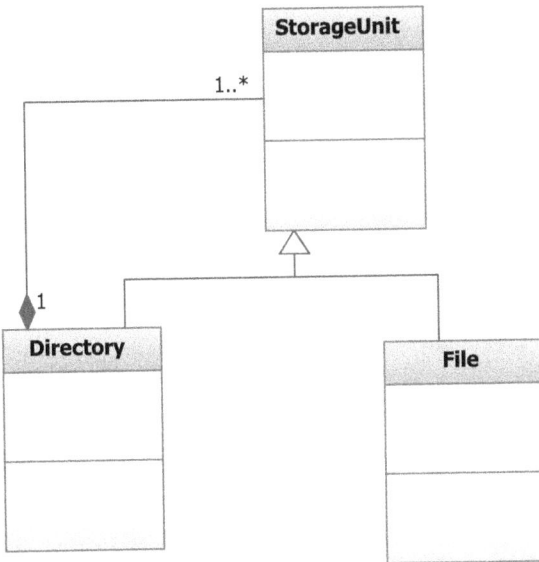

Figure 7.   File systems.

Example 5: A storage artifact includes directories (or folders) and files, while a directory may contain other artifacts such as sub-directories and/or files (Figure 7).

Example 6: In a society, both organizations and individuals are legal representatives, and an organization contains other legal representatives (Figure 8).

Example 7: Both a network and a computer are systems, and a network contains other systems such as sub-networks and computers (Figure 9).

The Representative Pattern: This pattern comes from the use of polymorphism to model exclusive associations. It may sometimes be called the party pattern (Hay 1996). Figure 10 shows two class diagrams using the pattern. Here, each bank account belongs to either an individual or an organization, but not both. Thus, we generalize Individual and Organization into Customer to be associated with Account. Each customer is the representative of an individual or an organization.

The relationship between claims and patients also follows this pattern. A claim may be either filed by an individual patient or by a clinic,

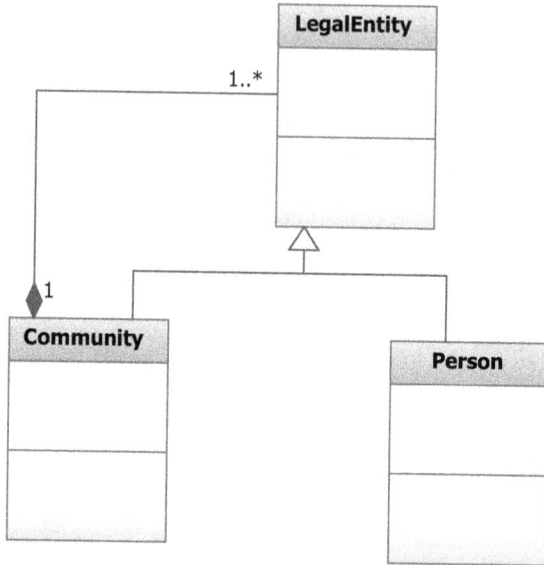

Figure 8.   Social systems.

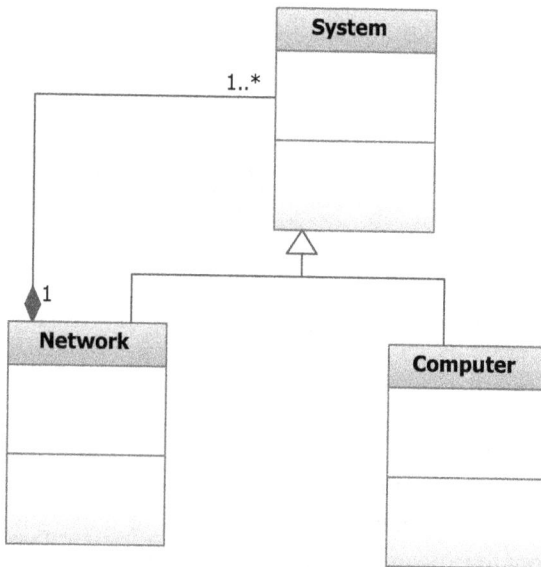

Figure 9.   Computer systems.

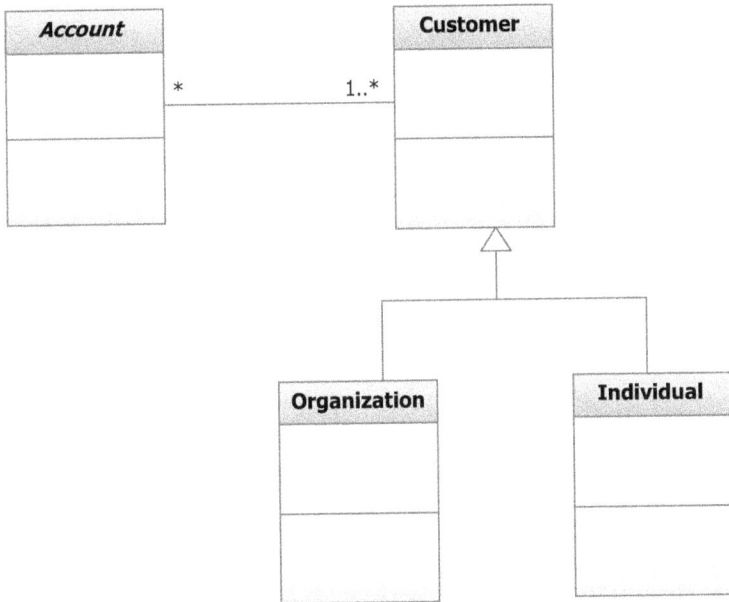

Figure 10.   The representative pattern.

but not both. Instead of using two exclusive associations — one between clinics and claims and one between patients and claims — we can create a representative, say `Filer`, which represents either a patient or a clinic to participate in an association with a claim. Figure 11 is the class diagram for the case.

Later in this book, we will find the representative patterns applicable to use case modeling, where a representative may be used to represent two or more actors who play the same role and two or more use cases that have almost identical sequence of interactions between an actor and the system. In fact, this pattern is widely used in practical class diagramming using phraseology: when we see a sentence that connects one or more subjects to one or more objects by a verb, the subjects and/or objects will often better be modeled by a representative. For example, to model the sentence "an office is assigned to either one full-time professor or multiple part-time instructors," we should find a representative of both full-time professors and part-time instructors to connect with offices.

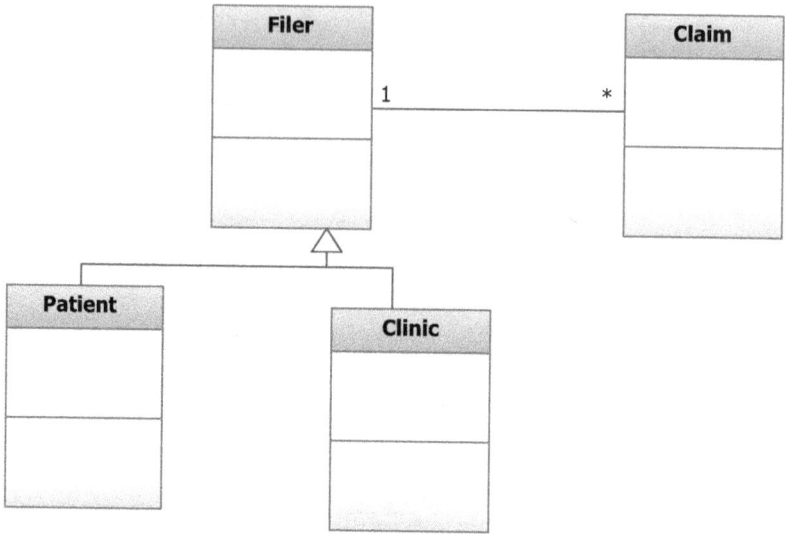

Figure 11. Insurance claims.

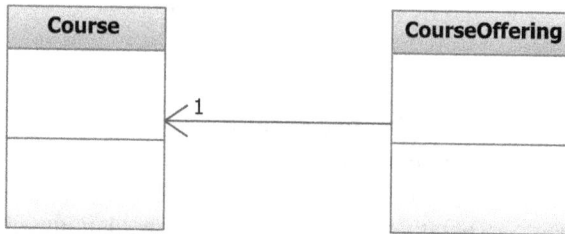

Figure 12. The manifestation pattern.

The Manifestation Pattern: What is difference between a course and a course offering? The former is a catalog entry listed in the college bulletin, whereas the latter is a concrete course section or a manifestation taught by an instructor and taken by students. Their relationships, shown in Figure 12, exhibit an interesting pattern that is followed by many other examples, which are as follows:

1. *Test*: Tests such as SAT and GRE are catalog entries with information such as test name and description. The actual tests are offered under catalog tests, scheduled on a specific date and place, and participated in by students.

2. *Flight*: Flights such as UA112 and UA111 are catalog entries with a flight number, a route, and days. Actual flights are those that passengers can book and are scheduled on a date.

3. *Video*: In a video rental store, videos are catalog entries with information such as ISBN, title, length, description, authors, and publisher. Actual video tapes are those that can be rented by customers and have information such as UPC bar code, whether it is rented out or not, etc.

Practical Skills for Identifying Objects and Relationships

Acquiring domain knowledge is the TAO to discovering objects. However, do not expect clients or users to tell you what objects they have or what you may need; they may not understand object-oriented concepts. Instead, what they can tell you are what they can see or touch, or their complaints with software or their work, or what they want to achieve, or how they do their work. Therefore, it is the business analyst's job to discover objects from what clients and users can offer.

Two common forms of what a user can offer are business forms/reports and problem statements. Business forms and reports tell what data are currently used. Domain objects are business entities that we need to keep or process data for. Thus, if we can find out what data an organization uses or is going to use, the holders or carriers of the data will be the domain objects to be identified. To find out data content, the best approach is to collect existing business documents, including sample forms and reports, in electronic or paper format. For example, Figure 13 shows a customer order form. According to the data shown on the form, there are three types of objects involved: customers, orders, and products. Within each order, each product has a separate order quantity and price, which can only be registered with the relationship between a product and an order. Therefore, we need to create an association class for the relationship between Order and Product, and Figure 14 shows the class diagram to capture the data on the form well.

Discovering objects and relationships using phraseology

Business-, user-, and software requirements are often expressed in textural descriptions. Phraseology is a technique to identify objects, attributes,

PVF Customer Order

Order No:	61384	Customer No:	1275
Order Date:	11/04/1997	Customer Name:	Computer Associates
Promised Date:	11/18/1997	Customer Address:	123 Wall Steet
		City-State-zip:	Austin, TX 73301

Product No	Description	Quantity	Unit Price	Sub Total
M123	Bookcase	4	$200	$800
B234	File Cabinet	2	$145	$290

Total: $1090.00

Figure 13.   Sample customer order form.

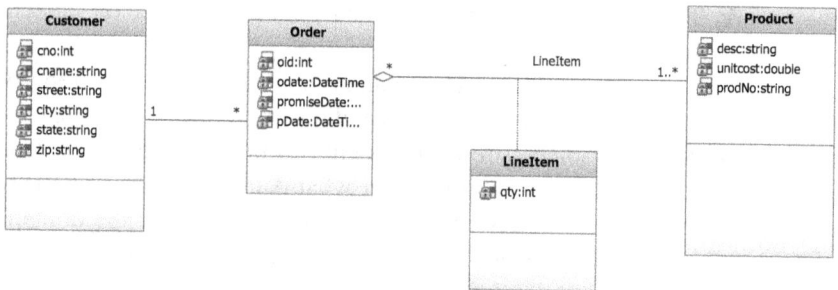

Figure 14.   Class diagram to capture data in the sample order form.

operations, and relationships by analyzing the grammatical structure of a written text. The technique focuses on the nouns contained in a text and decides whether each noun is relevant or not, and, if relevant, whether it should be modeled as an object or attribute. It then focuses on connection words like from, to, off, of, like, is, etc., that link one or more objects and decides whether each of such words establishes a relationship between the identified objects. Finally, it analyzes the verbs contained in the text, determines who carries out the action, and determines whether the verb is an action word that process data or not. If a data action is carried out by a business stakeholder, the verb will probably suggest a use case. If a data action is carried out by a domain object, the verb probably suggests an operation.

Action nouns sometimes indicate relationships. When being asked about what the users want to keep or process data, typically they describe things like enrollment, assignment, shipment, transaction, teaching, etc. These action words, most of time, indicate relationships between two or more objects. We may need to ask the user to describe the meaning of words to understand the role of various parties involved.

Of course, there will be many nouns, connection words, and verbs that are irrelevant to the problem domain as judged by use cases or requirements. They will be left out. In the following, we use two examples to demonstrate this technique.

Example 1: ABC University Business Office receives supplies from various vendors and checks out the items to internal departments. The actual cost of each item is billed to the departments who use the supplies. Internally, as a convention of organizing inventories, supplies are organized into categories. For each supply, the maximum and minimum inventory levels are kept so that when the stock of a part is below the minimum, a replacement order may be issued and sent to a vendor to get it refilled.

By reading the text, we identify the following nouns:

Business office, supplies, departments, actual cost, items, convention, inventories, categories, maximum inventory level, minimum inventory level, stock of a part, parts, vendors, and orders.

The nouns are then filtered in the following order:

1. Remove the objects that we do not want to keep data for: For example, since we are building a system for the business office, we do not need to keep and process data about the business office itself. Thus, the noun "business office" can be removed. Similarly, the noun "convention" does not belong to the domain and can be removed.
2. Remove the synonyms: For example, supplies, items, inventories, and parts are obviously synonyms. We can delete all but keep one, say supplies.
3. Differentiate objects from attributes: For each of the remaining nouns, some of them will be identified as objects while others are attributes. For example, supplies, vendors, categories, and orders are identified as objects. The nouns "maximum inventory level," "minimum inventory level," and "stock of a part" are attributes for supplies.

Thus, we identified the following objects with attributes:

Supplies (actual cost, stock, maximum inventory level, minimum inventory level), Departments, Categories, Vendors, and Orders.

As nouns indicate objects and attributes, connection words including verbs usually indicate relationships, especially when the object and the subject of a verb have both been identified as objects. For example, the sentence "supplies are organized into categories" indicates relationships between categories and items (synonym for supplies). Similarly, "Items are checked out to departments" indicates relationships between supplies and departments. For this purpose, let us underline all the connection words:

"ABC University Business Office <u>receives</u> supplies <u>from</u> various vendors and <u>check out</u> the items <u>to</u> internal departments. The actual cost <u>of</u> each item <u>is billed</u> <u>to</u> the departments who <u>use</u> the supplies. Internally, as a convention <u>of</u> organizing inventories, supplies <u>are organized</u> <u>into</u> categories. For each supply, the maximum <u>and</u> minimum inventory levels <u>are kept</u> so that when the stock <u>of</u> a part <u>is below</u> the minimum, a replacement order will <u>be issued and sent</u> <u>to</u> a vendor to <u>get</u> it refilled."

The underlined connection words are extracted into the following list: business offices receives supplies, supplies from vendors, business office checks out supplies, supplies are checked out to departments, departments use supplies, actual cost is of a supply, actual cost is billed (by Business Office), actual cost to departments, supplies are organized into categories, maximum and minimum inventory levels are kept (by Business Office), stock is of a supply, stock is below the minimum inventory level, order is issued and sent (by Business Office) to vendor, vendor fills order.

Now review the connections one by one in the following order:

1. Some of these connections are already known. For example, stock and actual cost are attributes of supplies. Remove them!
2. Some connections are linking objects with somebody or something not identified as objects. Remove them for now or consider them in use case modeling later. For example, "receive," "check out," "kept by," "billed by," and "issued and sent by" are the actions carried out by Business Office, which is outside the list. In fact, they indicate five use cases: receive supplies, check out items, bill departments, check inventory levels, and issue orders (see the next chapter).
3. Analyze the multiple connections between the same objects to see whether they are of different connections in terms of characteristics.

If they are the same, remove duplicates. For example, "order to vendor" and "vendor fills order" should be the same. Similarly, "supplies are checked out to departments," "departments use supplies," and "actual cost to departments" are identical.

After these steps, we come up with a short list as follows: supplies from vendors, supplies are checked out to departments, supplies are organized into categories, stock is below the minimum inventory level, and vendors fill orders. Among the five remaining connections, "stock is below the minimum inventory level" suggests an operation to compare stock levels. The other four connections suggest associations. For each association, we need to decide whether it is direct, or if it is established through another middleman object. The criterion is whether we need to keep data on the action verbs or not. If yes, the middleman object is used. Supplies are ordered from vendors, and supplies are checked out to departments. We need to keep data on the order and check out actions, for example, their dates, places, and amounts, etc., so that we can track the history of the actions. Thus, there is a middleman class, `ReplacementOrder`, between `Supply` and `Vendor`, and there is also `CheckOut` between `Supply` and `Department`. Also, by applying the transaction pattern to both cases, we identify two association classes. Finally, "Supplies are organized into categories" means that each supply belongs to a category. It is a direct association because we do not need to track the action of organizing supplies into categories.

Summarizing all the above findings, we come up with a class diagram as shown in Figure 15. Note that, after relationships are identified, the determination of mapping cardinality and relationship types (composite, aggregate, association, dependence, or generalization) are often not difficult.

In this example, all the objects and the relationships are identified using phraseology. This is not always the case. In particular, the relationships between objects are often not hinted by connection words. Instead, they must be identified through alternative routes. There are three alternative strategies we can use to determine whether there is a relationship between any objects.

1. *Navigational Test*: Given an object in one class, do we need to find the corresponding objects in the other class? If this test is positive, then there will be relationships between the two classes. For example, between `ReplacementOrder` and `Vendor`, given an order,

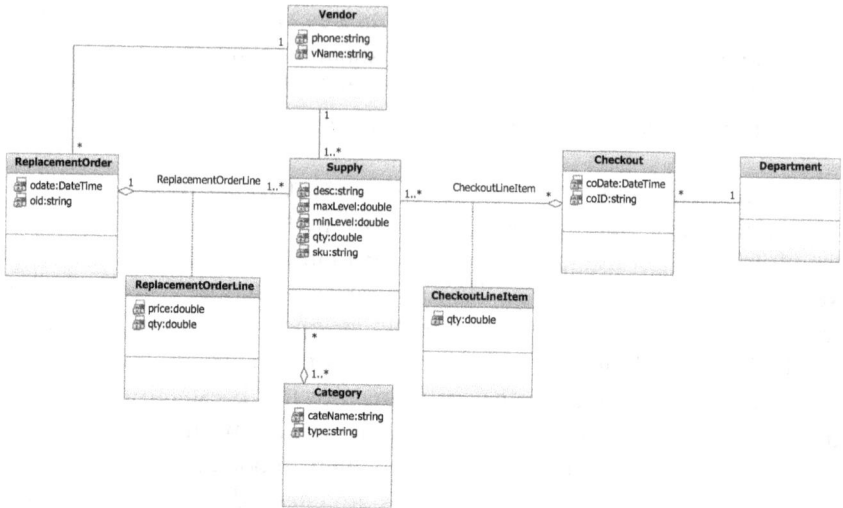

Figure 15.   Class diagram for Example 1.

do we need to know to whom we will send the order? Of course, the answer is positive. Therefore, we need to model relationships between `ReplacementOrder` and `Vendor`. For another example, between `ReplacementOrder` and `Supply`, given an order, do we need to know what is in the order? Yes, of course. Thus, there is an association between them.

2. *Responsibility Test:* If an object in one class has one or more operations that require the knowledge of and/or request services from an object in the other class, then these two classes should be connected. For example, we already know that to create a new replacement order if the stock is below the minimum inventory level, i.e., the "create order" operation in the `Order` class, requires knowledge about supplies. Thus, `ReplacementOrder` and `Supply` should be connected. In later chapters, we will see that after a use case is fully described, more such collaborations will become evident and can be used to identify the relationships.

3. *Domain Knowledge Test*: All the relationships must be consistent with domain knowledge. The navigational and responsibility tests may indicate high-level or indirect relationships. Domain knowledge will supplement them by providing detailed navigation and collaboration maps and direct relationships. For example, phraseology

indicates relationships between `Supply` and `Department`. But how is a supply related to a department? The domain knowledge tells us that a department makes checkout transactions and each transaction contains certain items. Therefore, we need to create a new class called `CheckOut` and then link `Department` to `CheckOut` and `CheckOut` to `Supply`.

Example 2: Insure-A-Person Inc. provides health insurance services to employees and their family members across America. Due to the need to promote its customer relations, the company has decided to open a web-based system for clinics and individual customers to be able to file claims on the Internet 24 hours a day and 7 days a week. The company has approached you to design the system for that purpose. According to the company, this is how the web-based system is supposed to work. Within 60 days of seeking treatments for himself or any of his family members, a customer needs to log on to the system and file a claim. First, you specify the name of a patient, the date and the place the service was provided, and the primary doctor providing the service. Then, you detail the procedures performed by the doctor. In the medical industry, all procedures have been standardized with fixed identification numbers and short descriptions. The insurance company will pay for the service based on all the procedures performed by the service.

Reading the text one round, we can identify the following nouns that may be relevant (irrelevant nouns are ignored):

> Employees, family members, customers, clinics, claims, 60 days, treatments, name, patient, date, place, service, primary doctor, procedures, identification numbers, descriptions.

Treatments and services are synonyms, and so are places and clinics. Employees and customers are also synonyms, and they are related to patients, which also include family members. Thus, we can eliminate the duplicates but keep patients as a generalization of employees and their family members. Nouns like 60 days, date, name, identification numbers, and descriptions are attributes. Thus, we identified the following objects along with the mentioned attributes:

> Patient (name), Employee, Dependent, Clinic, Claim, Treatments (date), Doctor, Procedure (identification numbers, description).

Next, let us analyze connection words. A few connection words like "of" indicate attributes of objects as shown above. Some are irrelevant such as "Insure-A-Person Inc. provides health insurance services." Ignoring all those, we come up with a short list to be analyzed:

> Clinics file claims, patients file claims, claims for treatments, treatments for patients, clinics provide treatments, doctors provide treatments, doctors perform procedures, treatments contain (perform) procedures.

Since a claim cannot be filed by both a patient and a clinic, the relationships are exclusive. To resolve the exclusiveness, we can abstract Patient and Clinic to an abstract representative called Filer (see the representative pattern). Procedures are described as a catalog entry, and actual procedures are performed by doctors. Thus, the manifestation pattern applies. Also, since doctors perform treatments and treatments contain actual procedures, the connection "doctors perform procedures" indicates an indirect relationship. Figure 16 shows the class diagram summarizing all the above findings.

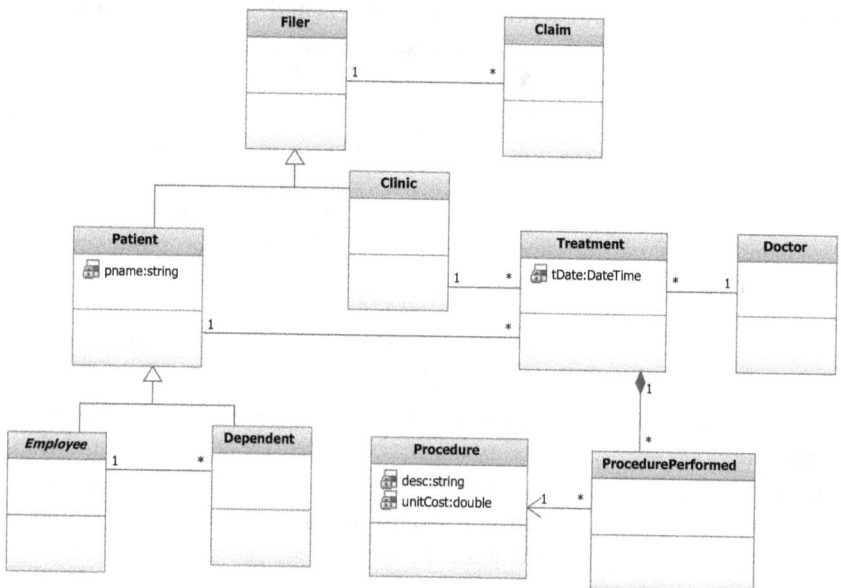

Figure 16.   Class diagram for Example 2.

Exercises

1. Give one of your own examples to apply the following patterns:
   a. Manifestation pattern.
   b. Transaction pattern.
   c. Representative pattern.
   d. Assembly pattern.
2. Construct a class diagram that reflects the data requirements for the data input form shown in Figure 17.
3. Construct a class diagram that reflects the data requirement from the data input form shown in Figure 18.
4. In Video Shack, customers are required to have a family membership card that is used mainly to ensure that they have a credit card, live in the neighborhood, and can be contacted in case they are late in returning their rentals. Video Shack has a varied stock of videos classified into such categories as comedy, adventure, children's, and romantic. Any title is obtainable from one distributor who owns the rights to it. Video Shack deals with about 25 distributors for different titles. It may carry many copies of a popular new title or only a single copy of some classics. Popular titles may have to be reordered. Most of the

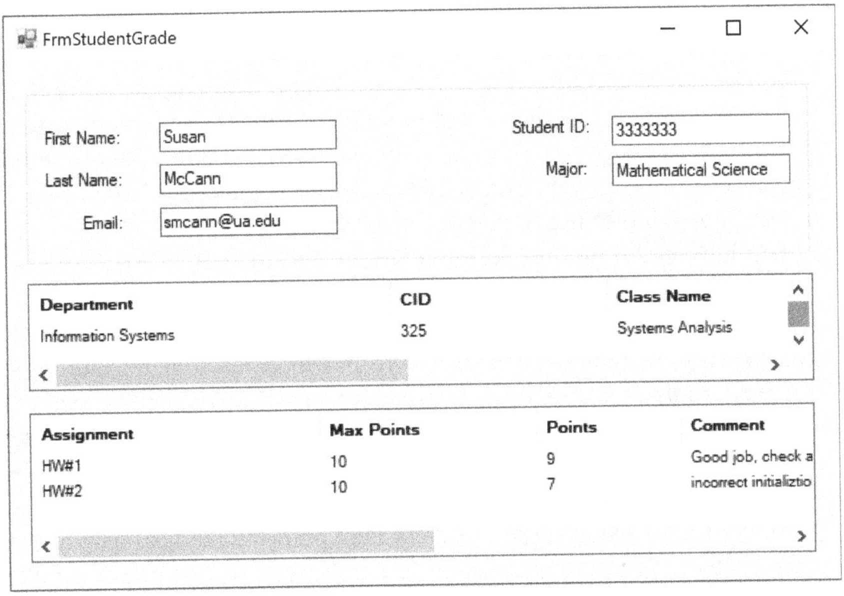

Figure 17.   A data input form for displaying student grades.

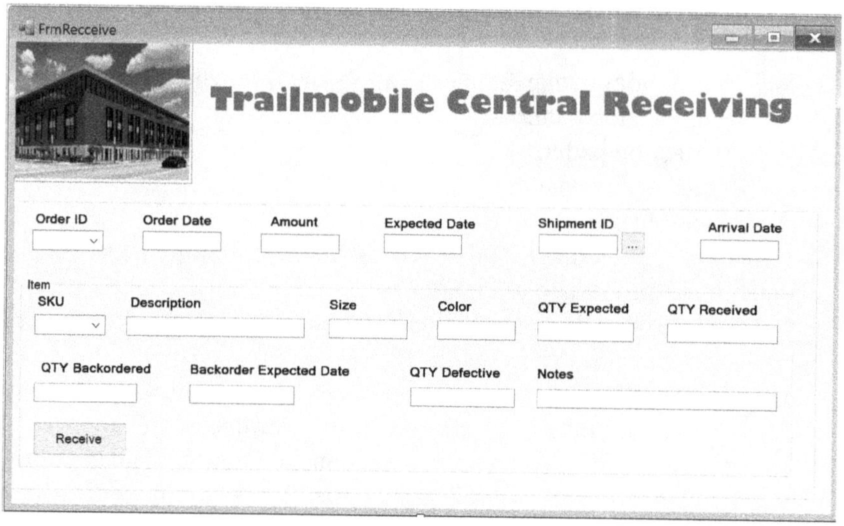

Figure 18.   The data input form for receiving shipments.

videos are rented for a standard price. However, there is sometimes a premium price for new releases. There is also a discount during weekdays. Customers agree to return rentals by noon of a set date, and they can reserve up to five videos in advance to ensure that they will be available when desired. Design a class diagram for a system that can be used to record purchases of videos from suppliers, record rentals and returns by customers, and produce a printed catalog of current holdings categorized by title and type. In addition, Video Shack would like to be able to get listings of how many copies they have of each title and how often each title has been rented.

5. You have been hired to design a system for a small healthcare organization. The clinic consists of several examining rooms and a few rooms for short-term critical-care patients. A core staff of seven physicians is supplemented by internists from a local teaching hospital. The clinic wants to computerize the patient records. All patient medical data is stored in a folder kept in a large central file cabinet. Arriving patients sign in at the front desk. A clerk checks the billing records, prints out a summary status sheet, and obtains the file number from the computerized system. The clerk then pulls the medical data folder and selects an examination room. After waiting for the physician, the clerk moves the data packet and the patient to the examination room.

A nurse records basic medical data (weight, blood pressure, etc.). The physician makes additional notes to both the medical and billing data and generally writes a prescription order, which is given to the patient and recorded on the charts. When the patient leaves, the clerk enters the new billing data into the system, collects any payments, and prints a list of charges and a receipt. The new billing data is forwarded to the appropriate insurance company. The medical data is returned to the filing cabinet. When the patient gets a prescription filled, the pharmacist calls the clinic for verification. A clerk retrieves the medical data, identifies the prescription, and verifies or corrects the order. Draw a class diagram to capture the requirements for an automated medical record system.

6. Create a class diagram to identify the primary business objects of the following business and show their relationships: A hardware store approaches you to design a new POS system to manage its inventories and sales. The store carries four different categories of products, including tools, lumber, plumbing items, and lawn-and-garden supplies. At first, you thought a simple sales system would work for this company. After all, each item has a cost, a description, and a list price. But after talking with the managers, you quickly learned that the items sold by this company have important differences.

   a. The tools have warranties, but the garden plants and lumber do not. Note that warranties have details including duration, description, and applicable conditions.

   b. Some items, like the plumbing supplies, have special attributes, and so managers often want to search based on those characteristics. For example, the plumbing department manager might want to know the inventory level of all 1/4-inch-interior-diameter pipes.

   c. All garden products have either a high temperature or low temperature or both — some plants must be moved inside if the temperature drops below a certain level.

   d. When clerks and managers check out items, they often must take different actions depending on which item is sold. For instance, when clerks ring up a sale for certain chemicals in the lawn-and-garden category, they are supposed to get the Federal pesticide license number of the customer.

   e. When the store orders certain electrical tools, it must send license and authorization numbers to the supplier.

f. The store also has two types of customers: Individuals and Contractors. Contractors are those who have registered with the store to obtain a lower price. Certain large contractors get additional discounts on some items like plumbing, electrical, and heating equipment. Discounts for these special contractors are negotiated individually.

g. A similar problem exists with suppliers — each one grants the store a different level of discounts. Each one also asks for different types of authorization numbers to get a discount on certain product, and some require the use of electronic data interchange (EDI) or touch-tone orders to get the best discounts. When managers order items, they must identify the supplier, find the best conditions, and follow the rules specified by that supplier.

7. Draw simple class diagrams for the objects referenced by each of the following sentences. Apply appropriate heuristics and design patterns and add a few appropriate data members and operations.

a. After each treatment, either patient or clinic can file insurance claims.

b. After each visit, the physician will create an order for additional tests or write a prescription.

c. In project management, of course, a project consists of one or more activities, and often a big project is made of many small projects.

d. Customers, including individuals and organizations, can order various products, some of which are perishable, with an expiration date, from the store.

e. A computer system is sometimes made of just one computer, but nowadays often made of one or more networks of computing devices that collaborate.

f. Vendors dispatch thousands of shipments, each of which contains one or more products, and track their status using RFID technology.

g. EMR systems keep not only static data on tests and medicines, but also detailed daily patient records such test results of the tests performed on each patient.

8. Draw simple class diagrams using the manifestation pattern for each of the following cases:

a. The video shack keeps multiple copies of each video title.

b. The hospital has thousands of procedurals in book but can perform only a handful per day.

c. Many courses are in book, but the department schedules less than 50 per term.

d. A patron can reserve up to three books at a time but can check out unlimited number of books.

9. Draw simple class diagrams using the representative pattern for each of the following cases:

a. An appointment may be made for a receptionist or a patient.

b. A customer can either buy or sell online with the web store.

c. A patient may come to the hospital for either an inpatient or outpatient visit.

d. The agent represents its clients or the employees of the clients in defending their cases in count.

10. Draw simple class diagrams using the assembly patterns for each of the following cases:

a. Army unit includes individual soldiers and smaller units.

b. An academic program is made of courses and other programs of education and training.

c. An ecosystem is made of individual species and smaller ecosystems.

11. Draw simple class diagrams using the transaction pattern for each of the following cases:

a. Each division may propose 0 or more bid request, each of which is sent to one or more vendors to request for bids. Some special instructions may be noted to different vendors.

b. Customers can make reservations, each of which can reserve one or more video titles, but the system needs to keep track of their urgency or priority.

c. A meal contains one or more foods, each of which uses one or more materials. One material can be used in 0 or more foods. For each meal, a specific instruction is noted for using each material.

Chapter 9

Use Case Modeling

Introduction

Use case analysis is a process of creating artifacts that represent dynamic system functions (processes) and procedures (operations), both of which we have studied in earlier chapters. It consists of two stages: use case diagramming and use case storyboarding. The deliverables, accordingly, include use case diagrams and use case descriptions along with user interfaces (UI). In this chapter, we will learn (1) the basic mechanics of use case modeling techniques and (2) how to identify use case model elements, including use cases and actors, from requirements.

Connections

We have learned how to draw class diagrams. There is no doubt that class diagrams are the most important deliverables of the systems analysis and design process, but remember that classes in a class diagram have both attributes and operations. Attributes are data about the objects and are often easy to identify (and the reader may have a separate Database Management course to cover the techniques extensively), but where and how can we come up with the operations?

In Chapters 3 and 4, we learned the concept of functions and how to represent them internally and externally as well as programmatically. The chapters also briefly discussed the methods of functional decomposition and data flow diagramming, which are the formal methods that the structured methodology uses to identify functionalities. These methods are meant to help us understand the concept of classes and objects, but they are not formal object-oriented methods for identifying functionalities.

In Chapters 5–8, we learned how to allocate functions, if captured, into the responsible objects. The chapters emphasized the heuristics and principles of data flow reduction and responsibility distribution. However, there are three problems to solve in order to complete the picture. The first is what major functions we should capture. The second is how to ensure that we will capture all the functions that are in support of the mission of the system, no more and no less. The third, which is somehow cynical, is how we can capture functionalities using a method that is consistent with the object-oriented methodology, i.e., obeying the principles of encapsulation, inheritance, and polymorphism.

The use case modeling resolves all these issues. From the point of supporting the mission of a system, we capture the *major functions that deliver values to the user*. These functions are called use cases. By describing each use case as a step-by-step sequence of interactions between the user and the system, we can capture all the sub-functions or operations needed to support the major function. In other words, we derive the subprocesses or operations by describing the sequence. Furthermore, by understanding that each action performed by the system is eventually performed by one or more objects, we can elaborate the use case description into a collaboration diagram that shows how participating objects collaborate with each other to perform a use case. Naturally, the actions performed by each object shall be those allocated to the object. Therefore, *the use case analysis discovers and formalizes the process requirements and enriches the class diagrams* by identifying and allocating operations into objects.

Use Case Diagramming Elements

Two basic constructs in use case diagrams are actor and use cases. The reader may be surprised to know that both actors and use cases are classes, and so use case diagrams are a special kind of class diagrams. Thus, actors and uses cases are connected through association and inheritance relationships.

Actors

An actor is a group of users that play the same role in using (or interacting with) the system to be developed. Users are instances of actors. Sounds familiar? Yes, an actor is a class, and a user is an instance or object. When understanding the concept and applying it to the identification of actors, we note the following four guidelines:

1. The user can be a person, an organization, or another system that communicates with the system. In general, it is an autonomous agent that is external to the system to be developed but interacts with it. So, we shall not restrict the concept of actors to physical human beings.
2. Sometimes there are many users who do the same thing, and sometimes one user may do different things with the system. In any case, we group all the users that play the same role into each group to make it an actor even though the group may have only one user. If a person plays different roles, then he or she belongs to different actors.
3. Not all the things that interact with the system are users or agents. A criterion to judge whether they are users or not is whether they have the capability of requesting services from or providing services to the system to be developed. For example, in an automatic teller machine (ATM), the credit card holder uses a credit card to interact with the ATM. Is the credit card a user? It is external to the system and interacts with the system. However, does it have ability to request services (or gain value) or provide service? The answer is no. Thus, credit cards should not be treated as an actor.
4. Actors must be logically outside the system to be developed. A system is made of one or more classes, including domain (or business) classes, control classes, and user interface classes. Sometimes, systems also include hardware components and databases. All these are not external to the system, and so they should not be considered as actors. Then what about employees who will use the system? Do we not already represent employees as the `Employee` class, which is a part of the system? A tricky question. Yes, we modeled employees into the system as the `Employee` class, but we did not actually teleport actual employees into the system. So, actual employees are still outside the system.

To recap, an actor represents a group of autonomous agents that provides services to and/or request services from the system and plays the same role in doing so. Depending on whether an actor can request services (and thus derive values) from the system or not, an actor may be classified into *primary and secondary*. A primary actor actively uses the system and gains value from the use, whereas a secondary actor plays a supporting role to serve the system.

Example 1 (ATM): Note that an ATM is a system consists of both hardware components and software objects, while most information systems are purely software. ATMs are a good example for us to learn the concepts of use case modeling. An ATM serves bank customers as well as other credit card holders, allowing them to withdraw money, deposit money, transfer money, and inquire about account balance. It requires services from machine operators to refill the cash dispenser, retrieve swallowed cards, and retrieve deposited checks. It also requires the bank system and the card networks to validate transactions. Thus, the actors of an ATM include Customer, Card Holder, Operator, Bank, and Card Network, with Customer, Card Holder, and Operator as primary actors. The values they derive from the ATM include serving their banking needs or fulfilling their job responsibilities. Bank and Card Network are secondary actors since they gain no value from the ATM.

In a use case diagram, actors are represented as stick human figures (see Figure 1). Some CASE tools use a rectangle box with a stereotype <<actor>> for non-human system actors. Indeed, an actor is a class, a stereotyped class. Rhapsody does not support the differentiation, and so we will use stick figures for all actors.

Actors are defined and identified based the role that their users play. Naturally, some users can play more roles than others, and so we may want to indicate that one actor is a special type of another actor based on the roles. For example, the role of Customer is a kind of Card Holder; as a card holder, a customer uses a debit card to withdraw cash. However, customers can do a lot more than be a card holder. For examples, customers

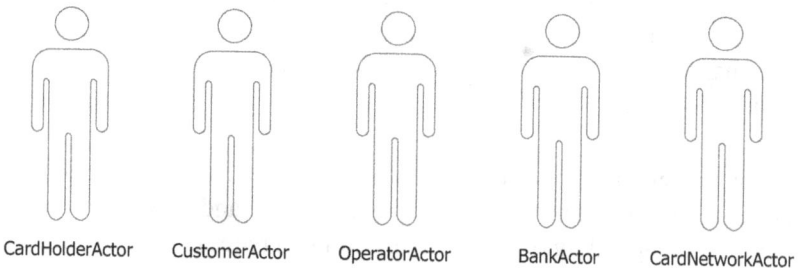

CardHolderActor   CustomerActor   OperatorActor   BankActor   CardNetworkActor

Figure 1.   Actors in the ATM example.

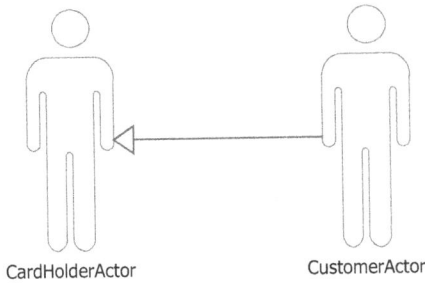

CardHolderActor   CustomerActor

Figure 2.   Role map.

can use the ATM to deposit money, transfer money, and inquire about account balances. Thus, we may model the Customer actor as a specialized or enriched Card Holder actor using an inheritance relationship (see Figure 2). A diagram that shows the roles of actors, like Figure 1, is called a role map. Remember that a specialized or child actor can do more, or perform more use cases, than its parent actor.

Note that there exists no association relationship between actors. By definition, actors are outside the boundary of the system to be developed. Thus, how they may interact with each other is outside our modeling scope, and thus the associations between actors are not our business. For example, in the ATM example, the Bank actor and the Card Network actor may have interactions; clearly, the bank lends money to a card holder, and it will contact the Card Network to get the money back. However, such *interactions between actors are not modeled* in a use case diagram.

Use cases

A *use case*, aka *system use case*, captures a major function to be provided by the system. Conceptually, it is a class — a stereotyped class. The concept may be defined differently from two perspectives. The first is based on usage instances, each involving instances, or users, of one primary actor who uses the system to derive a value from the system. Example usage instances include: a student using the registration system to register for a course offering, a customer using an ATM to withdraw cash, etc. *A use case is a group of usage instances that deliver the same type of value to the same primary actor.* Thus, a use case is a class.

Note that some usage instances may produce results of no observable value due to failures and due to incompleteness, whereas other may deliver more than one observable result of value. A use case is a class of usage instances, each of which delivers one observable result of value to the primary actor. The reader shall note these three key points here: (1) all usage instances of a use case serve the same purpose or deliver the same type of value to the primary actor, (2) all usage instances result in one and only one observable result of value, and (3) all usage instances are performed by the users of the same primary actor. In brief, *a use case must deliver one observable result of value to one primary actor.* Usages that result in no observable value to the primary actor or two or more values will not be considered as instances of a use case.

The second perspective is to define use cases based on the interactions between a user and the system. Let us contemplate a sequence of interactions. Imagine a card holder — who may be a passenger in an airport, a customer in a grocery store, or a patient in a hospital — wants to withdraw cash from an ATM. The typical sequence of interactions between the user and the system is as follows:

1. Card Holder inserts a card into ATM.
2. ATM validates the card.
3. ATM asks for a pin number.
4. Card Holder enters a pin number.
5. ATM validates the pin number.
6. ATM asks for a withdrawal amount.
7. Card Holder enters a withdrawal amount.
8. ATM authorizes the withdrawal amount.
9. ATM dispenses cash.
10. Card Holder takes cash.
11. ATM prints a receipt.
12. ATM records the transaction.
13. ATM releases the card.
14. Card Holder takes the card.
15. ATM goes idle.

Note that the above sequence of interactions does not refer to a specific card holder, a specific teller machine, a specific card, or a specific withdraw amount. By varying all these parameter values, the sequence represents a group of all usage instances between a card holder and the system for the purpose of withdrawing money. Therefore, *a use case may be defined as a sequence of interactions that leads to one observable result of value to the primary actor.*

This second definition is practical and may be more appealing to structured systems developers because it essentially makes a use case equivalent to a function, which, by definition, is made of a sequence of activities that work together to perform a high-level task. In practice, this mindset does not cause any trouble to business analysts, with two exceptions. One exception is that a use case model shows associations between actors and use cases, and thus it is a stereotyped class diagram. In contrast, in the structured development, a process model shows data or work flows between functions.

Another exception is that a use case must deliver an observable result of value to the user whereas a function does not have to. Note that the definition of use cases emphasizes one primary actor and one purpose in terms of one observable result of value to the primary actor. *Not all functions performed by the system are use cases.* For example, the ATM validates cards, validates pin numbers, prints receipts, etc. Since these functions do not lead to any observable result to the user, they are not use cases. By the same token, any incomplete sequence of interactions is not a use case.

Not all complete sequences are use cases either. Here is one example:

1. Card Holder inserts a card into ATM.
2. ATM validates the card.
3. ATM displays "invalid card".
4. ATM releases the card.
5. Card Holder removes the card.
6. ATM goes idle.

Apparently, a card holder wants to use the ATM for some purpose but encounters failure. Since the sequence does not lead to any observable result of value, this is not a use case either.

Here is another example. A card holder, who happens to be a customer, uses an ATM provided by the customer's bank. The customer first deposits a check into her account and then moves on to withdraw some cash. The sequence of interactions obviously serves two purposes, deposit money and withdraw money, and it will lead to two observable results of value to the user. Such a sequence is also not a use case. A sequence that delivers multiple observable results of value may have to be split into atomic ones. Otherwise, there will be too many different combinations of sequences to be captured as use cases.

In sum, an incomplete sequence of interactions is not a use case, and a complete sequence that does not end with any observable result or ends with two or more results is also not a use case.

This delineates a dramatic difference between structured and object-oriented development: the former is feature-focused while the latter is value-focused. In structured development, we try to capture any desirable functionality or feature — simple or complex, creative or destructive — as a function. Functions are arranged in a layered structure via function decomposition, where simple lower-level functions support a complex higher-level one. In contrast, in object-oriented development, we capture only those functionalities that provide a value to a primary actor as use cases. Validating cards and printing receipts are necessary functions that an ATM must possess, but they are not use cases because no user uses the ATM merely for validating a card or printing a receipt. Similarly, playing video games and streaming movies may be desirable features, but they are not use cases because they do not serve any ATM users.

Here is value-focused thinking. To capture use cases, we need to first capture who the users or primary actors are. Then we ask the users what they want from the system or what values they can obtain from the system. In the ATM example, Card Holder, Customer, and Operator are three primary actors. Card holders use the system to withdraw cash. Bank customers use the system to deposit money and inquire about account balance. They also withdraw cash like other card holders. An operator uses an ATM for maintenance purpose, including refilling cash dispenser and printer and retrieving deposits and swallowed cards. Therefore, there are four use cases for the ATM system: withdraw money, deposit money, inquire balance, and perform maintenance. Since nobody uses an ATM for validating cards or streaming videos, these features will not be captured as use cases.

In use case diagrams, use cases are shown as oval circles (see Figure 3), and *each use case is named with a verb,* similar to the way in which we name functions in data flow diagrams or actions and activities in activity diagrams.

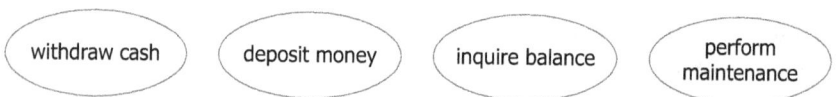

Figure 3.   Use cases for ATM.

Use Case Diagrams

A *use case diagram* is a graphical representation of how actors are associated with system use cases. The primary components of a use case diagram include actors, use cases, associations between an actor and a use case, a role map to show the role generalization/specialization relationships between actors, and relationships between use cases such as generalization/specialization, inclusions, and extensions (see Chapter 11). Figure 4 shows the use cases we have identified so far and their associations with the primary actors in the ATM example.

Since a use case is a class, and so is an actor, their relationship is an association, representing that a user in the role of the actor uses or performs an instance of the use case. A use case diagram is a special kind of class diagrams. Thus, it follows the syntax rules of class diagramming. For example, an association may be marked with cardinalities, quantifying how many users are involved with one usage instance of a use case and how many usage instances of a use case are associated with one user. The cardinality is useful to model problems like game playing, where several game players may be involved in one gaming session, and video streaming,

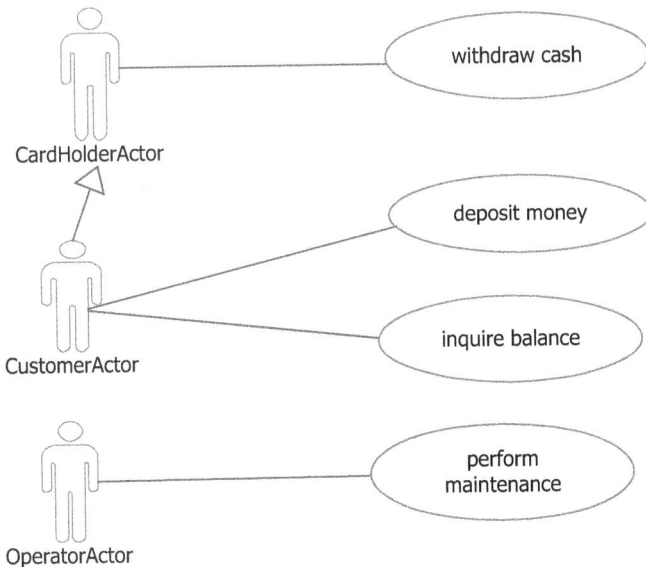

Figure 4.   Use cases and associations with actors.

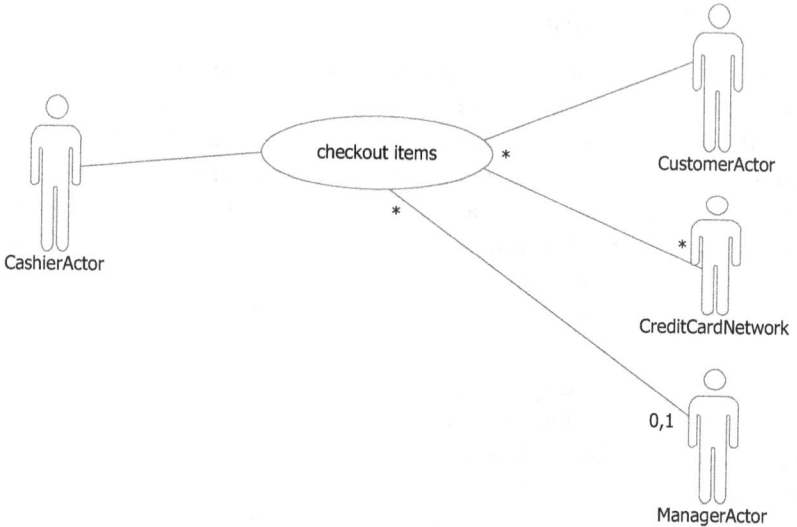

Figure 5.   "Checkout Items" use case.

where one user downloads two or more videos at the same time. It also allows an actor to be optional in an association with a use case as in the case where a manager user is optional in checking out items in a store, for example, to override prices (see Figure 5). Note that, by default, a use case diagram does not show cardinalities. In Rhapsody, we can right click on an association and go to Display Options dialog box to check Multiplicity in order to view cardinalities.

Unlike a regular class diagram, a use case diagram has one special syntax rule: *each use case is associated with one and only one primary actor* in accordance with the requirement that each use case serves one and only one primary actor. However, each use case may be associated with multiple secondary actors. For example, the "checkout items" use case in a point of sale (POS) system is associated with one primary actor — Cashier — and three secondary actors — Customer, Manager, and Credit Card Network (see Figure 5).

The reader may notice in Figure 5 that the primary actor is drawn on the left-hand side of a use case whereas secondary actors are placed on the right-hand side. This is a convention. Indeed, secondary actors play only supporting roles, and the focus of a use case diagram is on the primary actor.

Adding secondary actors to Figure 4, we have an updated use case diagram for the ATM example (see Figure 6). Here, all the use cases involving bank accounts must interact with the Bank for authorization and account updates. For a card holder, the "Withdraw Cash" use case must interact with a card network for authorization. Of course, the card network will need to contact the bank that issued the card to gain authorization and transfer funds, but that is not a concern when designing the ATM.

Note that Figure 6 uses role maps to address two special needs. First, both bank customers and card holders perform the "Withdraw Cash" use case, in which they play the same role. Since a customer does more than just withdraw cash, we invoke an inheritance relationship to model Customer to be a special kind of Card Holder, and thus Customer will inherit the association between Card Holder and the "Withdraw Cash" use case. By doing so, we conform to the rule that each use case is associated with one and only one primary actor.

Second, the "Withdraw Cash" use case involves either Card Network or Bank as a secondary actor for authorization. We may use two separate associations to connect the use case to both Card Network and Bank actors since there is no restriction on how many secondary actors can be associated with a use case. Note that these two associations are exclusive. When withdrawing money from a bank account, there is no need to involve a card network, and vice versa. How can we model such a situation? One

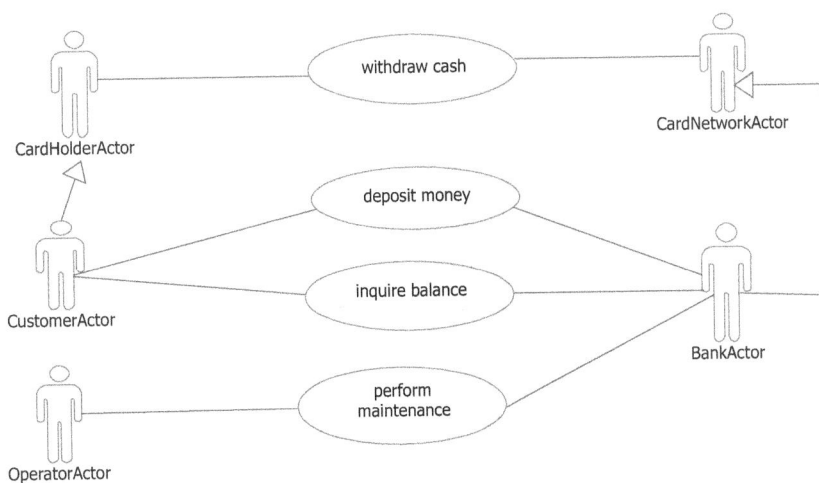

Figure 6.   Use case diagram for ATM example.

solution is to use {OR} constraint, linking the associations to indicate that the concerned associations are exclusive (see Chapter 7). The second solution is to split the use case into two separate ones: one for card holders and one for bank customers. When using the second solution, we will not need to use the inheritance relationship between Card Holder and Customer actors. Both solutions are undesirable because they create duplicate artifacts in a use case diagram as well as redundancies and complexities in describing the use cases (see the next chapter).

The third solution is to use the representative pattern and model Bank as a special kind of Card Network, as in Figure 6. Alternatively, we can create Authorization actor as a representative to both Card Network and Bank and use a role map to show that Card Network and Bank are special kinds of Authorization actors and then associate the "Withdraw Cash" use case to Authorization and other use cases to Bank (see Figure 7).

We will have other alternative solutions to the above problem. In Chapter 11, we will learn three use case optimization techniques to optimize use case models in order to reduce the redundancies and complexities of use case description. These techniques include factorization, extension, and inheritance. A brief definition of these is provided in the following:

1. *Factorization*: We may factorize a portion of a use case into a separate use case, called inclusion use case, and use <<include>> relationship between the base and inclusion use case to show the factorization.

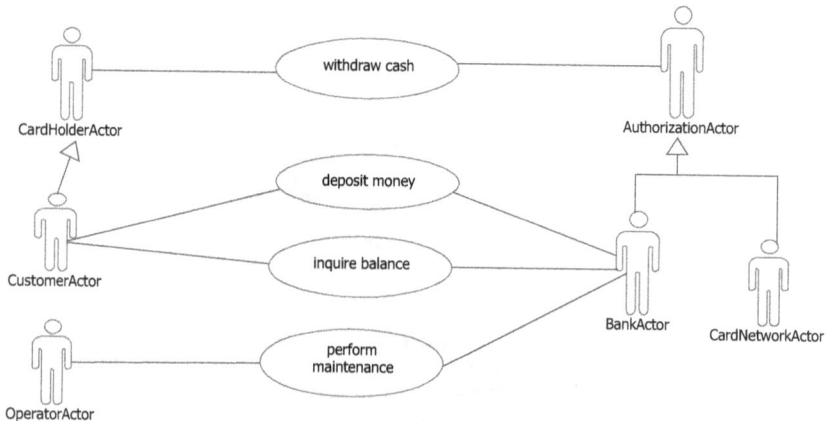

Figure 7.   Alternative use case diagram for ATM example.

2. *Extension*: We may optionally insert a sequence of interactions into a use case to enrich or extend its value. We model the sequence to be optionally inserted as an extension use case and use the <<extend>> relationship to connect it to the use case to be enriched.
3. *Generalization*: If one use case is essentially the same as the other except for the differences in a few steps in their sequences of interactions, we may invoke the representative pattern and model them as child use cases of a representative one.

Example 1: The Board of Watson Town Memorial Hospital has recently decided to develop a new information system to manage their patient admissions and discharges. The hospital handles two types of patients: outpatient and resident patient. As typical, each time a new patient comes, the data about his/her identification, address, phone, and issuance carriers are recorded. If a patient is a resident, he/she will be assigned to a bed and an admission date is recorded. After the treatment, a nurse must sign off the discharge card. For an outpatient, the nurse will set a check-back time after each treatment. Identify actors and use cases based on the text and develop a use case diagram.

The primary users of this system include receptionists, nurses, and patients, where receptionists and nurses are primary actors who actively request services while patients play a secondary role in assisting

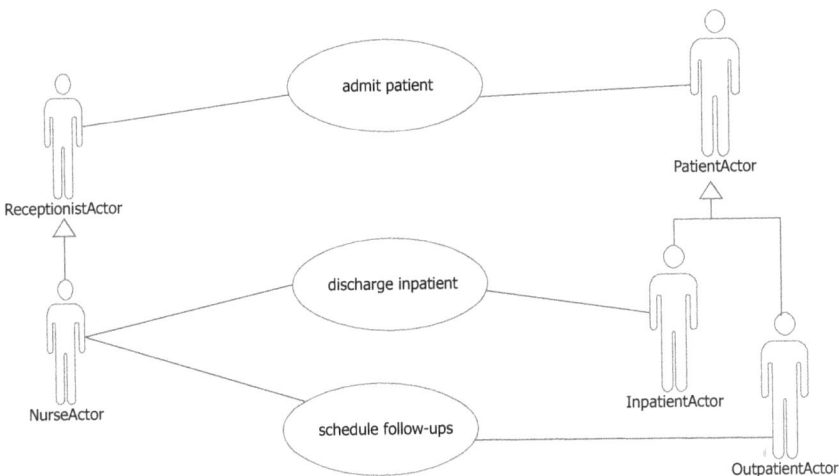

Figure 8.   Use case diagram for patient admission system.

receptionists and nurses to fulfill their functions. The use cases include admit patient, discharge inpatient, and schedule follow-ups for outpatients. The use case diagram is shown in Figure 8.

Example 2 (Student Registration System): Many users will use a registration system: students use it to search for courses and manage registrations, departments use it to schedule and change offerings, registrar uses it to print transcripts, advisors use it to override prerequisites,

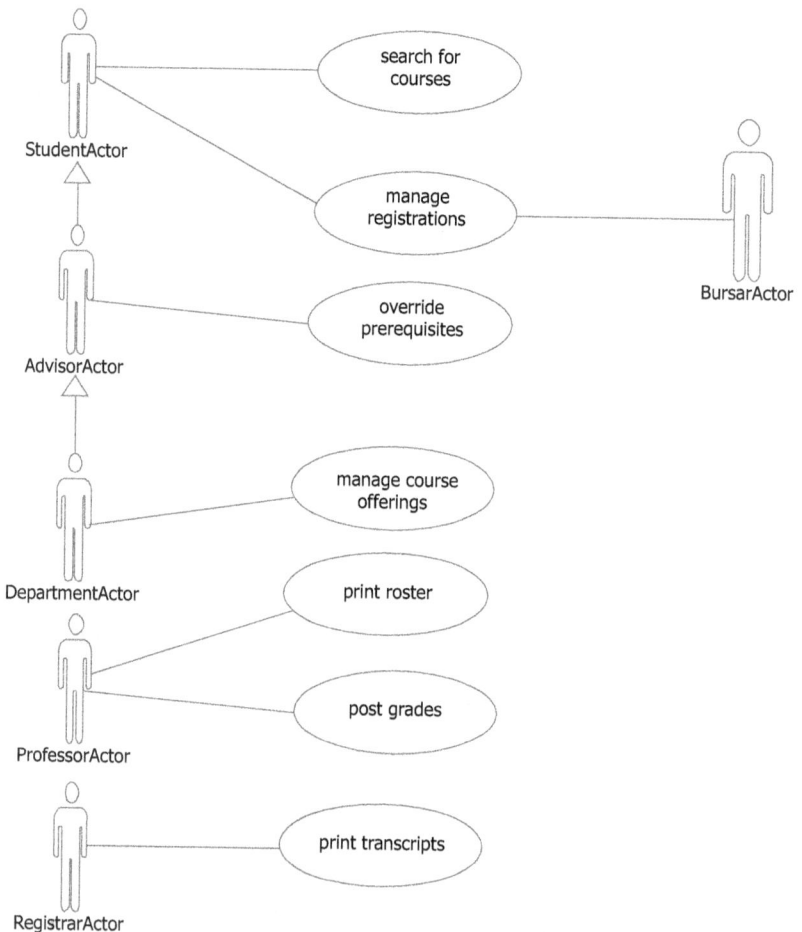

Figure 9.   Use case diagram for student registration system.

and professors use it to print rosters and post grades. Identify primary and secondary actors, identify use cases, and draw a basic use case diagram.

Grouping the users by their roles, we can identify five primary actors: Student, Professor, Advisor, Department, and Registrar. We identify six use cases as follows: search for courses, mange registrations, override prerequisites, manage course offerings, print rosters, post grades, and print transcripts. We identify Bursar as a secondary actor because the "Manage Registration" use case will need to update the bursar with updated credit hours for assessing tuitions and fees and inquire about payment status in order to allow students to register for a new course.

Students, advisors, and departments can all enroll students into courses. However, an advisor can override prerequisites besides adding or dropping offerings, and a department can manage offerings beyond what students and advisors can do. Thus, we use a three-layer role map to show that Advisor is a child of Student, and Department is a child of Advisor (see Figure 9).

Exercises

1. Using the example of a point-of-sale system for a small retail store in a mall, list relevant actors and use cases.
2. If you are a structured systems developer, think of two example functions that you have developed wherein one is a use case and the other is not. Use the concept of use cases to explain why they are so.
3. In the ATM example, is the "play game" a good use case? Why or why not? What about "steel money"? Is it a use case for the actor "theft"?
4. How would you handle the situation of two or more users using the same use case? (Hint: You need to consider whether the users are playing the same role and whether the users are performing the same use case simultaneously).
5. (Restaurant): When a customer walks into a restaurant, a waitress comes and greets him and takes his order. The system will then convert the order bill into a kitchen order so that the cook can make the food. Finally, the customer will come to the front desk to pay for the food and service using the system. Periodically, the restaurant owner consults with the system for revenues, profits, and inventories. The

actors of this system will include Customer, Waitress, Owner, and Cook. Identify use cases and draw a basic use case diagram.

6. (Inventory System): An inventory system is responsible for generating orders if the actual stock falls below the minimum re-order level and pay invoices to suppliers. It is responsible for updating inventory added if a new order is received and updating inventory used based on the inventory decrement data generated from the food ordering system. It should also allow the manager to query inventory levels. Identify actors and use cases and create a use case diagram for the inventory system.

7. (Student Club): A student club wants to have a database to manage its data on members. The club assigns members to its numerous committees. It is possible that one member can serve in more than one committee and will chair at most one committee. Each year, the club organizes many events. Each time, the club assigns one committee to oversee an event. Identify actors and use cases and create a use case diagram for the club management system.

8. (Facility Management): University A has several meeting facilities that can be reserved for individuals and departments, called customers. Here are the ongoing tasks for the facility manager: make and change reservations; print out the list of customers who will be using each facility for each day; and if a customer calls in, she can query the facility the customer reserved. In addition, after the facility is used as scheduled, its reservation status will be updated, and an invoice will be sent to the customer for payments. Identify actors and use cases and draw a use case diagram for the reservation manager.

Chapter 10

Use Case Storyboarding

Introduction

Use case modeling is to discover and capture functional requirements, whereas use case storyboarding is to procedurally and logically describe each captured use case as a sequence of interactions along with flows to handle exceptions and alternatives.

There are several models for procedure modeling, including activity diagrams, sequence diagrams, communication diagrams, state transition diagrams, and structured text descriptions. We have learned how to use activity diagrams to model the internal logic of functions in Chapter 3. The same model, of course, can be used to describe use cases too. However, the convention is to use structured English along with an established template. Structured text has the advantages of being expressive and free from interpretation errors if done accurately and precisely. In fact, storyboarding is the primary method for modeling the internal logics of a use case because *a well-developed and validated use case description often acts as a reference contract, in case of disputes, between business users and programmers in the systems development process*. Of course, textual description comes with some disadvantages. For example, it does not give the developer a quick visual view of how the sequences of control move along. The programmer must read it carefully before grasping the big picture.

Use case storyboarding is a process to capture detailed functional requirements to fulfill a use case. Through storyboarding, we understand what low-level or elementary functions that the system needs to perform in order to deliver the observable result of value to the primary actor.

Storyboarding serves the same purpose as function decomposition in structured development but has many advantages over function decomposition. First, storyboarding ensures that all the derived functions are necessary and essential to fulfill the use case and that all the derived functions work together to deliver value to the user. We will not create fancy but useless functions, and we will not miss essential ones.

Second, storyboarding allows one to see how the derived functions logically follow each other to bring a use case to success. Function decomposition does not render such a logic, and a different model such as a structured chart may be needed as supplement.

Third, storyboarding renders requirements modeling a creative process. In story writing, a novelist can creatively design a plot to resolve conflicts. Similarly, in use case storyboarding, an analyst can creatively re-engineer how the users may perform a use case by designing better graphical user interfaces (GUIs), re-arranging the sequence of some interactions, plotting a different sequence to bring a use case to the resolution, etc.

Concepts and Templates

A use case description is *a structured text via a list of interactions between the primary actor and the system with a clearly identifiable beginning and end.* It exposes and develops the plot of a use case based on one typical instance of the actor, i.e., a user, and one typical instance of the use case, i.e., a usage, as the representative instances of their respective classes.

In real life, typical users follow a common flow to the completion of the use case, but occasionally some users may encounter obstacles due to running conditions and must adopt a workaround. It is important that we describe a sequence of interactions not only in the most common scenario but also in alternate scenarios that end up with success through an alternate route or exceptional scenarios that may result in a failure. This is done through the distinction of the *basic flow* for the most common success scenario used by typical users, *alternate flows* for alternate scenarios due to different run-time conditions, and *exceptional flows* for error scenarios that will lead to the failure of the use case. For example, when describing the "withdraw cash" use case in the ATM problem, the basic flow will list the interactions that lead to the successful withdrawal of cash without hiccups. However, what if the inserted card cannot be validated? What if the user enters pin codes incorrectly? What if the account has exceeded the maximum daily withdrawal limit? What if the user forgets to take his card from the ATM? Some of these scenarios may have workarounds to eventually lead to success and will be described as alternate flows. Others may be doomed to failure and will be described by exceptional flows.

To improve the readability of use case stories, organizations often follow a template to describe use cases. A template typically splits a use case story into four categories: (1) A required identification and brief summary of the use case, (2) required lists or flow of interactions in all scenarios, (3) optional graphical user interface requirements, and (4) optional non-functional requirements. Some templates, example, the Business Requirement Document template in Howard Podeswa (2005), are more comprehensive and include additional categories such as pre- and post-conditions, class diagrams, activity diagrams, message and prompts, business rules, and external interfaces. The additional entries are either supplementary to the textual description or common references by the description. Class diagrams can provide background information about the system, and activity diagrams augment the use case description with logic details. Business rules and user interfaces are the common reference sections to which the description may point. Prompts and messages are simplified user interfaces meant for the system to solicit simple data from human users and to display simple feedback.

This book follows a streamlined template consisting of five sections: Use Case Overview, Flow of Events, GUIs, Business Rules, and Prompts and Messages. User case overview lists the name, the ID, the purpose, the primary actor of a use case, and a use case diagram to show the context of the interactions to be described. Flow of Events is the main section housing basic, alternate, and exceptional flows. The other three sections are optional; they exist if necessary.

A template also prescribes a system to label the parts of a use case description. Note that, after being drafted, a use case story may have to go through several stages of evaluation before being ratified. It is important that there is a label system that can identify each elementary part of a story, such as a step in a flow of events, a business rule, a user interface, etc. by a unique numerical label. This book suggests the following system. In describing the flow of events, the steps in the basic flows are labeled as 1, 2, 3, etc. An alternate or exceptional flow is labeled based on the label of the step from which the alternate or exceptional scenario is created.

For example, the following is a snapshot of the basic flow for the "withdraw cash" use case:

4. Card Holder enters a pin number
5. ATM validates the pin number
6. ATM asks for a withdraw amount

There are two possible scenarios or run time errors that may come out of Step 5: Pin may be incorrect, and account may be on hold. In the first case, the ATM shall allow the user to re-enter a pin, eventually leading to success. It should be described as an alternate flow. Under the second scenario, the use case will fail indefinitely, and it should be described as an exceptional flow. These two flows may be labeled, respectively, as 5a and 5b, with the steps in each alternate or exceptional flow labeled as .1, .2, .3, etc.

Alternate Flows:

> 5a: invalid pin:
>
> > .1 ATM displays "invalid pin" message (PM1)
> > .2 ATM verifies the number of error pin entries is less than five
> > .3 Go to Step 3

Exceptional Flows:

> 5b: account on hold:
>
> > .1 ATM displays "account on hold" message (PM2)
> > .2 ATM ejects the card
> > .3 Card Holder takes the card
> > .4 ATM goes idle

If there are further alternate or exceptional scenarios out of alternate flows, these scenarios are labeled according to the new step label. For example, out of the second step of the "5a: invalid pin" flow, i.e., 5a.2, there is a scenario wherein the card holder enters wrong pin codes five times, in which event the system shall kill the use case for security purposes. The scenario will lead to a new exceptional flow as follows:

Exceptional Flows:

5a.2a: too many pin entry errors:

.1 ATM displays "too many pin entry errors" message (PM3)
.2 ATM ejects the card
.3 Card Holder takes the card
.4 ATM goes idle

The label 5a.2a symbolizes the first exceptional scenario out of the second step of the alternate flow 5a. Additional scenarios, if any, may be labeled as 5a.2b, 5a.2c, etc., accordingly.

Using the above labeling scheme, each flow and each step have a unique label. For example, labels such as 5a, 7b, etc. identify alternate or exceptional flows, whereas 5a.2a.3 and 5a.1, respectively, identify Step 3 in the flow 5a.2a and Step 1 in the flow 5a.

Business rules, user interfaces, and prompts and messages are labeled sequentially. For example, BI2 refers the second business rule, UI5 the fifth user interface, and PM2 the second prompt or message. For example, the following is a snapshot of the Prompt and Message section, where PM1, PM2, and PM3 are referenced in the above sample flows.

Prompts and Messages:

PM1 (invalid pin): Your pin entry does not match. Please try again.
PM2 (account on hold): Your account is on hold. Please call (800) 111-1111 for details.
PM3 (too many pin entry errors): Your account is locked due to pin entry errors.

Flow of Events

The essential part of a use case story is the "Flow of Events" section. A few guidelines are outlined here on how to write the events or interactions in a flow:

1. *Describe each interaction using a short concise sentence in the active voice, and do not use ambiguous non-measurable adjectives.*

For example, we should say "Card Holder inserts a card" instead of "A card is inserted by Card Holder." For another example, we should say "ATM displays 'Invalid Pin' message for 3 seconds" instead of "ATM displays 'Invalid Pin' for a while."

2. *Use consistent names for subjects and objects.* For example, if we call the user Card Holder, we should carry the name through and should not call the user "Card Holder" in one place and "Customer" in another.

3. *Follow the logical order of events and remember that there must be a response following a request.* For example, when an ATM asks for a pin number, the next logical step should be a response from the user to enter a pin number.

4. *Assume a positive running condition in the current flow and use alternate or exceptional flows to handle negative ones.* Avoid using if-else statements inside a sentence. For example, after the sentence "ATM validates the card," we assume the card is valid in the basic flow but use an exceptional flow to document the opposite scenario.

5. Understand what constitutes the system and do not describe inter-actions among actors and any other entities that are not a part of the system. For example, an ATM consists of hardware components and interfaces, and thus we document events like "Card Holder inserts a card." In contrast, a POS system is made of software objects and a barcode scanner. Shopping basket and conveyor belts are not a part of the system, and so actions like "Customer places items on the belt" are not interactions between the user and the system. Cashier and Customer are both actors, and so actions like "Cashier greets Customer" are not interactions between the user and the system either. These actions should not be described in a flow of events.

In this section, we will provide two complete examples of story-boarding: Withdraw cash using an ATM, and checkout items using a POS system. The examples demonstrate the differences in describing two types of systems: an ATM is made of mostly hardware components and interfaces, whereas a POS mostly software objects. The reader should also pay attention to the design of GUIs to appreciate how an effective GUI may help simplify storyboarding.

Figure 1. The context diagram for the "Withdraw Cash" use case.

Storyboarding via examples: Withdraw cash

The "withdraw cash" use case allows a card holder to withdraw cash using an ATM. The primary actor is Card Holder and the secondary actor is Card Network. We assume that, as a precondition, the ATM is idle and has enough cash to begin with. The postcondition includes: (1) the amount of cash is equal to the pretransaction amount minus the withdrawal amount, (2) the slot reader is empty or else swallows the forgotten card. Figure 1 shows the portion of the use case diagram that acts as the context diagram for storyboarding.

We will omit the use case summary section, but the reader can fill in the blank based on the above problem description. The following is a snapshot of the "Flow of Events" section:

Flow of Events
  Basic Flow:

1. Card Holder inserts a card
2. ATM validates the card
3. ATM asks for a pin (UI1)
4. Card Holder enters a pin (UI1)
5. ATM asks Card Network for authorization
6. Card Network authorizes the request
7. ATM asks for a withdrawal amount (UI2)
8. Card Holder enters a withdrawal amount
9. ATM checks and validates the withdrawal amount
10. ATM dispenses cash
11. Card Holder takes cash
12. ATM prints a receipt (UI3)
13. ATM ejects the card
14. Card Holder takes the card
15. ATM records the transaction
16. ATM goes idle

Alternate Flows:

   6a. invalid pin:

   .1 ATM displays "invalid pin" message (PM1)
   .2 ATM checks if pin entries exceed the limit
   .3 Go to Step 3

   9a. withdraw limit exceeded:

   .1 ATM displays "withdraw limit exceeded" message (PM4)
   .2 Go to Step 7

Exception Flows:

   2a. invalid card:

   .1 ATM displays the "invalid card" message (PM6)
   .2 ATM ejects the card
   .3 Card Holder takes the card
   .4 ATM goes idle

   6b. account on hold:

   .1 ATM displays "account on hold" message (PM2)
   .2 ATM ejects the card
   .3 Card Holder takes the card
   .3 ATM goes idle

   6a.2a. too many pin entry errors:

   .1 ATM displays "too many pin entry errors" message (PM3)
   .2 ATM ejects the card
   .3 Card Holder takes the card
   .4 ATM goes idle

   11a. cash forgotten:

   .1 ATM takes cash back
   .2 ATM displays "forgotten cash" message (PM5)
   .3 ATM swallows the card
   .4 ATM goes idle

User Interfaces

UI1. Pin Entry UI

Enter Pin Code:
xxxxxx|

1	2	3	Cancel
4	5	6	Clear
7	8	9	Enter
.	0	+	

UI2. Withdraw Amount UI

Enter Withdraw Amount:
xxx0.00|

1	2	3	Cancel
4	5	6	Clear
7	8	9	Enter
.	0	+	

UI3. Receipt UI

Transaction ID: 42333
ATM: Montrose of OH 44335
Account No: XXX3455
Withdraw: $50.00
Transaction Date: 1/23/2009 14:29PM
New Balance: $2406.33

Thank You!

Prompts and Messages

PM1 (invalid pin): Your pin entry does not match. Please try again.
PM2 (account on hold): Your account is on hold. Please call (800) 111-1111 for details.

PM3 (too many pin entry errors): Your account is locked due to pin entry errors.
PM4 (withdraw limit exceeded): The amount exceeds the withdrawal limit.
PM5 (cash forgotten): ATM took cash back, and transaction was voided.
PM6 (invalid card): The card is not recognized.

In this example, the reader shall note that a reference item such as a user interface, a business rule, or a message may be referenced by multiple steps. For example, UI1 is referenced in Steps 3 and 4.

Changing textual description can be very tedious due to how we label offspring alternate and exceptional flows. For example, if you add or delete a step, the whole sequence number changes, and so do the labels for alternate and exceptional flows.

Note also that all alternate flows will route back to the basic flow, whereas exceptional ones will not. If all flows must end with the same system state, we use a postcondition to stipulate the state rather than write a flow to reach the state. For example, what should we do if a card holder forgets to remove his card? It may happen in basic flow and exceptional flows, and so it is not clear whether we should handle it as an alternate or exceptional flow. In the basic flow, since cash is withdrawn, the use case succeeds, and we should not handle it as an exceptional flow. On the other hand, an exceptional flow will lead to the failure of the use case, and so we cannot handle the forgotten card scenario as an alternate flow. The best option is probably to state a postcondition that the ATM swallows the forgotten card if the card reader is not empty.

Storyboarding via examples: Checkout items

For a POS, the most important use case is probably to check out items for customers. It is performed by a cashier with the help of a customer. A credit card network may be involved to pay for the transaction. Thus, Cashier is the primary actor, and Customer and Credit Card Network are secondary actors. Note that a manager may need to provide overrides in case the price cannot be retrieved or the discount cannot be applied. This override may be another sequence of interactions that can be optionally included into the "checkout items" use case. We will handle optional sequences in the next chapter. Figure 2 shows the context diagram for the use case.

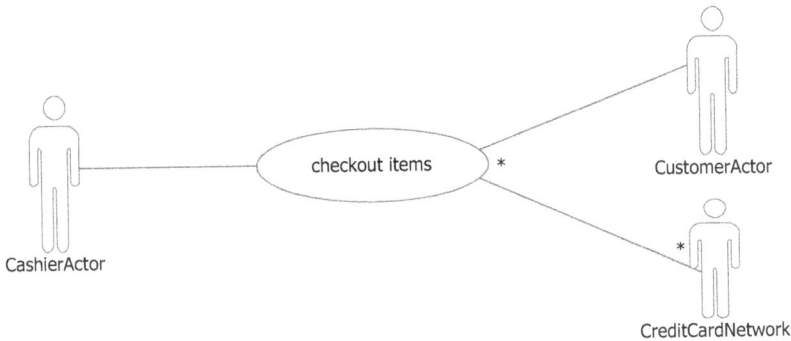

Figure 2.   The context diagram for the "Checkout Items" use case.

In the following description, we assume that a customer commonly swipes a credit card for a full payment of the purchase, but the system is designed to allow multiple payments to be made against one transaction.

Flow of Events
Basic Flow:

1. Cashier starts a new transaction

   [Steps 2–4 repeat for each item in the basket]

2. Cashier scans an item
3. The system retrieves the item description and price (UI1)
4. The system computes the subtotal and tax

   [Steps 5–11 repeat for each payment until zero balance]

5. Customer swipes a credit card
6. The system displays credit card payment option (UI1)
7. The system retrieves credit card number, expiration, and security code (UI1)
8. Cashier verifies the credit card
9. Cashier presses Pay button (UI1)
10. The system requests payment authorization from Credit Card Network
11. The system computes remaining balance (UI1)
12. The system prints a receipt
13. The system goes idle

Alternate Flows:

2a. produce lookup:

   .1 Cashier selects produce lookup (UI1)

    .2 The system gets the weight

    .3 Continue to Step 3

  2b. bad bar code:

    .1 The cashier enters bar code (UI1)

    .2 Continue to Step 3

  5a. check payment:

    .1 Cashier presses check payment screen (UI1)

    .2 Cashier enters check number, routing number, and amount (UI1)

    .3 Cashier stamps the check

    .4 Cashier presses Pay button (UI1)

    .5 The system opens Cashier drawer

    .6 Cashier inserts check and closes drawer

    .7 Continue to Step 11

  5b. cash payment:

    .1 Cashier presses cash payment screen (UI1)

    .2 Cashier enters tendered amount (UI1)

    .3 The system computes change

    .4 The system opens Cashier drawer

    .5 Cashier takes change and closes drawer

    .6 Continue to Step 11

  5c. credit card payment with amount limit:

    .1 Cashier presses credit card payment screen (UI1)

    .2 The customer chooses payment amount

    .3 Customer swipes credit card

    .4 Continue to Step 7

Exception Flows:

  1–4a. customer cancels:

    .1 Cashier presses Cancel button (UI1)

    .2 The system prompts "Cancellation Confirmation" message (MP1)

    .3 Cashier presses Yes button

    .4 The system goes idle

Prompts and Messages

PM1. Cancellation Confirmation — "Are you sure to cancel? Press Yes to continue."

User Interfaces

UI1. Checkout Screen

This example illustrates the power of user interfaces: a well-designed graphical user interface reduces the number of interactions, improves the efficiency of performing a use case, and make the description of the use case clearer and easier to understand. As we see in UI1, by using a pictured combo box, the cashier can select a produce or fruit to weigh instead of opening another form with a portfolio of produce to select. By embedding tabbed controls for payment options, it avoids the need to go to another screen for handling payments. The same checkout screen can even handle tax exempt and coupon options, which may be optional sequences to be plugged into the "checkout item" use case (see Chapter 11 for extension use cases).

GUI Design

User interfaces, including screens and reports, not only support use case storyboarding but also serve as the device for business analysts to communicate with end users to elicit their implicit knowledge about the procedures of conducting use cases. User interfaces include software as well as hardware components that the users use to interact with the system. For example, the ATM machine has a card reader, a cash dispenser, a receipt printer, a numerical keyboard to enter pin number and amounts, and keys for confirming, canceling, or going back. The ATM must also have a screen to display messages and prompts. These are hardware user interfaces.

The reader should not confuse user interfaces with external interfaces, which are hardware and software components through which the system interacts with external systems. For example, the modem used by an ATM to communicate data with a credit card network and the bank system are external interfaces, and these may be listed under the External Interfaces section of a use case story.

In modern windowing and web-based systems, user interfaces are mostly graphical. GUIs play a critical role in use case storyboarding. While a story creatively delineates how the process will be used, GUIs creatively imagine the look and feel of the system. These two aspects of a use case story are intertwined: a creative GUI can lead to the creation of a new plot for a use case, and an effective GUI can also help improve the effectiveness of its description.

There are many books discussing the criteria, principles, or guidelines for GUI designs. Most of the guidelines are common sense, while some are based on psychological research. This book does not have room for detailed exposition of these guidelines but summarizes the principles in three words: Character, Control, and Cognition.

Character refers to the goal of achieving aesthetic characteristics, i.e., GUIs must look good. This principle embodies at least two sub-aspects. First, each form or report (or web page) must be neat and clean with appropriate grouping of related information. It does not waste valuable screen spaces and, at the same time, does not look too crowded. Second, across forms or reports (or web pages), there should be consistency in sizes, looks, and locations of controls such as labels, text boxes, buttons, etc.

Control refers to the goal of giving users a sense of self-control, i.e., the users feel that they are in control of the system rather than being controlled by the system. This principle implies many sub-principles. For example, after each user action, the system must provide feedback through words and signs, and users must be given an option to reverse their action to go back to the previous state, if possible. Otherwise, enough warnings must be present before an irrevocable action.

Cognition refers to the goal of helping users to overcome their limited cognitive capabilities. For example, users do not want to memorize and do not want to think, and so we should not design GUIs that count on their short-term memory or mental math. For another example, making a decision is hard, and so we should avoid too many user choices and present a clear task flow for users to follow instead. For yet another example, people have perceptual or behavioral biases, and so we should design GUIs to take advantage of their psychology to prevent errors and improve efficacies.

Prototyping in Visual Studio

User interfaces are often created using rapid application development (RAD) tools such as Microsoft Access, Microsoft Visual Studio, Eclipse, Oracle Developer, etc. In the following, we will learn how to use Visual Studio to create the checkout screen seen earlier in this chapter.

When creating a new project using C# language and Windows Application template, Visual Studio will create a form with a blank canvas for us to draw controls on. We can add addition forms using Project → Add Windows Form menu. Visual Studio has a rich set of graphical controls in the toolbox (see Figure 3). The common controls include Button, Label, TextBox, CheckBox, ComboBox, RadioButton, and MenuStrip. A label is for displaying a text that cannot be changed by the user, and a text box for a text that can be changed. If an entry has only two to five possible values, e.g., a gender is either male or female, radio buttons or check boxes may be better alternatives than a text box for data entry. Radio buttons allow one value to be entered, but check boxes allow multiple values to be selected. If an attribute has tens of possible values, e.g., music genre can be rock, jazz, blues, etc., a list box or a combo box may be used. A combo box allows one value to be selected, but a list box allows multiple values to be chosen. Buttons are for executing commands. If there are many commands, they should be grouped into menu strips.

The use of each control amounts to dragging the icon from the toolbox, dropping it onto the canvas, and setting its properties using the Properties Window, which can be opened by using the View → Properties Window menu if not opened by default. The two basic properties of each control are name and text: the name is just like that for a variable and follows the same naming conventions (see Chapter 2), and the text is the caption of the control. A list box and a combo box have an Items property that allows us to enter a list of values. To attach a command to a control, click on the

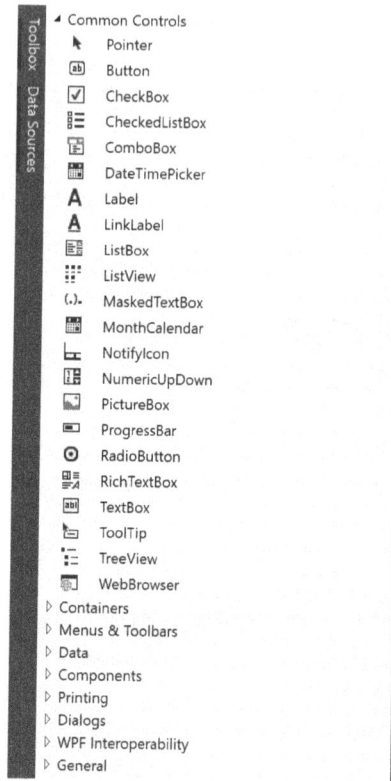

Figure 3.   The Visual Studio toolbox for GUIs.

Events tab of Properties Window and double click on the name of the event, which will trigger the execution of the command. For example, for a command button or a menu item in a menu strip, the typical event to trigger a command is Click.

The checkout screen in the earlier section used three advanced features. The first is using a label to show a live clock. This is achieved by the following procedure:

1. Add Label to the form and rename it lblClock.
2. Add a timer object to the form and rename it timerCO.
3. In the properties window of timerCO, set Enabled to true, Interval to 1,000 (milliseconds).

4. Double click on timerCO, and add the following code to timerCO_ Tick() event:

```
private void timerCO_Tick(object sender, EventArgs e)
{
    lblClock.Text = DateTime.Now.ToLongDateString();
    lblClock += " " + DateTime.Now.ToLongTimeString();
}
```

The second feature is to use a DataGridView with button columns to show checkout items by using the following procedure and code:

1. Add DataGridView control on the form and rename it dgvItems.
2. Add two button columns to the lgvItems and label them as Delete and Edit by clicking on the little triangle of the grid control and selecting Add Columns menu.
3. Double click on any blank spot of the form, and add the following code to the form load event:

```
DataTable dt = new DataTable();
dt.Columns.Add("SKU", typeof(string));
dt.Columns.Add("Description", typeof(string));
dt.Columns.Add("Price", typeof(double));
dt.Columns.Add("QTY", typeof(double));
dt.Columns.Add("Sub Total", typeof(double));

DataRow dr = dt.NewRow();
dr["SKU"] = "01-23456";
dr["Description"] = "Apple Iphone X";
dr["Price"] = 999.99;
dr["QTY"] = 1;
dr["Sub Total"] = 999.99;
dt.Rows.Add(dr);

dr = dt.NewRow();
dr["SKU"] = "01-23423";
dr["Description"] = "Apple Macbook Pro 256GB 13.3";
dr["Price"] = 1999.00;
dr["QTY"] = 2;
dr["Sub Total"] = 3998.00;
dt.Rows.Add(dr);

dgvItems.DataSource = dt;
```

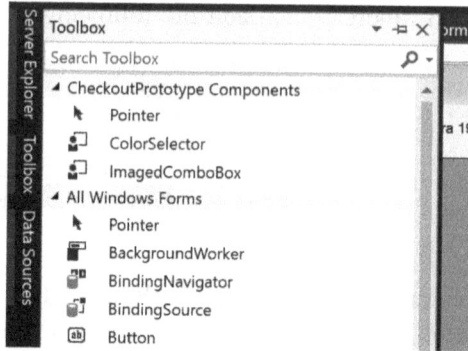

Figure 4.   The ToolBox with ImagedComboBox.

The third feature is to use a combo box to show a list of produce and fruits. This cannot be done using the standard ComboBox control. Thus, we need to extend its capability to draw pictures in the list items. This can be done in many ways. The Appendix in the end of this chapter shows two such extensions. The following procedures show how to use the extensions to create the pictured combo box in UI1:

1. Add a new class called ComboBoxExtension and copy the code given in the Appendix to replace everything in the ComboBoxExtension.cs file.
2. Build the project using Build → Build Solution menu.
3. Go to the checkout form in design view and open the ToolBox. You will see two custom controls added: ColorSelector and ImagedComboBox (see Figure 4).
4. Drag ImagedComboBox to the form in design view and rename it imageCBOProduce.
5. Add photos for some fruits and produce such as apple.png, grapes. png, and carrot.png into the project bin\debug folder.
6. Write the following code inside the form load event:

```
imageCBOProduce.Items.Add(new ComboBoxExtension.
ComboBoxItem("Apple", Image.FromFile("apple.png")));

imageCBOProduce.Items.Add(new ComboBoxExtension.
ComboBoxItem("Grapes", Image.FromFile("grapes.png")));

imageCBOProduce.Items.Add(new ComboBoxExtension.
ComboBoxItem("Carrot", Image.FromFile("carrot.png")));

imageCBOProduce.SelectedIndex = 0;
```

That is, after compiling or running the project, we will see a combo box filled with a list of photos for apples, grapes, and carrots beside the fruit names.

Exercises

1. (Student Registration System): It is perceivable that many people use a registration system. For example, students use it to enroll into courses, departments use it to schedule courses, advisors use it to enforce prerequisites and enroll students into a class under special conditions, professors and registrars use it to print rosters, the school business system uses it to assess payments, etc. Grouping these agents by their roles, we can identify actors as follows: Student, Professor, Registrar, Department, Advisor, and Business System. Here, Business System is a secondary actor and all others are primary actors. Both professors and registrar print rosters, but conceivably there are some differences. For example, a professor can print rosters for his or her own course whereas the registrar can do it for all courses. Identify primary and secondary actors, identify use cases, and draw a basic use case diagram. Then describe the use case "drop a class" using the streamlined template. Make sure to include all scenarios as well as GUIs.

2. (Restaurant): When a customer walks into a restaurant, a waitress comes and greets him and takes his order. The system will then convert the order bill into a kitchen order so that the cook can make the food. Finally, the customer will come to the front desk to pay for the food and service using the system. Periodically, the restaurant owner consults with the system for revenues, profits, and inventories. The actors of this system will include Customer, Waitress, Owner, and Cook. Identify primary and secondary actors, identify use cases, and draw a basic use case diagram. Then describe the use case "take food order" using the streamlined template. Make sure include all scenarios as well as GUIs.

3. (Video Rental): In Video Shack, customers are required to have a family membership card that is used mainly to ensure that they have a credit card, live in the neighborhood, and can be contacted in case they are late in returning their rentals. Video Shack has a varied stock of videos classified into such categories as comedy, adventure, children's, and romantic. Any particular title is obtainable from one

distributor who owns the rights to it. Video Shack deals with about 25 distributors for different titles. It may carry many copies of a popular new title or only a single copy of some classics. Popular titles may have to be reordered. Most of the videos are rented for a standard price. However, there is sometimes a premium price for new releases. There is also a discount during weekdays. Customers agree to return rentals by noon of a set date, and they can reserve up to five videos in advance to ensure that they will be available when desired. Design a use case diagram for a system that can be used to record purchases of videos from suppliers, record rentals and returns by customers, and produce a printed catalog of current holdings categorized by title and type. In addition, Video Shack would like to be able to get listings of how many copies they have of each title and how often each title has been rented. Describe the use case "Reserve Video".

4. (Electronic Medical Records): You have been hired to design a system for a small health care organization. The clinic consists of several examining rooms and a few rooms for short-term critical-care patients. A core staff of seven physicians is supplemented by internists from a local teaching hospital. The clinic wants to computerize the patient records. All patient medical data is stored in a folder kept in a large central file cabinet. Arriving patients sign in at the front desk. A clerk checks the billing records, prints out a summary status sheet, and obtains the file number from the computerized system. The clerk then pulls the medical data folder and selects an examination room. After waiting for the physician, the clerk moves the data packet and the patient to the examination room. A nurse records basic medical data (weight, blood pressure, etc.). The physician makes additional notes to both the medical and billing data and generally writes a prescription order, which is given to the patient and recorded on the charts. When the patient leaves, the clerk enters the new billing data into the system, collects any payments, and prints a list of charges and a receipt. The new billing data is forwarded to the appropriate insurance company. The medical data is returned to the filing cabinet. When the patient gets a prescription filled, the pharmacist calls the clinic for verification. A clerk retrieves the medical data, identifies the prescription, and verifies or corrects the order. Draw a use case diagram to capture the functional requirements for an automated medical record system.

5. Design a graphic user interface for airline agents to book flights for a customer. We assume that one booking is for one trip from one airport

to one destination, and the customer can buy several tickets for several passengers.

6. Design a graphical user interface for a video rental store to do check-out. Note that, besides renting, the store also sells used videotapes and other miscellaneous items. For rentals, the cashier needs to scan a membership card and the screen should display the member information and if there is an outstanding balance due to late returns or damages.

7. Design a graphic user interface for a small restaurant to take customer orders. Note that the waiter depends on the order information to remember the exact table and seat to deliver foods and drinks and send the final bill.

8. Design a graphic user interface for the receiving dock employee to check in shipments. When a shipment has arrived, a shipping slip is used to retrieve the order information. The employee will check each ordered item to make sure the quality and quantity are correct. If error occurs, e.g., missing items or damaged items, the employee will note the problem on the screen, and the system can then generate a notice to the vendor and create a backorder when the check-in job is finished.

9. Design a graphic user interface for student enrollment. The screen should have the capability to search for offerings and add offerings to a basket. The user must be informed of whether a student has met the prerequisites to take an offering and whether she has time conflicts with another course she has enrolled in or is about to enroll in.

10. Design a graphic user interface for creating questions in a question library. Each question is of True/False, Multiple Choice, Select-All-That-Apply, or short answer type. It has the main question text along with 0–5 possible answers. Occasionally, some question text or answers may contain image, video, music, or document files.

Appendix: Combo Box Extensions in C#

The following code is by Bassam Alugili for a tutorial in codeproject.com. It is reproduced here for convenience since the website requires an account to download the code.

```
using System;
using System.Drawing;
using System.Windows.Forms;
```

```csharp
using System.ComponentModel;
using System.Linq;
using System.Collections;

namespace ComboBoxExtension
{
    public sealed class ColorSelector : ComboBox
    {
        public ColorSelector()
        {
            DrawMode = DrawMode.OwnerDrawFixed;
            DropDownStyle = ComboBoxStyle.DropDownList;
        }

        protected override void
            OnDrawItem(DrawItemEventArgs e)
        {
            e.DrawBackground();

            e.DrawFocusRectangle();

            if (e.Index >= 0 && e.Index < Items.Count)
            {
                DropDownItem item = (DropDownItem)
                Items[e.Index];

                e.Graphics.DrawImage(item.Image,
                e.Bounds.Left, e.Bounds.Top);

                e.Graphics.DrawString(item.Value, e.Font,
                    new SolidBrush(e.ForeColor),
                    e.Bounds.Left + item.Image.Width,
                    e.Bounds.Top + 2);
            }
            base.OnDrawItem(e);
        }
    }
}

public sealed class DropDownItem
{
    public string Value { get; set; }

    public Image Image { get; set; }

    public DropDownItem()
        : this("")
```

```csharp
    { }

    public DropDownItem(string val)
    {
        Value = val;
        Image = new Bitmap(16, 16);
        using (Graphics g = Graphics.FromImage(Image))
        {
            using (Brush b = new SolidBrush(Color.
                FromName(val)))
            {
                g.DrawRectangle(Pens.White, 0, 0, Image.
                    Width, Image.Height);
                g.FillRectangle(b, 1, 1, Image.Width -
                    1, Image.Height - 1);
            }
        }
    }

    public DropDownItem(string val, Color color)
    {
        Value = val;
        Image = new Bitmap(16, 16);
        using (Graphics g = Graphics.FromImage(Image))
        {
            using (Brush b = new SolidBrush(color))
            {
                g.DrawRectangle(Pens.White, 0, 0, Image.
                    Width, Image.Height);
                g.FillRectangle(b, 1, 1, Image.Width -
                    1, Image.Height - 1);
            }
        }
    }

    public override string ToString()
    {
        return Value;
    }
}


[Serializable]
public class ComboBoxItem
{
    private object _value;
```

```
private Image _image;
public object Value
{
    get
    {
        return _value;
    }
    set
    {
        _value = value;
    }
}


public Image Image
{
    get
    {
        return _image;
    }
    set
    {
        _image = value;
    }
}


public ComboBoxItem()
{
    _value = String.Empty;
    _image = new Bitmap(1, 1);
}


public ComboBoxItem(object value)
{
    _value = value;
    _image = new Bitmap(1, 1);

}


public ComboBoxItem(object value, Image image)
{
```

```
            _value = value;
            _image = image;
        }


        public override string ToString()
        {
            return _value.ToString();
        }
    }
}


public class ComboCollection<TComboBoxItem> : CollectionBase
{

    public EventHandler UpdateItems;
    public ComboBox.ObjectCollection ItemsBase { get; set; }

    public ComboBoxItem this[int index]
    {
        get
        {
            return ((ComboBoxItem)ItemsBase[index]);
        }
        set
        {
            ItemsBase[index] = value;
        }
    }

    public int Add(ComboBoxItem value)
    {
        var result = ItemsBase.Add(value);
        UpdateItems.Invoke(this, null);
        return result;
    }

    public int IndexOf(ComboBoxItem value)
    {
        return (ItemsBase.IndexOf(value));
    }

    public void Insert(int index, ComboBoxItem value)
    {
        ItemsBase.Insert(index, value);
```

```
        UpdateItems.Invoke(this, null);
    }

    public void Remove(ComboBoxItem value)
    {
        ItemsBase.Remove(value);
        UpdateItems.Invoke(this, null);
    }

    public bool Contains(ComboBoxItem value)
    {
        return (ItemsBase.Contains(value));
    }
}


public class ImagedComboBox : ComboBox
{
    private ComboCollection<ComboBoxItem> _items;

    [DesignerSerializationVisibility(DesignerSerialization
        Visibility.Hidden)]
    public new ComboCollection<ComboBoxItem> Items
    {
        get { return _items; }
        set { _items = value; }
    }

    public ImagedComboBox()
    {
        DropDownStyle = ComboBoxStyle.DropDownList;
        DrawMode = DrawMode.OwnerDrawVariable;
        DrawItem += ComboBoxDrawItemEvent;
        MeasureItem += ComboBox1_MeasureItem;
    }

    protected override ControlCollection
        CreateControlsInstance()
    {
        _items = new ComboCollection<ComboBoxItem>
        {
            ItemsBase = base.Items
        };
```

```csharp
        _items.UpdateItems += UpdateItems;

        return base.CreateControlsInstance();
    }

    private void ComboBox1_MeasureItem(object sender,
        MeasureItemEventArgs e)
    {
        var g = CreateGraphics();
        var maxWidth = 0;
        foreach (var width in
            Items.ItemsBase.Cast<object>().Select(element =>
                (int)g.MeasureString(element.ToString(),
                Font).Width).Where(width => width > maxWidth))
        {
            maxWidth = width;
        }
        DropDownWidth = maxWidth + 20;
    }


    private void ComboBoxDrawItemEvent(object sender,
        DrawItemEventArgs e)
    {
        e.DrawBackground();
        if (e.Index != -1)
        {
            var comboboxItem = Items[e.Index];
            e.Graphics.DrawImage(comboboxItem.Image,
                e.Bounds.X, e.Bounds.Y, ItemHeight,
                    ItemHeight);

            e.Graphics.DrawString(Items[e.Index].Value.
                ToString(),
            Font,Brushes.Black,new RectangleF(e.Bounds.X
                + ItemHeight,
            e.Bounds.Y, DropDownWidth, ItemHeight));
        }
        e.DrawFocusRectangle();
    }
}
}
```

Chapter 11

Use Case Optimization

Introduction

In the previous chapter, we learned how to describe a use case textually. Storyboarding is time consuming and error prone, and it is important that use cases are parsimonious, i.e., captured use cases do not have redundant sequences in need of descriptions. In this chapter, we will learn three advanced use case modeling techniques on how to optimize use case models. Particularly, we will learn how to use two special dependency relationships — include and extend — and inheritance relationships between use cases to reduce redundant storyboarding.

Use Case Factorization

Often, we find that the descriptions of different use cases overlap. In the last chapter, for example, we detailed the sequence of how to perform the "withdraw cash" use case. If we had worked out other use cases such as "inquire balance" and "deposit money," we would realize that these use cases have a few common subsequences of interactions. For example, in the basic flow for the "withdraw cash" use case in the last chapter, Steps 1 to 6 are required for all these use cases, and so are all alternate and exceptional flows coming out of those steps. How could we optimize these use cases in order to remove or reduce duplicates?

If two or more use cases share a common sub-sequence of interactions, a standard technique is to factor this common sequence out into a separate use case, called *inclusion use case*, and then add the inclusion use case to the *base use cases* from which the sub-sequence was taken out. For example, Steps 1 to 6 may be factored out into a use case that does nothing but authorize transactions. We anticipate that all other use cases require similar authorizations for them to be performed successfully. Therefore, we create "authorize transactions" as an inclusion use case and use a dashed line labeled with <<include>> pointing to the "authorize transactions" use case (see Figure 1). Such a dashed line denotes an *inclusion relationship*.

Conceptually, an inclusion relationship is a stereotyped dependency, a special kind of dependency between classes. Here, it means the base use case depends on the inclusion use case. This can be understood in the sense that the base use case depends on the inclusion use case to perform a part of its task, and any change in the inclusion use case affects how the base case works.

Since an inclusion use case is a sub-sequence, it does perform a complete task to produce an observable result of value to the primary actor. Thus, it is not a use case by itself.

It is possible that base use cases have several segments of common sub-sequences to be factored out. For example, in a POS, both "checkout items" and "return orders" use cases involve a common sequence to process payments, which may be factored out as an inclusion use case "process payments." Besides, both use cases also involve another common sub-sequence that deals with item scanning. Thus, we may factor "scan items" as another inclusion use case (see Figure 2).

The reader should be aware that the factorization technique should not be overused, i.e., we should not create an inclusion use case if a common

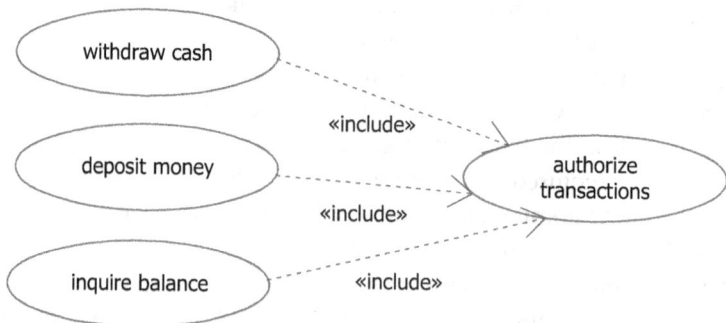

Figure 1.   Inclusion use case "Authorize Transactions" in the ATM.

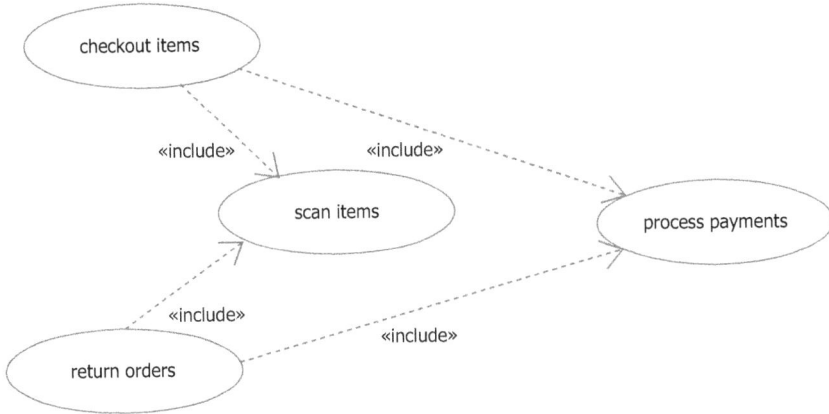

Figure 2.   Inclusion use cases for a POS.

sequence is short. Otherwise, we achieve a small marginal reduction of the redundancy in use case description at the cost of making a use case model unnecessarily complex. We may end up with too many small use cases, and the management of these cases becomes difficult.

Operationally, an inclusion use case is inserted into a step in the description of the base use case. For example, we may have "include (authorize transactions)" as a step in the description of the "withdraw cash" use case. If the inclusion use case has hiccups, we document possible scenarios of inclusion failure as alternate or exceptional flows just like we do for any other interaction steps. The following is a snapshot of the updated description of the "withdraw cash" use case.

Use Case: withdraw cash
Type: base use case

Flow of Events
   Basic Flow:
                1. ATM checks authorization status
                2. ATM asks for a withdrawal amount
                3. Card Holder enters a withdrawal amount
                …..

   Alternate Flows:
             1a. not authorized:
                  .1 Include (authorize transactions)
                  .2 Continue to Step 2

Exception Flows:
      1a.1a. authorization failed:
          .1 ATM displays "authorization failed" message
          .2 ATM ejects the card
          .3 ATM goes idle

Note that in the above description, we include "authorize transactions" use case in an alternate flow rather than in the basic flow. This helps to prevent the inclusion use case from being executed repeatedly if the user may need to perform multiple use cases that all require the inclusion use case as the predecessor. A customer may deposit a check and then move on to withdraw cash. We do not want the customer to exit the current session after finishing the "deposit money" use case and then do another round of authorization for performing the "withdraw cash" use case. Instead, the ATM should check if the customer has been authorized first and then require authorization if not.

Since the inclusion relationship is unidirectional, an inclusion use case is described without reference to the base use case; one inclusion use case may be included into multiple base use cases. The following is a complete description of the "authorize transactions" inclusion use case:

Use Case: authorize transactions

Type: inclusion use case

Flow of Events
    Basic Flow:
          1. Card Holder inserts a card
          2. ATM validates the card
          3. ATM asks for a pin (UI1)
          4. Card Holder enters a pin (UI1)
          5. ATM asks Card Network for authorization
          6. Card Network authorizes the request

    Alternate Flows:
        6a. invalid pin:
          .1 ATM displays "*invalid pin*" message (PM1)
          .2 ATM checks if pin entries exceed the limit
          .3 go to Step 3

    Exception Flows:
        2a. invalid card:
          .1 ATM displays "invalid card" message (PM4)
          .2 ATM ejects the card

.3 Card Holder takes the card

.4 ATM goes idle

6a.2a. too many pin entry errors:

   .1 ATM displays "too many pin entry errors" message (PM3)

   .2 ATM rejects the card

   .3 Card Holder takes the card

   .4 ATM goes idle

6b. account on hold:

   .1 ATM displays "account on hold" message (PM2)

   .2 ATM ejects the card

   .3 Card Holder takes the card

   .4 ATM goes idle

User Interfaces

UI1. Pin Entry UI

Prompts and Messages

PM1 (invalid pin): Your pin entry does not match. Please try again.

PM2 (account on hold): Your account is on hold. Please call (800) 111-1111 for details.

PM3 (too many pin entry errors): Your account is locked due to pin entry errors.

PM4 (invalid card): The card is not recognized.

Use Case Extension

Factorization optimizes use case descriptions by factoring out a required common sequence, whereas extension lets the base use case optionally plug in another sequence of interactions, called an *extension use case*, to be enriched.

During performing the "withdraw cash" use case, at the point when the ATM asks for a withdrawal amount, a customer may want to know his or her account balance. Instead of canceling the ongoing use case to check balance, isn't it better that the "withdraw cash" use case has the option to allow the customer to inquire about balance?

It might be tempting to create a new "withdraw cash with checking balance" use case to have a built-in sequence to check balance. However, the new use case will duplicate the behavior of the existing ones "withdraw cash" and "inquire balance." Of course, most customers do not bother to check balance when withdrawing cash, and nobody wants to perform "withdraw cash" in order to check balance. Thus, we cannot leave out the existing "inquire balance" and "withdraw cash" use cases.

A better alternative is to model the "inquire balance" sequence as an optional extension to the "withdraw cash" use case. The optional on-demand insertion of a use case is modeled using a dashed line with the <<extend>> label, called an *extension relationship*. The use case to be plugged in is called an extension use case, whereas the use case that receives the plug-in is called the base use case. The extension relationship is pointing to the base use case; the extension use case can augment, extend, or enrich the behavior of the base use case. Like the inclusion relationship, the extension relationship is conceptually a stereotyped dependency between classes. Here, it means that the extension use case depends on the base use case, and changes in the base use case affect the extension use case but not vice versa.

Extension makes a base use case more powerful; it enables the base use case to be more efficient and more capable. It allows the base use case to handle more varieties of applicable situations while allowing some users to take shortcut routes. Although it does not change the value to be delivered to the primary actor, it does make some users more satisfied.

When using extension relationships, the base use case may indicate the extension points at which the extension use case may be inserted. The extension use case may also specify the condition under which the extension will be triggered. For example, Figure 3 shows an extension relationship, where "withdraw cash" is a base use case and "inquire balance" is an extension use case. The extension point is when the customer enters a withdrawal amount. Note that some CASE tools will automatically create a compartment for extension points in a base use case when an extension relationship is drawn. Unfortunately, Rhapsody does not have this feature yet. The condition of an extension relationship as well as the references to

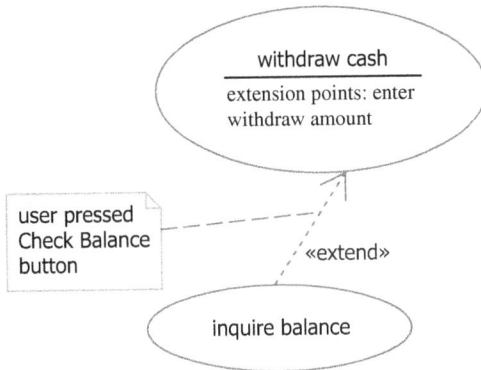

Figure 3.   Extension of "Withdraw Cash" by "Inquire Balance."

the extension points are optionally shown in a comment note attached to the corresponding extension relationship.

Situations involving extension use cases are widely prevalent. The following are sample scenarios that occur to me on a typical day. At this time of writing this book or performing the "write books" use case, I occasionally check spellings and change fonts or insert the extension use cases "check spellings," "change font," etc. While taking a break, I go to an online bookstore to perform the "make orders" use case. Realizing that I have a coupon, I request the "apply coupons" extension use case. On my way home, I pass by a video rental store to borrow video tapes. While performing the "rent videos" use case, the cashier realizes that I am not a member yet, and so she must insert the "create memberships" use case as an extension to the "rent videos" use case so that she does not have to void the transaction and start all over again. Figure 4 shows the above extension relationships.

Note that one base use case may have multiple extension use cases, and one extension use case may be inserted into multiple places, or extension points, of a base use case. For example, at any point of performing the "write books" use case, I may apply "check spelling" and "change fonts" use cases.

Operationally, this is how extensions work. The base use case is complete on its own, and the extension use case interrupts the flow of the base use case at extension points. While carrying out the flow of the base use case, when it reaches the first extension point, the condition attached to the extension use case is evaluated once to decide whether the flow should be interrupted. Once the system confirms that the extension applies, the extension will be invoked at all the succeeding extension points without re-evaluation of the condition.

(a) Write books

(b) Checkout items

(c) Rent videos

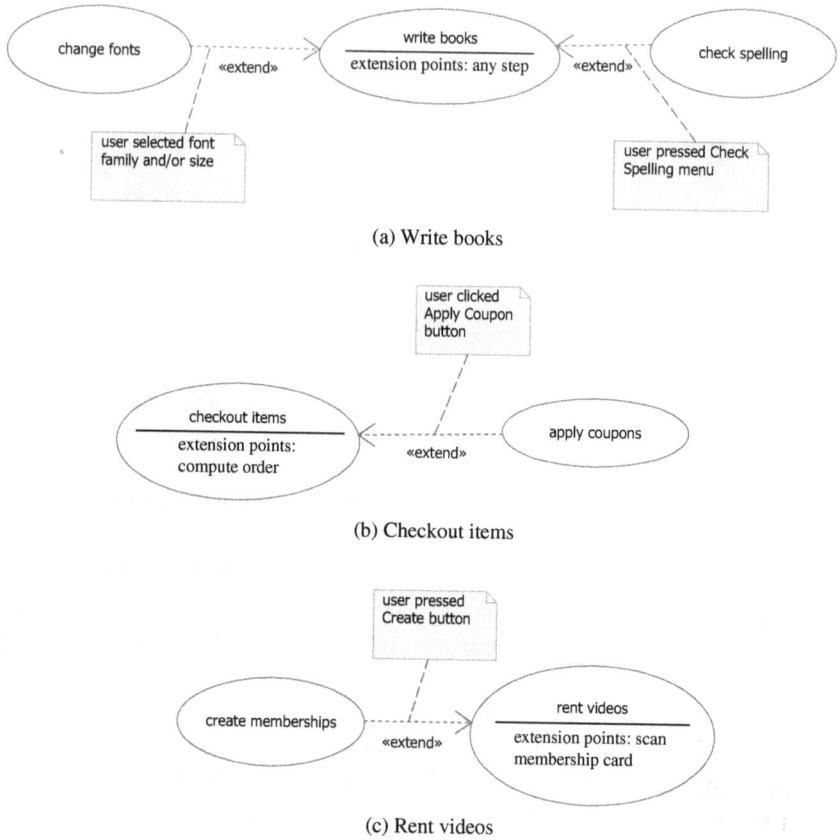

Figure 4.   Examples of extension use cases.

There are two different approaches to the description of an extension use case. A simple approach is to use an alternate flow, with the execution condition as the flow scenario, to describe the extension use case inside the base one. Indeed, an extension is an interruption to the basic flow but will flow back to the basic flow. The following is a snapshot to describe the "apply coupons" extension with the graphical user interface UI2 and message PM5 not included here:

Flow of Events
    Basic Flow:

        ...

      4. The system computes the order total
      5. Customer swipes a credit card

        ...

Alternate Flows:
   4a. user clicked Coupon button:
      .1 The system displays the coupon dialog box (UI2)
      .2 Cashier enters a coupon code
      .3 Cashier presses Apply button
      .4 The system validates the coupon
      .5 go to Step 4

   4a.4a. coupon not valid:
      .1 The system displays "coupon not valid" message (PM5)
      .2 go to Step 4a.2

The simple approach is applicable if the extension use case does not deliver an observable result of value to any primary actor as in the case of "apply coupons," "check spellings," and "change fonts" in Figure 4. However, when an extension use case is itself a base use case, using alternate flows is not an option. In this case, we will use the second approach. First, the base use case explicitly marks the extension points at the beginning of the flow of events. The following shows how to describe the "withdraw cash" use case along with the "inquire balance" extension:

Use Case: withdraw cash
Type: base use case
Flow of Events
   Extension point: enter withdrawal amount (Step 3)
   Basic Flow:
      1. ATM checks authorization status
      2. ATM asks for a withdrawal amount
      3. Card Holder enters a withdrawal amount
          .....

   Alternate Flows:
      1a. not authorized:
         .1 Include (authorize transactions)
         .2 Continue to Step 2

   Exception Flow:
      1a.1a. authorization failed:
         .1 ATM displays "authorization failed" message
         .2 ATM ejects the card
         .3 ATM goes idle

Then, an extension use case is described as usual. It does not have to reference which base use case is to be extended but may optionally document the conditions in which the extension use case is activated.

For example, the "inquire balance" extension may be triggered when the user presses a three-dot button next the withdrawal amount or the "check balance" menu item.

> **Use Case:** inquire balance
> **Type**: extension use case
> **Flow of Events**
> > *Condition: the user pressed "check balance" button*
> > *Basic Flow*:
> > > 1. Check authorization status
> > > 2. ATM asks Bank for account balance
> > > 3. Bank responds with account balance
> > > ...
> > *Alternate Flow*:
> > > 1a. not authorized:
> > > > .1 Include (authorize transactions)
> > > > .2 Continue to Step 2
> > *Exception Flow*:
> > > 1a.1a. authorization failed:
> > > > .1 ATM displays "authorization failed" message
> > > > .2 ATM ejects the card
> > > > .3 ATM goes idle

Both inclusion and extension involve the insertion of one sequence of interactions into another. They have some important differences. First, an inclusion use case is a required sequence by the base use case, while an extension use case is optional. Second, the directionality of the two relationships is opposite: the inclusion relationship is pointing away from the base use case, while the extension is pointing toward the base use case. Third, an inclusion use case is not a complete sequence to deliver an observable result of value to the primary actor, while an extension use case may itself be a base use case.

Use Case Generalization

If the descriptions of two or more use cases are similar overall but do have differences here or there in some steps, we can generalize these use cases using a representative one. For example, the two use cases of deposit check and deposit cash are very similar in their descriptions. The difference is that, for a cash deposit, an ATM counts cash and updates available account balance, while for a check deposit, a customer enters a deposit amount,

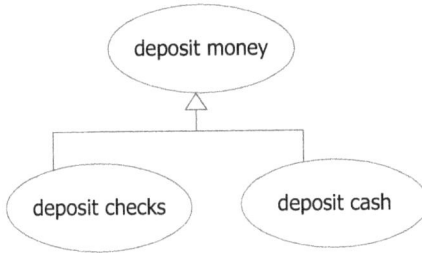

Figure 5.   Generalization of "deposit checks" and "deposit cash."

and the system updates pending account balance. Figure 5 shows the use case diagram for the generalization of "deposit checks" and "deposit cash" into the representative one, "deposit money," using an inheritance relationship.

Generalization applies not only to base use cases but also inclusion and extension ones. When it is applied to inclusion or extension use cases, in fact, it can not only optimize use case descriptions but also offer a device to model exclusive inclusion or extension relationships. For example, the inclusion use case "authorize transactions" deals with two different authorizations, one with Card Network for Card Holder and one with Bank for Customer. Earlier, we proposed treating the Bank actor as a special kind of the Card Network actor. We could model the situation by splitting "authorize transactions" into two separate inclusion use cases: "authorize with Card Network" and "authorize with Bank" (see Figure 6). By doing so, we will have a paradox here: which one of them should be the inclusion use case for the "withdraw money" use case. On the one hand, the "withdraw money" use case selectively includes "authorize with Card Network" for card holders and "authorize with Bank" for bank customers. On the other hand, by definition, an inclusion use case is not optional and must always be included into the base use case. How do we handle such a paradox?

The solution lies in the notion of polymorphism. Remember the representative pattern? If an object has an exclusive association with two or more other objects, we generalize the two or more objects into a representative one. Similarly, if a base use case has an exclusive inclusion relationship with two or more inclusion use cases, we generalize the inclusion use cases into a representative inclusion use case. So, the solution is to create a generalized inclusion use case called "authorize transactions" that can take two forms: "authorize with Card Network" for card holders and

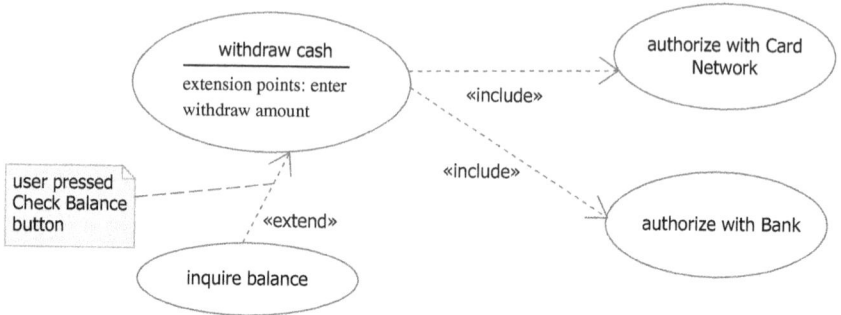

Figure 6.   Which inclusion use to include?

"authorize with Bank" for bank customers. This is a better solution than the earlier ones. It not only resolves the above paradox but also clarifies the relationships with other use cases and the two secondary actors of the ATM system: (1) the "authorize with Card Network" use case is associated with the Card Network actor only, (2) the "Authorize with Bank" use case is associated with the Bank actor only, and (3) the "authorize with Bank" use case will be the only inclusion use case for the "deposit money" and "inquire balance" use cases (see Figure 7).

Generalization is often used when two or more use cases achieve the same goal via different technologies or achieve different goals via the same business process. The following are additional examples of use cases that can be generalized:

- "take order via Internet" and "take order over the phone" can be generalized into "take order" use case
- "check prerequisite with finished courses," "check prerequisite with equivalent experience," and "check prerequisite with instructor's consent" can be generalized into the "check prerequisite" use case
- "book a flight" and "register training courses" may be generalized into an abstract "make reservation" use case
- "file case documents" and "file regulatory documents" may be generalized into "file documents" use case

Here, the first two examples are use cases of achieving the same user goal via alternate routes, whereas the last two examples are use cases of achieving different goals using the same process.

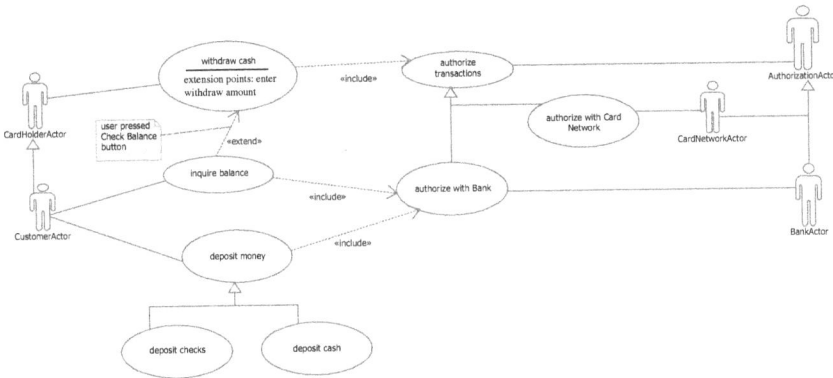

Figure 7.   The use case model for ATM.

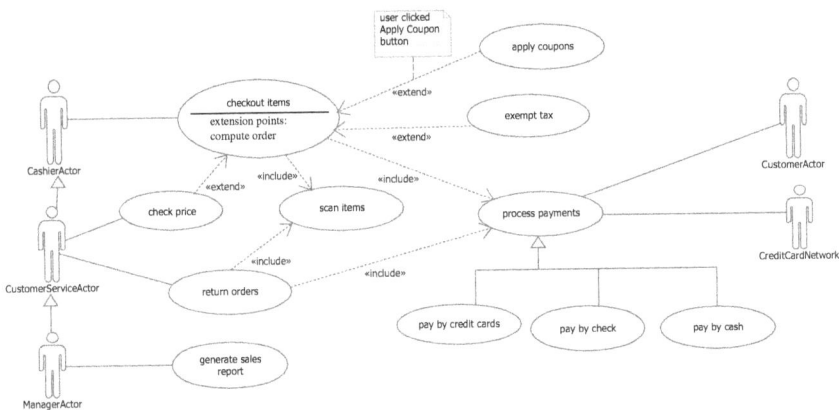

Figure 8.   The use case model for POS.

Figure 8 shows an optimized use case model for the POS, where a cashier performs "checkout items" use case, which is extended by "apply coupon," "exempt taxes," and "check price." The use case includes "process payments" as an inclusion use case. Since there are three different techniques to process payments, a generalization technique is employed to deal with their similarities.

Use case generalization allows one to abstract similar use cases into a common requirement while simultaneously retaining the possibility of describing the differences at the sequence level. It is much like the counterpart for regular objects, where child objects inherit from parent objects all data and operations while retaining the possibility of adding specialized attributes and operations and overriding the behavior of inherited operations.

Therefore, to describe similar use cases, we first fully describe their representative one and then partially describe the different interactions pertaining to each sub–use case to override the interactions in the representative use case. For example, the basic flow of the "authorize transactions" use case is as follows, where the reference UI1 was shown in the previous chapter but omitted here.

Use Case: authorize transactions
Type: inclusion use case
Flow of Events
   Basic Flow:
      1. Card Holder inserts a card
      2. ATM validates the Card
      3. ATM asks for a pin (UI1)
      4. Card Holder enters a pin (UI1)
      5. ATM asks for an authorization center for authorization
      6. The authorization center replies with authorization

Then, the child use cases can modify the above generic sequence slightly into the following two overridden flows:

Use Case: authorize with Card Network
Type: sub use case
Flow of Events
   Basic Flow:
      5. ATM asks Card Network for authorization
      6. Card Network replies with authorization

Use Case: authorize with Bank
Type: sub–use case
Flow of Events
   Basic Flow:
      5. ATM asks Bank for authorization
      6. Bank replies with authorization

At the sequence level, both use case factorization and generalization deal with duplicate interactions. However, the two techniques have differences. First, the use cases to be generalized are similar in the sense that they deliver the same value to the primary actor via different technologies or different values to different actors via the same flows. In contrast, the use cases from which we extract an inclusion use case may just happen to have a common segment in their descriptions. Second, use cases to be generalized look almost identical except for a few steps here or there, i.e., sporadically. They may have no common sub-sequence to be extracted as an inclusion use case, or they may have too many, but each common segment has very few interactions to be justified to be a separate use case.

Practical Use Case Modeling

Discovering use cases from business requirements documentation is an essential skill that the business analyst needs to learn. No matter whether it is a vision statement or a business use case (see the next chapter), a requirement document describes some aspects of the system such as system interactions, functionalities, or constraints. These verbal descriptions often carry a lot of message about business objects and system use cases. In an earlier chapter, we learned how to uncover business objects for class diagramming. In this section, we demonstrate how to uncover system use cases from verbal descriptions using an example.

Example 1: ABC University Business Office receives supplies from various vendors and checks out the items to internal departments. The actual cost of each item is billed to the departments that use the supplies. Internally, as a convention of organizing inventories, supplies are organized into categories. For each supply, the maximum and minimum inventory levels are kept so that when the stock of a part is below the minimum, a replacement order may be issued and sent to a vendor to get it refilled.

To discover use cases, we first underline all the verbs while reading the passage because some verbs indicate interactions with the system. Next, we identify the nouns that are the names of business stakeholders or existing systems. Those nouns will likely give clues on the identification of actors. Then, for each underlined verb, we determine whether the action is executed by a business stakeholder and whether the action needs the system to be carried out. If the answers to both tests are positive, then the verb suggests a system use case. Finally, after all system use cases are identified, one or more optimization techniques may be applied to reduce model complexity and description redundancy.

Following the above procedure, let us first underline all the verbs:

ABC University Business Office <u>receives</u> supplies from various vendors and <u>check out</u> the items *to* internal departments. The actual cost of each item <u>is billed</u> to the departments that <u>use</u> the supplies. Internally, as a convention of organizing inventories, supplies <u>are organized</u> into categories. For each supply, the maximum and minimum inventory levels <u>are kept</u> so that when the stock of a part <u>is below</u> the minimum, a replacement order will <u>be issued and sent</u> to a vendor to <u>get</u> it <u>refilled</u>.

The names of business stakeholders include Business Office, Internal Department, and Vendors. The underlined verbs suggest interactions carried out by each of the business stakeholders and are grouped as follows:

- **Business Office:** receive [supplies], check out [supplies], bill [departments], organize [supplies into categories], keep [minimum and maximum inventory], check [inventory level], issue [replacement order], send [replacement order]
- **Department:** use [supplies]
- **Vendor:** fulfill [orders]

Are departments and vendors going to interact with the system? Probably not. Thus, we identified Business Office as the only primary actor. Look at each action performed by Business Office: keep inventory and check inventory indicate the same action; issue and send orders are related interactions, which can be represented in one use case; organize inventory into categories is probably an action that does not need the system to do anything. Thus, we come up with five use cases: receive supplies, check out items, bill departments, check inventory levels, and issue orders. Among them, "bill department" is something the Business Office will have to do when checking out supplies to departments. Thus, we can model it as an inclusion use case. "Issue and send order" may be optionally executed when the inventory level is low. Thus, we may model it as an extension use case to enrich the behavior of checking inventories. Figure 9 shows the final use case diagram for ABC Business Office.

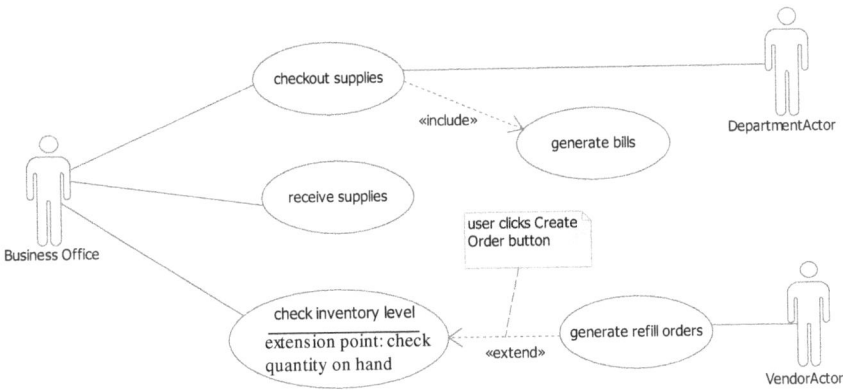

Figure 9.   The use case diagram for ABC Business Office.

Packaging Use Cases

Packaging means to re-structure the use cases into packages for a simplified high-level view of use cases and their dependence. For diagrams like Figures 7 and 8, packaging is probably unnecessary. However, if a use case diagram contains tens or hundreds of use cases, it is going to be messy and overwhelming instead of being informational. Packaging then becomes essential.

We may use either top-down or bottom-up approaches to develop packages. In the top-down approach, we start with packages or sub-systems and then develop a use case diagram for each package. In the bottom-up approach, we start with discovering individual use cases and then regroup them into packages.

There are several strategies for regrouping use cases into packages. These include packaging by actors, packaging by functional concerns, packaging by deployment locations, packaging by changes, etc. Based on the actors, it makes sense for the ATM system to separate the use cases associated with the machine operator from use cases that are associated with customers and card holders. Based on functional concerns, it makes sense to separate all transaction use cases from support ones. For example, we may group the three use cases related to authorization into one package and the other use cases, including "withdraw cash," "deposit money," and "inquire balances" into another package. Finally, if some use cases are anticipated to be more frequently changed than others, it would make sense to single them out.

(a) Package diagram

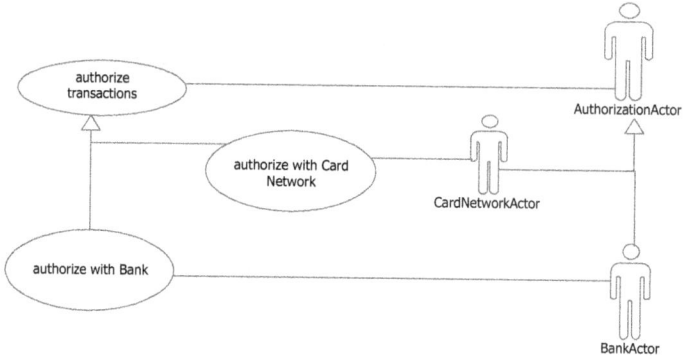

(b) Authorization services package

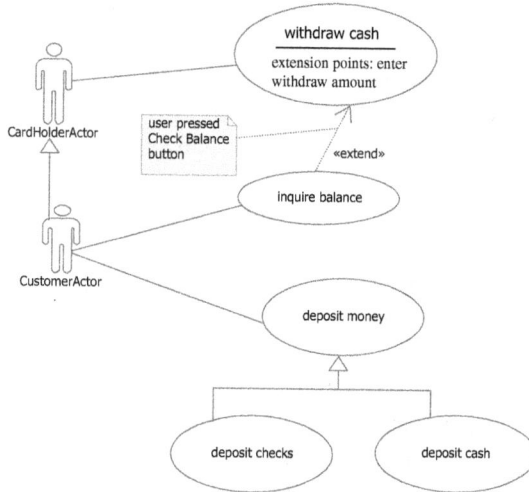

(c) Transactional services package

(d) Operational services package

Figure 10.   Use case packages for the ATM system.

The most important criterion for a successful packaging is that the resulting packages are loosely coupled, i.e., they are not heavily dependent on each other. Such dependencies are represented by directed dash lines pointing from the depending package to the dependent one. Figure 10 shows the package diagram for the ATM system and associated use case diagrams. Note that in the package diagram, the transaction service package depends on the authorization services package.

Review Questions

1. Think of an example wherein two or more use cases contain a common sequence of interactions that may be factored out into a separate inclusion use case.
2. Think of an example of two or more similar use cases that may be generalized into a representative use case.
3. Refer the textual descriptions for "authorize transaction," "authorize with Card Network" and "authorize with Bank," and then think about whether you can use <<include>> to replace a generalization relationship.
4. If two or more primary actors use a same use case, how could you make the use case to be associated with only one primary actor?
5. If one use case exclusively includes one of a few inclusion use cases, how could you handle the UML modeling difficulty?
6. Draw an extend use case diagram by considering the use case "shrink balloon" and the possibility that a balloon gets shrunk automatically over time while flying.

Exercises

1. In the purchasing department, each purchase request is assigned to a caseworker within the department. This caseworker follows the purchase request through the entire purchasing process and acts as the sole contact person with the person or unit buying the goods or services. The department refers to its fellow employees buying goods and services as "customers." The purchasing process is such that purchase requests over 1500 must be out for bid to vendors, and the associated request for bids for these large requests must be approved by the department. If the purchase is under 1500, the product or service

can simply be bought from any approved vendor, but the purchase request must still be approved by the department and they must issue a purchase order. Create a use case diagram with appropriate application of optimization techniques and create a textual description for the use case "process purchase requests."

2. Create a use case diagram with appropriate application of optimization techniques for a standard point of sale system and develop a textual description of the use case "process sale." Here is the business use case description. A customer arrives at the checkout to pay for her selected items. The cashier scans each item's bar code and records quantity, if it is greater than one. The cash register displays the price of each item, its description, and quantity. When all the items are entered, the cashier indicates the end of sale. The cash register displays the total cost of the purchase including tax. Occasionally, a customer may have tax exempt status, and so the cashier must check the certificate and remove the sales tax. Very often, a customer may come with a special coupon that the cashier may need to scan or record in order to apply discounts. The customer may select one of the three following methods to pay for the transaction:

   a. **Cash:** the cashier takes the money from the customer and puts it into the cash register, and the cash register indicates how much change is due to the customer.
   b. **Check:** the cashier verifies that the customer is in good standing by sending a request to an authorization center via the cash register.
   c. **Credit card:** the customer slides her credit card and the cash register sends a request for authorization to an authorization center.

   After the payment, the cash register records the sale and prints a receipt, which the cashier then hands to the customer.

3. BizbyOrder Books is specialized in ordering books for two types of customers: individuals and businesses in lower Manhattan. This is how these two customers are different. When an individual customer orders books, he or she has to pay 20% downpayment. A business customer can establish a credit line with BizbyOrder and pays 50% downpayment if and only if the order amount exceeds its credit limit.

The bookstore orders its books through five national distributors. Because of various special agreements in the book industry, each publisher sells its books exclusively to one distributor. This is how the bookstore runs its daily business. Each time a customer comes in to buy a book, the bookstore uses its database system to find the title and locate the distributor that sells the book. Then the customer will leave their contact information and make a downpayment (if needed) for BizbyOrder to send the order to a distributor. When an ordered book comes in, the customer will be contacted to pick up and pay the rest of the balance.

   a.   Please draw a use case diagram to capture the functional requirements by BizbyOrder,

   b.   Storyboard use case "process order."

4. Here are some descriptions of the OMCA club operations. When members join OMCA health club, they pay a fee for a certain length of time. Most memberships are for 1 year, but memberships for short periods are available. Due to various promotions throughout the year, it is common for members to pay different amounts for the same length of membership. The club wants to mail out reminder letters to members to ask them to renew their memberships one month before their memberships expire. Some members have been angry when asked to renew at a much higher rate than their original membership contract. So, the club needs to keep track of the price paid so that the managers can override the regular prices with special prices when members are asked to renew. The system must keep track of these new prices so that renewals can be processed accurately. One of the problems in the health club industry is the high turnover rate of members. Although some members remain active for many years, about half of the members do not renew their memberships. This is a major problem because the club spends a lot in advertising to attract each new member. The manager wants to track each time a member comes into the club. The system will identify heavy users and generate a report so the manger can ask them to renew their memberships early at a reduced rate. Likewise, the system should identify those who do not come to the club often so that the manager can call them and attempt to attract them to the club.

   a.   Create a use case diagram with appropriate application of optimization techniques for the OMCA health club.

   b.   Describe the use case "renew membership."

5. Here is a high-level procedural description of how ecourse.org handles student enrollment. To enroll a student into a class, the registration system must check whether the student has all the prerequisites taken, whether the class is still open, and whether the total number of credit hours the student registers is not beyond the maximum allowed. After a student finishes her registration, she will need to pick up a printed confirmation that shows all the courses she has registered for, the date/time, section number, credit hours, ecourse.org access code, and instructor for each class. Also, the confirmation paper shows the student status, state of residence, the total number of credit hours, and the total amount to be paid to the college. The student will bring the confirmation to the business office and make a deposit, which is equivalent to 20% of the total amount, to reserve her registration. If she fails to do so within 10 days, her registration will be canceled. The system also actively monitors the number of students signed up for each class. Three days before the class starts, if the number of registered students for a class is less than 15, the class will be canceled. The registered students will be informed to find alternative classes. To better serve the students and departments, the system has functionality for students to make course requests for future terms. The requests will be summarized and sent to departments so that they can make informed decisions on what is to be offered in the future.
   a. What is the scope of the system to be designed?
   b. Please draw a use case diagram and apply the optimization techniques if necessary.
   c. Package use cases into a package diagram.
   d. Design the printed confirmation after registration.
   e. Who is the actor for the use case "cancel registration" due to failed payment, and describe the use case in English with appropriate graphical user interfaces?

6. Professor Bizmind does a lot of consulting in his life. He used to use FastBook to manage his bills and payments. Now he feels that the software cannot be customized to fit his needs. In particular, he would like his clients to be able to make job requests using the Internet. The client can get feedback immediately if the requested time conflicts with his existing schedule. The request can then be modified for another time, canceled, or sent regardless. Bizmind then looks at all the requests every day. If the requested time can be honored, he will

update his schedule and send a confirmation to the client. Otherwise, he will talk to the client using email or phone to set up another time. Then he updates the schedule on the agreed date/time and sends an automatic confirmation. In terms of request details, the professor has itemized a list of standard activities such as Data Analysis, Systems Administration, IT planning, etc. A client can simply select one or more activities when making a a job request. The professor also has a standard unit fee associated with each activity. He may give discounts based on the quantity (e.g., number of hours) performed on an activity. A bill will be sent after each job is finished and at the beginning of each month, if a client has outstanding balance. A minimum payment and a due date will be specified on the bill. Late fee may be accessed if a payment is overdue.

   a.   Please draw a use case diagram, apply the optimization techniques if necessary.

   b.   Package use cases into a package diagram.

   c.   Describe the use case "make job request" with appropriate graphical user interfaces.

7.   Insure-A-Person Inc. provides health insurance services to employees and their family members across America. Due to the need to promote its customer relations, the company has decided to develop a web-based system for clinics and individual customers to be able to file claims on the Internet 24 hours a day and 7 days a week. The company has approached you to design a relational database for that purpose. According to the company, this is how the web-based system is supposed to work. Within 60 days of seeking treatments for himself or any of his family members, a customer needs to log on to the system and file a claim. First, you specify the name of a patient, the date and the place the service was provided, and the primary doctor providing the service. Then, you detail the procedures performed by the doctor. In medical industry, all procedures have been standardized with fixed identification numbers and short descriptions. The insurance company will pay for the service based on all the procedures performed by the service.

   a.   Create a use case diagram and apply optimization techniques if necessary.

   b.   Package the above use case diagram into a package diagram.

   c.   Describe the use case "File Claim" with appropriate graphical user interfaces.

8. The Board of Watson Town Memorial Hospital has recently decided to develop a new information system to manage their patient admissions and discharges. The hospital handles two types of patients: outpatient and resident patient. As typical, each time a new patient comes, the data about his/her identification, address, phone, and insurance carriers are recorded. If a patient is a resident, he/she will be assigned to a bed and an admission date recorded. After the treatment, a nurse has to sign off the discharge card. For an outpatient, the nurse will set a check-back time after each treatment.
    a.   Create a use case diagram with appropriate optimization techniques.
    b.   Storyboard the use case "check in patient" including appropriate graphical user interfaces.

9. Use the appropriate optimization technique to draw use case diagrams for the following use case(s):
    a.   While checking out videos, a cashier in BlockBuster Video may optionally create a new membership for a customer who does not have an account yet or search for membership ID if a customer does not bring his membership card.
    b.   While checking out in a grocery store, some customers may present coupons for discounts.
    c.   Overall, for the POS, checking out items and returning items are pretty much the same except that for returns, a receipt is scanned first before scanning each item and at the end, the payment is typically a refund. How do you optimize these use cases?
    d.   In a hospital registration system, checking in inpatient and outpatient processes are almost identical except that for inpatients a bed is assigned.
    e.   In a student registration system, to be able to enroll into a class, the system must make sure the prerequisites are met. In some other processes, such as plan for future courses, the system must also consider prerequisites.
    f.   For an ATM, it seems that in all transactions, the final steps will be abut printing receipts, recording transactions, and dispensing the card. How do you optimize the use cases for an ATM?

Chapter 12

Requirements Documentation

Introduction

Requirements discovery and development proceeds object and use case modeling, but we postpone it to this later chapter because we need to first understand the concepts of use cases and objects before we can see how requirements connect to the concepts. In this chapter, we will go back to the initial stage of the development process and learn the techniques on how to document requirements. In detail, we will learn how to develop vision statements, how to specify project scopes, how to develop major features, how to develop business rules, how to develop business use cases, and how to develop software requirements.

Requirements

Requirements, functional and non-functional, may be classified into three levels: business requirements in the blue sky level, user requirements at the sea level, and software requirements in the deep ocean level (see Table 1). Different elemental requirements can be classified into one or more of these categories for documentation. When deciding where to include what, besides considering the scope differences of the three categories, we may also consider two other criteria: (1) contractual vs. informational; and (2) sensitive vs. public. In this chapter, we will describe the elements by scope only.

Table 1.   Requirements classification.

	Functional	Non-functional
Business	Background, business objectives, risks, strategic alignment, value chain integration	
	vision, scope, feature, business use cases	*business rules*
User	User goals, tasks, resources including usability, usefulness, and quality	
	system use cases	*information, interfaces*
Software	Measurable software specifications	
	behavioral statements	*PIECES: performance, integrity, efficiency, control, economy, service (reliable, flexible, accessible, scalable, etc.)*

Business requirements express high-level expectations by an organization on the system to be developed. They specify how the mission of the system should be aligned with the mission of the organization. Business requirements describe the primary benefits that the system will provide to all stakeholders, including customers, employees, users, and sponsors. They provide background on the rationale and historical or organizational context of the project; describe the business problems to be solved, business opportunities to be taken, and business threats to be addressed by the project; outline business objectives and success criteria that the project will help to achieve; and summarize the business risk associated with developing or not developing the system.

User requirements are specifications of what values the system must bring to its users. They are the derivatives of business requirements and are expressed as system use cases.

Software requirements, expressed as classic "shall" statements, are the most detailed specifications of how the system shall look and feel, shall answer user requests, shall enforce business rules, and shall respond to environmental events. They also include other non-functional requirements on performance, control, security, usability, integrity, availability, reliability, compatibility, etc.

Three levels of requirements represent three different perspectives in specification. Their boundary lines are often fuzzy, in part because they are all deliverables in the requirements development stage and are produced by requirements engineers with the assistance of business stakeholders. They are fuzzy also because some requirements or related artifacts are cross-boundary in nature, e.g., (1) business objects and data are referenced and specified in all requirements, (2) business rules are referenced in user requirements and enforced in software requirements, and (3) business use cases (or business processes) belong to either business requirements or user requirements.

Vision Statements

Business requirements provide the foundation for the development of user and software requirements. As an important part of business requirements, *a vision statement* presents the long-term purpose of the system and an idealistic picture of how the system aligns with business objectives and satisfies the needs of all business stakeholders. Essentially, a vision statement synthesizes and summarizes business requirements and provides a clear vision for the entire software development process. Thus, it guides user and software requirements development. Because of this nature, the vision statement may be included into either the business requirements document or the user requirements document, if an organization desires to have separate documents for different requirements.

Often a vision statement may be phrased using a standard template. For example, one such template, suggested by Geoffrey A. Moore (1991), is as follows:

> *For [target customer]* **who** *[statement of need or opportunity],* **the** *[product name]* **is** *[a product category]* **that** *[key benefits, compelling reasons to buy or use].* **Unlike** *[primary competitive products, current system, or current business process],* **our product** *[statement of primary differentiation and advantages]*

The following are two example vision statements following the template. The first is to envision a food order system targeted at cafeteria and restaurant owners, and the second is a centralized electronic medical record system.

> **The Vision of E-Servant:** For restaurants and cafeteria who desire to improve service quality and reduce service staffing costs, the e-servant

is an electronic food order system that allows customers to make orders without using a waiter or waitress. The system allows the customers to browse specials of the day, to custom food flavor or taste, and to format and dispatch kitchen orders automatically. It also allows the customer to check out with credit card without waiting for a cashier. Unlike the current manual ordering system, our product will cut customer wait time by 90%, reduce food wastes by 30%, and reduce staffing by 80%. Our product will also generate all sales reports and submit sales tax in compliance with all local government regulations.

The Vision of Electronic Medical Records: For all healthcare providers and consumers who want a single point of access to all patient information, the Electronic Medical Records is an information system that provides a central repository of patient medical data and a suite of tools to retrieve and update the data. The system will contain information such as medical history of the patient, past diagnosis, test results, prescriptions, and communications between doctors and patients. It will also act as a legal record of the care and provides information to public health, epidemiological studies, and clinical research. Unlike the current manual or paper records, our product will defragment separate medical records into a single universal patient record, which improves data accessibility and data quality and makes it easy for physicians to perform effective and efficient diagnosis and exchange patient information across clinics and practices. It will also reduce the workload for managing patient data and cut the clinic staffing costs by 60%.

Scope

To elaborate a bit more about a vision statement, business requirements should include a system scope, which delineates a set of software features and business tasks (or business use cases) that the system enables and delimits certain capabilities that the system will not include. Unlike a vision, a scope is often tied to a specific timeframe or release version due to resource limitations at a certain point in time. The scope may change over time. The scope description sets the boundary of the system to be developed and the context the system lives in. It specifies what is included and what is not. Sometimes, a graphic model such as a *context diagram* provides an alternative to a scope statement by showing the system in the context of the universe of actors or stimuli. A scope description may be included into either the business requirements document or the user requirements document.

Depending on how we view the relationship between a system and its context, there may be four different kinds of context diagrams to circumscribe a scope. The first type, focusing on interactions between the system and its users, shows the system instances as a use case and the connections to the context as associations. Figure 1 depicts such an interaction context diagram for the ATM.

An interaction context diagram circumscribes the system boundary by showing who external actors are and how their instances are associated with a system instance. It is a high-level use case diagram from a bird's-eye view. As such, it may omit a lot of technical details such as role maps among actors, dependencies among associations, and multiplicities of the associations. Figure 1 explicitly displays multiplicities to show that, at any moment, there is 0 or 1 instance of each primary actor connected to one system instance for the purpose of understanding the nature of concurrent interactions and the requirements for concurrency control.

An interaction context diagram may serve as a starting point to develop use case diagrams for the system. Yet it does not convey much information

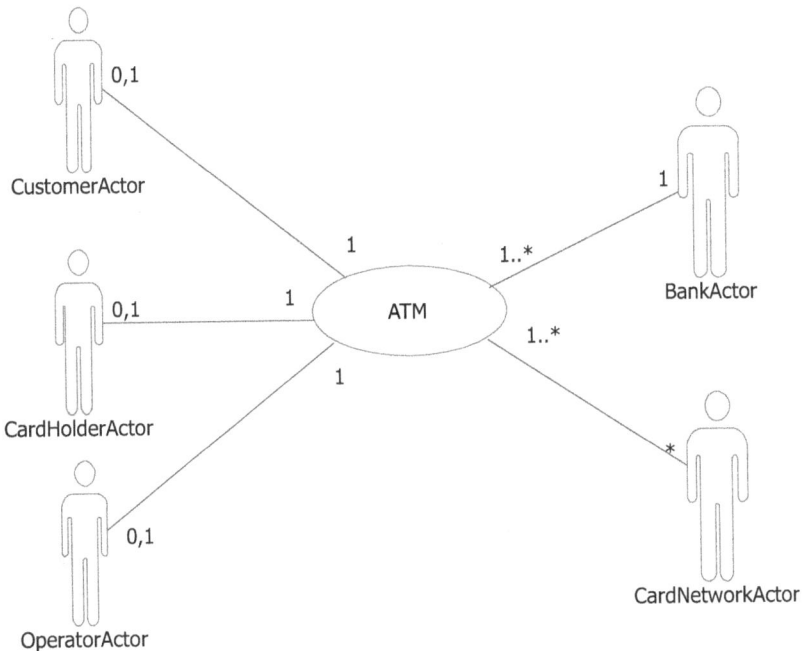

Figure 1. An interaction context diagram for an ATM.

on the functional requirements for analysts to discover use cases except that the role of each actor implies a system use case to be executed by the actor.

The second type of context diagrams, focusing on data flows between the system and external entities, shows the system as one function (or data processor) and its connections to the context as data flows. These data flows are either initiated from an external entity or delivered to an external entity. They are often mission-critical, justifying the purpose or mission of the system in the context. Figure 2 shows the context diagram for a food order system. This type of context diagram is an important deliverable for structured development; data inputs and outputs implicitly convey what the system is supposed to do, and therefore may give hints for discovering functional requirements. For example, the food order system in Figure 2 can take food orders and customer payments, generate customer receipts, update the inventory system with inventory being used, generate formatted kitchen orders, and print sales reports for managers. Clearly, it conveys the related functional requirements, some of which may become use cases.

The third type of context diagrams addresses event–response relationships between the system and its context. The diagram shows the system as the responder of events and the context as the dispatcher of events. Unlike the other types of context diagrams, here, the context may not be described as a set of actors, external entities, or agents, and event–response relationships may not be represented as associations, data flows, or dependencies. Figure 3 shows a gate control system that responds to the change of light, time, and weather as well as the presence of objects

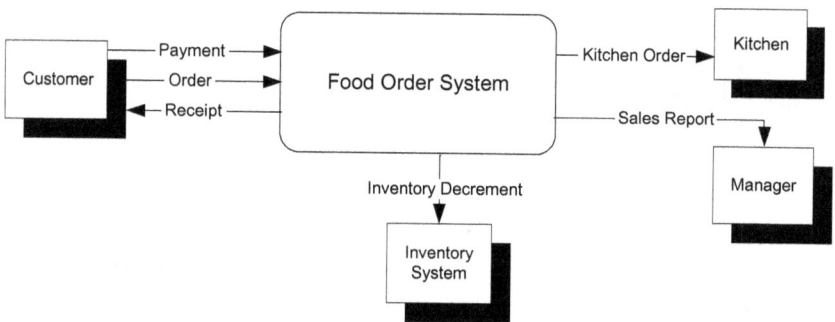

Figure 2.   A data-flow context diagram for a food order system.

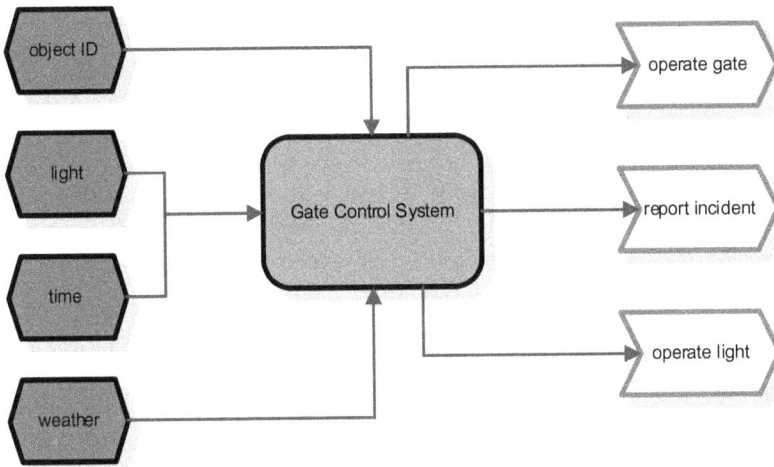

Figure 3.   Event context diagram for a gate control system.

by invoking appropriate responses including "open gate," "close gate," "report incident," "turn on lights," "close lights," etc.

An event context diagram is important for developing real-time event-driven systems. A gate control system, for example, does not have any particular users. Instead, its use cases are triggered by external events. During the evening or weekend, the system will not respond to any events and keep the gate and light off. During working hours, events like bad weather, low light, and the presence of an object will trigger the light to turn on. The object will be scanned for verification and let in if recognized. Otherwise, the system will ask for an identity and a permission from the management in order to open the gate. In the event of false attempt, the incident will be reported to police.

Event context diagrams may be useful for discovering functional requirements and business rules in developing management information systems. For example, many systems periodically send invoices or statements to customers. When a payment is overdue, it sends an automatic reminder email or phone call. These use cases are executed in response to timing events. As another example, some manufacturers actively monitor the inventory status of their products in retail locations and automatically create replenishment orders when the level is low. Thus, the use case "create orders" is triggered by inventory status rather than being executed by any actor.

The last type of context diagrams, focusing on the dependencies between the system and external agents, shows the system as one agent and its connections to external agents as dependencies. It is a simplified version of the so-called *strategic dependency model* based on the *i** techniques proposed by Eric Yu (1995). The basic idea is that all agents in the context have intentional properties such as goals, and they depend on each other to fulfill their goals. A strategic dependency model consists of a set of nodes for agents, including the system, and links among them. Dependency links capture the motivation and the rationale of agents. The *i** technique distinguishes four types of dependencies: *goal dependency* (oval shape), *resource dependency (rectangle shape), task dependency (hexagon shape)*, and *soft goal dependency (cloud shape)*. In a goal dependency, an agent depends on another to fulfill a goal. In resource dependency, an agent depends on another agent to provide physical or informational resources. In task dependency, an agent depends on another to carry out a task. A soft goal dependency is like a goal dependency except that a soft goal is not precisely defined and is often associated with non-functional requirements.

Figure 4 shows a dependency context diagram for *SmartCD* order system (Alencar *et al.*, 2000) along with two external agents: *Client* and *Store*. *Client* depends on *Store* to buy CDs (resource dependency) and wishes the services to be of good quality (soft goal). *Client* depends on *SmartCD* to take orders and receive notifications when the ordered CD arrives (task dependencies). *Client* expects the access to *SmartCD* to be secure (soft goal). *Store* relies on the system to process internet order (goal) and to update inventories (task).

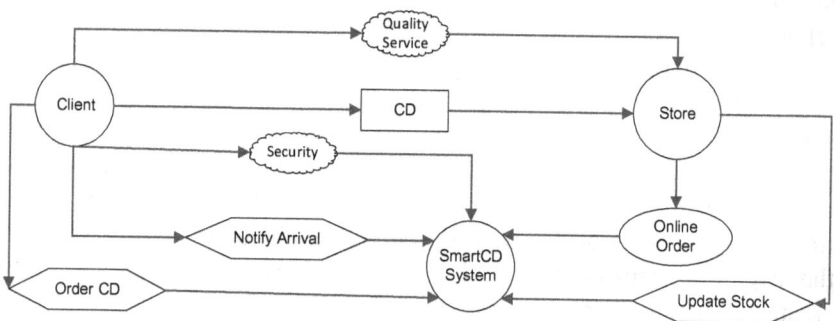

Figure 4.   Dependency context diagram for SmartCD.

Four types of context diagrams represent four different views of a system boundary. The utility of each may vary depending on the nature of the system, and they may complement each other. A dependency context diagram seems to include the first two types of context diagrams as special cases, where a data flow is equivalent to information resource dependency and interactions are equivalent to tasks. Since it also represents goal and soft goal dependencies, it carries more information on requirements, both functional and non-functional, than other types of context diagrams.

Major Features

Besides a scope statement of what is included and what is not, or one or more context diagrams visualizing the relationships between the system and its context, other important artifacts that may be included in the business requirement document are major features and prioritization plans. Major features are high-level system capability statements. Each major feature is typically labeled with a unique ID such MF1, MF2, etc. These features will be referenced throughout all requirement documents. If incremental releases of the system are planned, major features may be scheduled for incremental implementations at various stages with various priorities. To this end, a prioritization plan may be documented as a table detailing what major features are not to be implemented, partially implemented, or fully implemented, in which release and which major feature has low, medium, or high priority.

A major feature is derived from a vision or scope statement and may be expressed as the "the system shall be capable of ..." statement. All the major features, if combined, must be in alignment with or in support of the vision statement and must be within the system scope as described or pictured. As an example, the following is a list of major features within the scope of the system as pictured in Figure 2:

MF1: The system shall be capable of taking customer orders.
MF2: The system shall be capable of processing customer payments.
MF3: The system shall be capable of formatting and submitting kitchen orders.
MF4: The system shall be capable of printing sales report.
MF5: The system shall update inventory upon taking an order successfully.

Similarly, the following is a list of major features implied by the context diagram in Figure 3:

MF1: The system shall be capable of controlling the gate operation in response to time changes, presence of vehicles, and the knowledge of vehicles.

MF2: The system shall be capable of controlling the light in response to time change, weather change, light change, and the presence of vehicles.

MF3: The system shall be capable of reporting an incident in the event of failure to recognize and validate a vehicle.

Business Use Cases

A business use case is an end-to-end business process that delivers an observable result of value to a business or its stakeholders. It is called a use case because it is a sequence of interactions. However, here, the interactions are among business stakeholders, and some interactions may be of physical nature, i.e., non-data activities, and may take an extended period of time.

Before developing a business use case, stakeholder profiles may be developed for reference. A stakeholder profile may be expressed as a table detailing all business stakeholders involved in the project, their values or benefits due to the project, their likely attitudes toward the project, their interests to be considered, and their constraints to be accommodated.

A business use case is an end-to-end business process, and a software project is often proposed to reengineer the whole process or automate one or more activities in it. When developing a business use case, we need to explicate whether it is a current business case or a future one. Lower-level requirements will be derived from a future business use case.

The following is a business use case for a hospital. It details a sequence of interactions among patients, doctors, receptionists, nurses, and insurance companies. Each interaction involves one activity, data or non-data, and engages one or more business workers, partners, and systems. The whole sequence describes an end-to-end business process of handling patient visits.

A patient arrives at the hospital for a treatment or a general checkup. Irrespective of whether the patient is new or not, he has to fill out a patient form with basic information, like name, SSN, allergies, reason for the visit, etc. Once the form is completed, a receptionist pulls his file, verifies the information, and hands the file over to a nurse. The nurse will take initial tests and acquire preliminary medical information such as the patient's blood pressure, allergies, current medications, if any, etc. Once done, a doctor sees the patient. The doctor will go through the initial results obtained by the nurse and also the patient's file before conducting formal diagnosis. The doctor may recommend additional tests. Once the reports of all tests are reviewed, the doctor will prescribe medication. If needed, the patient is admitted to the hospital for inpatient treatments, and the doctor and nurses go for routine visits till the patient is discharged. For any treatment, the bill will be sent to the patient if the patient does not have a medical insurance. Otherwise, the bill will be sent to the insurance company, which will then deal with the patient henceforth for settling the bill.

Diagrams and/or structured descriptions may be used in lieu of an unstructured textual description of a business use case. For example, we may treat a business use case as a system use case and any stakeholder as an actor and then model their interactions using a use case diagram. Figure 5 shows a business use case diagram with three business use cases where the "manage hospitalization" extends the "manage patient visit" use case, and both the "manage patient visit" and "mange hospitalization" use cases include the "mange payments" use case.

A business often has multiple ongoing processes, all or some of which may be captured as business use cases. For example, a hospital may have a process for employee hiring, a process for patient appointments, a process for emergency response, a process for supply procurement, etc. The business use cases to be captured and modeled must, of course, reflect the vision and the scope of the system.

As for system use cases, after a use case diagram is created, a procedure model may be produced. For example, for each business use case, we may use a textual description or a graphical model to describe it.

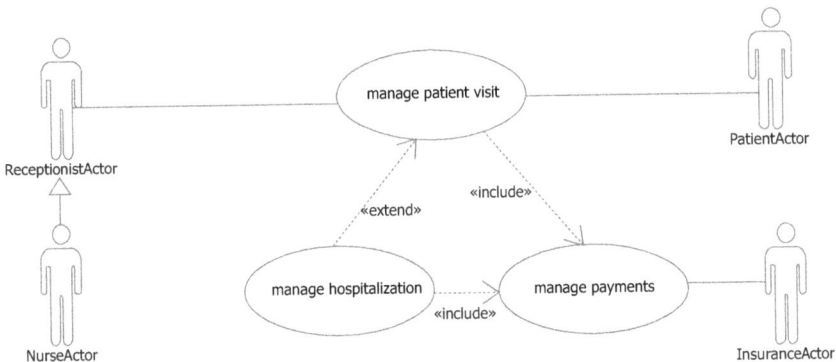

Figure 5.   A business use case diagram for patient visits.

Business use case via examples: Relocation order

Figure 6 shows a business use case diagram for a moving company followed by textual descriptions of two use cases. The company wants to develop an information system capable of handling all moving-related activities, from initial estimates, to packing and moving, and to the collection of payments. However, due to the complexity in the initial stages of the process, the company has an urgent need to create a system that is capable of handling packing household goods and initial customer payments. Therefore, we show two business use cases: one for packing household goods and one for handling payments.

The following are textual descriptions of the two business use cases in Figure 6. The format of describing a business use case follows that of a system use case, but with a less rigorous structure. First, alternate or exceptional flows may be embedded into the basic flow wherever hiccups occur. Second, longer and more complex unstructured sentences may be

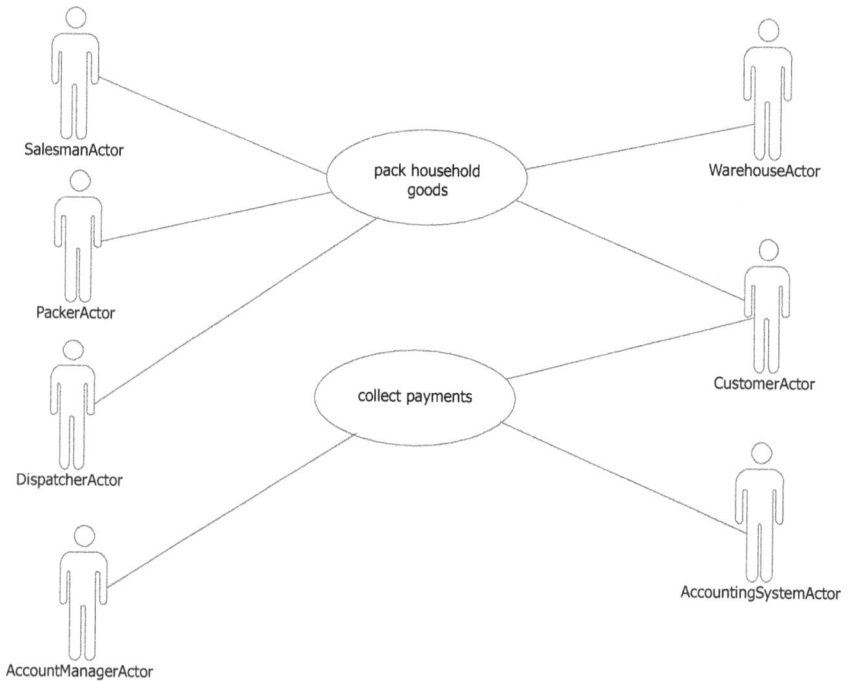

Figure 6.   A business use case diagram for a moving company.

used to describe each event. The end result is a semi-structured description that outlines the logical flow of interactions among stakeholders.

Business Use Case: Pack Household Goods (HHG) for Relocation

Precondition: A mover is ready to take relocation orders.

Postconditions on success: The customer's household has been packed by the mover and is ready for relocation, or the customer has purchased packing material to perform self-packing service.

Flow:
1. A customer calls the moving company and schedules a time for a salesman to go to his home and give him a cost estimate for HHG relocation.
2. A dispatcher schedules the time and date of the estimate and a salesman on the job.
3. The salesman completes an in-house quote when that day arrives.
4. After the salesman gives the quote, he collects the customer's information, creates a quote number, and gives the number to the customer.
5. The salesman also gives a hard copy of the estimate to the dispatcher.
6. When the customer decides that he wants to use the mover, he contacts the salesman and gives him the relocation date as well as other origin and destination information.
   6.1. If the customer chooses not to schedule the packing job, the estimate is disposed after one year.
7. The dispatcher receives customer orders from each salesman and schedules packing services.
   7.1. If the customer wants to perform self-packing service, the dispatcher creates the material order and sends it to the warehouse for fulfillment and shipping. Customer pays at the time of delivery.
   7.2. If the customer wants the mover to perform the packing service, the dispatcher enters the customer's name and packing date into the schedule book and collects the quoted amount in advance.
8. One day prior to the packing date, the dispatcher assigns one or more packers who will perform the packing service. Then he notifies the customer of the estimated time that the crew will arrive.
9. On the day of the packing job, the crew obtains the needed material from the warehouse and proceeds to the customer's home.
10. Once the packers complete the service, they return to the office with the signed paperwork that indicates that the service has been completed. The form also specifies how many cartons have been packed and what the sizes those cartons are. Warehouseman restocks unused cartons.

Business Use Case: *Collect Payment*

Precondition: A sale for packing material and/or services has been completed.

Postconditions on success: A payment has been made to the company and to the salesperson involved. Revenue and expense accounts have been updated.

Flow:
1. The dispatcher sends a copy of the completed packing forms to the accounting department. The accounting manager verifies the actual amount of the sale against the estimated amount that was collected in advance.
    1.1. If the amount collected is more than the actual amount, the account manager refunds the customer via check or by credit issued to their card.
    1.2. If the amount collected is less than the actual amount of the service, then the accounting manager charges the balance due to the customer's credit card or he contacts the customer to obtain a cashier's check for the balance.
2. Account manager enters payment, revenue, and expense information into the existing accounting information system. (The existing accounting information system will be used to record payments to the proper accounts, pay refunds, maintain collections, and pay commissions to the sales staff and to record payroll expenses that are related to the sales commission).

Business Rules

A business rule defines or regulates certain aspect of a business. It includes corporate policies, laws, principles, conventions, standards, etc. Like business use cases, business rules are of high-level requirements or their determinants. They may be included in a business requirement document and should be cross-referenced in other requirements when applicable. For systems analysis, business rules are incorporated into conceptual or business models. They may also form the constraints or references that all business use cases need to observe and are thus included in use case descriptions. For systems development, business rules are incorporated into logical class diagrams and the procedural descriptions of object operations. Since business rules reflect the domain knowledge of a business, their utility as references or constraints are often beyond one or two projects. Thus, organizations may develop central rule repositories for storing, organizing, and managing business rules so that multiple software projects can access and share a common set of rules.

A business rule, depending on what it may impact on, may be classified into three broad categories: structural rules, algorithmic rules, and behavioral rules.

Structural rules

A structural rule defines or regulates objects, object attributes, and object relationships. Its direct impact is on the development of object models such as class diagrams. For example, a structural rule defines or regulates how objects are composed and related. Here, an object includes not only a conceptual one like a business entity but also a logical one like a user interface or control object. The following are some examples of structural rules:

SR1: A customer address is the physical location where a customer primarily resides.

SR2: An airline has one or more planes.

SR3: A course is an educational product that delivers knowledge of specific breadth and depth to a receiver upon its completion.

SR4: Each lab order includes a unique identifier, the date the order was created, the date the test was done, and the physician who performs the test.

SR5: All correspondence regarding an order shall disclose the last four digits of a credit number while hiding the rest of the digits.

Implicitly, a structured rule specifies what object operations are needed in support of its definition or regulation and how objects shall be in collaboration to realize a use case. Therefore, a structural rule may constrain one or more interactions in use cases. For example, rule SR5 specifies a functional requirement in performing activities related to creating customer correspondence.

Algorithmic rules

An algorithmic rule defines, regulates, or derives possible computational results under certain conditions. The simplest manifestation is a formula or logic for arriving at a result. A generalized manifestation is a constraint that delimits a possible set of results. In any case, its impact is on object states, attribute values, and computational results. When impacting object states or attribute values, an algorithmic rule implies actions to reset an object state or attribute or restrain the choices of the state or attribute. When impacting computational results, it determines how certain activities in a use case are executed and results calculated or selected. The following are a few examples of algorithmic rules:

AR1: A section with the number of students less than its cap is considered open.

AR2: If a payment is not received within 30 days after the due date, the account will be delinquent.

AR3: A customer who made 2 orders in the last five years is considered active.

AR4: An order quantity of an item is no larger than the quantity-on-hand of the item.

AR5: A customer must be 18 years or older to purchase alcohol products.

AR6: A professor can check out a reference book for 6 months.

AR7: A grade takes on five possible values including A, B, C, D, and F.

AR8: The discount rate is determined by order amount as follows: 5% for order of $500 or more, 8% for orders of $1,000 or more, and 15% for orders of $2,000 or more.

AR9: The amount payable is computed based on the formula: amount = (price * quantity) (1 − discount rate) + tax + shipping and handling.

An algorithm rule is typically stated as a reference in use case descriptions. However, if an algorithmic rule specifies possible results with no conditions, like AR7, the rule may be expressed as an object attribute or state and is captured in an object model. Otherwise, the rule is better expressed as a decision table, decision tree, formula, or algorithm and included into a use case or operational description. For example, AR8 may be expressed as a contingency or decision table as in Table 2.

Table 2. Discount rates.

Sales Amount	Discount
$500.00–$999.99	5%
$1,000.00–$1,999.99	8%
Over $2,000	15%

Behavioral rules

A behavioral rule prescribes or regulates actions or activities in response to certain conditions. Like an algorithmic rule, it may be expressed as a decision table or tree. However, it leads to actions instead of object states, attribute values, or computational results. Like some algorithmic rules that act as a mechanism to trigger the change of object states or attribute values, a behavioral rule acts as the mechanism to trigger actions or activities. The following are a few examples of behavioral rules:

BR1: Each order must be verified and acknowledged within 2 business days of receiving it from a customer.

BR2: At the end of each month, generate a billing statement if an account balance is greater than $1.00.

BR3: If the number of students enrolled in a section is 80% full within one week of opening, notify the department of the enrollment status.

BR4: If the card reader detects a card, verify the validity of the card.

BR5: When an object is detected, the sight is dark, and it is during working hours, turn on the light.

BR6: If a debit card is inserted, display all choices that a customer may choose. Otherwise, display "withdraw money" choice only.

A behavioral rule may be captured in various requirements. If the resulting actions or activities are interaction steps of a use case or operation (e.g., BR3, BR4, and BR6), it may be documented in the user requirement document, for example, in use case or operation descriptions. If the resulting activities are use cases (e.g., BR1, BR2, and BR5), the rule may be captured in the business requirement document.

Behavioral rules can be alternatively expressed using decision trees and contingency/decision tables in lieu of verbal statements. Figure 7 shows a decision tree that describes the decision to turn on or off light in response to three interrelated factors: time, presence of an object, and light condition (see BR5). In a decision tree, we use a circle to represent a condition variable and a rectangle to represent a decision. We use a branch out of a condition variable to represent one possible value of the condition variable.

A decision tree may be equivalently represented as a decision table. To develop a multidimensional decision table, we first need to decide on the number of possible value combinations of the conditional variables. A decision table shows the responses under each of these possible combinations. In our current example, there are three condition variables, and

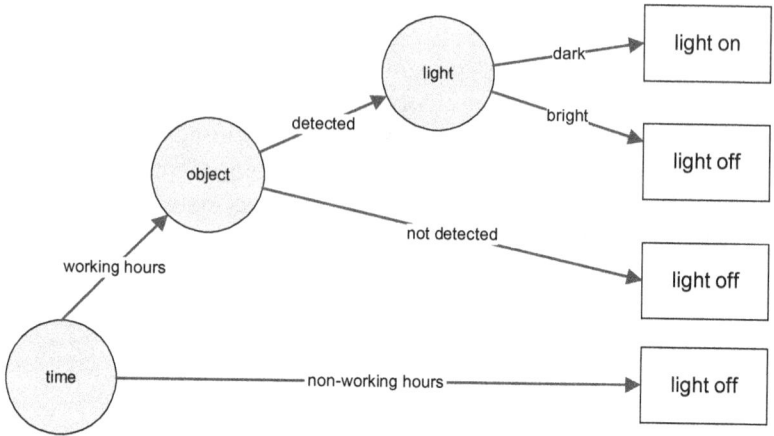

Figure 7.   Decision tree for light control.

Table 3.   Decision table for light control.

	1	2	3	4	5	6	7	8
Time	W	W	W	W	NW	NW	NW	NW
Vehicle	P	P	NP	NP	P	P	NP	NP
Light	B	D	B	D	B	D	B	D
Light Control	Off	On	Off	Off	Off	Off	Off	Off

Note: W: working hours; NW: non-working hours; P: presence; NP: non-presence; B: bright; D: dim.

each takes on two values. Thus, there are eight possible combinations. We create a table with 8 + 1 columns (one extra column for the names of the variables), and with all condition variables, one variable in each row in the top portion of the table, and the decision variables, one in each row in the bottom part of the table (see Table 3).

If a behavioral rule implies the capability of the system in responding to conditions or events, it may be captured as a major feature in the business requirement document or a functional requirement in the software requirement document. For example, the major feature "The system shall be capable of controlling the light in response to time change, weather change, light change, and the presence of vehicles" reflects BR5. For another example, BR3 may be translated into a functional requirement as "The system shall verify the validity of the card when its reader detects a card."

Functional Software Requirements

Software requirements are derived from and must be in support of business and user requirements. They can be classified into two categories: functional and non-functional.

A functional requirement, like a major feature, may be expressed as "the system shall be capable of ..." statement. The difference is that a major feature states a high-level system capability that may have to be realized via one or more system use cases, whereas a functional requirement states the system capability at the interaction level. For example, a major feature of the food ordering system may be stated as "The food ordering system shall be capable of taking customer orders." In contrast, a functional software requirement at a detailed level may be stated as "The system shall confirm that an order has been successfully submitted."

Like any other requirements, functional requirements are uniquely labeled. Since these requirements are often hierarchical in nature, we typically label them as sections, paragraphs, sentences, etc. There are many ways in which one can group functional requirements into hierarchies. The choices include grouping by major features, grouping by use cases, grouping by operational mode, grouping by actors, grouping by objects, and grouping by events or responses. Different templates may be adopted depending on the choice of the strategy. For example, if one chooses grouping by major features, the following sample section of requirements is related to the major feature of taking order and may serve as a template:

1.1: The system shall let a service person who is logged in to the food ordering system create an order of one or more meals for a customer
   1.1.1: The system shall confirm that the service person is logged in a service person
      1.1.1.1: If the service person is not logged in, the system shall give him or her option to login and continue to create an order
      1.1.1.2: If the service person is logged in not as a service person, the system shall give him or her choice to log out
   1.1.2: The system shall display the list of all meals and their availability

1.1.2.1: The system shall allow the service person to filter the list using a full or partial category name or meal name

1.1.2.2: The system shall allow the service person to select or deselect a meal in the list by placing or removing a check mark at the beginning of each meal

1.1.2.3: If the service person checks a meal that is unavailable, the system shall display a message "the selected food is not available"

If requirements are organized by use cases, one may label each requirement by referring to the use case ID and flow step ID. The following text presents a sample section of a use case and the corresponding functional requirements:

Use Case (UC1): Withdraw cash

Basic Flow:

1. Card Holder inserts a card
2. ATM validates the card
3. ATM asks for a pin
4. Card Holder enters a pin

....

Functional Requirements:

UC1.1: When it is idle, the system shall let a card holder insert a card
.a: The system shall display the welcome screen (UI1.2)
.b: The card reader shall be empty

UC1.2: The system shall be capable of validating a card if detected
.a: The card reader shall be capable of scanning the magnetic bar for account number
.b: The card reader shall inform the system if card scan succeeds or not
.a: The card reader shall reject the card if card scan fails
.b: The card reader shall pass account number to the system if card scan succeeds
.c: The system shall confirm if the card scan fails

UC1.3: If card scan succeeds, the system shall display "enter pin" prompt for the user to enter a pin for 10 seconds or until the user presses "enter" key (UI1.3)
.a: If the user does not enter press "enter" key after 10 seconds, the system shall display "more time" prompt

asking the user if he or she needs more time for 2 seconds and indicate the choice by pressing Yes or No key (UI1.4)

.a: If the user presses "Yes" key within 2 seconds, the system dismisses the "more time" prompt and add 10 seconds to continue "enter pin" prompt

.b: If the user presses "No" key within 2 seconds or does not make a choice after 2 seconds, the system dismisses "more time" prompt, dismisses "enter pin" prompt, rejects the card, and displays "remove card" prompt for 10 seconds (UI1.5)

    .a: If the card is removed within 10 seconds, the system dismisses "remove card" prompt and enters the idle state

    .b: If the card is not removed after 10 seconds, the system dismisses "remove card" prompt, swallows the card, and enters the idle state

In the above statements, labels like UI1.2, UI1.3, etc., are referring to user interface prototypes, which may be in one section of the user requirement document for references.

If functional requirements are grouped by responsible objects, they may be labeled as `CardReader`, `CardReader.Scan`, `CardReader.Scan.Fail`, `EnterPinScreen.Display`, `EnterPinScreen.Display.TimeExceed`, etc. The following are two example statements:

`EnterPinScreen.Display`: If card scan succeeds, the system shall display "enter pin" prompt for the user to enter a pin number for 10 seconds or until the user presses "enter" key (UI1.3)

`EnterPinScreen.Display.TimeExceed`: If the user does not enter press "enter" key after 10 seconds, the system shall display "more time" prompt asking the user if he or she needs more time for 2 seconds and indicate the choice by pressing a Yes or No key (UI1.3)

Like writing textual descriptions for use cases, functional requirements must also be written accurately and precisely. In particular, they should be written using the active voice, complete sentences, consistent terms, and simple grammatical structures. Try to avoid ambiguous terms that are not measurable or verifiable such as fast, efficient, robust, user-friendly, simple, easy, flexible, adequate, sufficient, state-of-art, etc.

Try to avoid non-specific quantities or options such as at least, at most, including, not limited to, between, several, multiple, as much as possible, etc. Try to avoid words that are impossible to implement such as instantaneously, immediately, seamlessly, transparent, etc. Try to avoid words that do not pinpoint to specific actions or activities such as enable, support, ensure process, etc. The following are a few bad examples with explanations:

> *Upon a card being inserted, the system shall instantaneously validate the card.*

This is a bad statement since no system can perform activity without taking time. Thus, this requirement cannot be implemented. It may be changed to something verifiable like this: Upon a card being inserted, the system shall validate the card in less than 0.1 second.

> *If card scan succeeds, the system shall display "enter pin" prompt for at least 10 seconds.*

This is a bad statement because it is confusing on several levels. First, it doesn't say what happens if the user finishes the pin entry in less than 10 seconds. Second, it doesn't say what the system should do if the user does not enter a pin after 10 seconds. Does it stay on forever?

> *If possible, the system shall still enable a customer to withdraw cash even though the system cannot print a receipt.*

This is a bad example. First, what does it mean by "if possible"? Does it mean that the system shall allow a user to withdraw cash regardless of whether the bank authorizes it or not. Second, the word "enable" is confusing. How can the system enable a customer? What the statement really means, if revised, is that the system shall still allow a bank customer to withdraw money in the event the machine cannot print a receipt.

Non-Functional Software Requirements

Non-functional requirements are anything but functional. Clearly, they form a broad category. Note that a non-functional requirement is by no means a secondary or unimportant requirement. In fact, within this category are two sub-categories of requirements that are essential to any project: interface requirements and informational requirements. Particularly for management information systems, interface and information requirements are equally or probably more important than functional requirements. As has been mentioned earlier, information requirements are captured by conceptual object models, which can be converted into logical data models for database implementation.

Interface requirements describe the logical characteristics of all interfaces to ensure proper communications between the system and externals. There are four types of interfaces: user interface, hardware interface, software interface, and communication interface. Therefore, there will be four sub-categories of interface requirements. Among them, user interfaces are probably the most critical because they ensure the usability and acceptability of the system. User interface requirements specify the common characteristics of all user interface designs such as corporate GUI standard for fonts, labels, buttons, images, color schemes, layouts, menus, hot keys, and message displays. Note that user interface prototypes are the references for use case descriptions and artifacts to discover user requirements. They are different from user interface requirements and should not be included in the software requirements document.

Besides information and interface requirements, other non-functional requirements include specifications on system performance, security, accessibility, control, scalability, currency, economy, efficiency, integrity, and other software quality attributes. The following are a few examples:

Security:
ATM shall hide a pin when the user enters it into the system
ATM screen shall not be visible from 3 feet distance or 45-degree angled view
Performance:
ATM screen response time shall not exceed 2 seconds
Each transaction shall not exceed 2 minutes
Availability:
ATM shall be accessible 24/7
A lack of paper for printing receipts shall not prevent a card holder from using the system
Integrity:
ATM shall be sturdy to avoid vandalism by any impact force less than 3000 Foot/CBs

Review Questions

1. Compare the utilities of four types of context diagrams in developing business, user, and functional requirements.
2. What elements should be included in business requirements? Create a template of the business requirements document based on your understanding of the elements.
3. What requirements are cross-referenced in all requirements?
4. Develop a list of major features based on the context diagram in Figure 3.
5. What are non-functional requirements? Give an example statement in each sub-category.
6. What is wrong with the functional requirement "The system shall attempt to get all necessary information to validate a user"? Please revise the statement into a good functional requirement.
7. Think of a business process, describe its background, and develop business objectives. Then write a vision statement to summarize the business requirements.
8. What are the strategies one may choose to organize functional requirements?
9. How are business rules captured in business requirements, user requirements, and software requirements?

Exercises

1. Develop a business use case and write all business rules based on the following description of an end-to-end business process. In the purchasing department, each purchase request is assigned to a caseworker within the department. This caseworker follows the purchase request through the entire purchasing process and acts as the sole contact person with the person or unit buying the goods or services. The department refers to its fellow employees buying goods and services as "customers." The purchasing process is such that purchase requests over 1,500 must be sent out for bid to vendors and the associated request for bids for these large requests must be approved by the department. If the purchase is under 1,500, the product or service can simply be bought from any approved vendor, but the purchase request must still be approved by the department and they must issue a purchase order.

2. Create a context diagram and list major features for a point-of-sale system. Here is the business use case description. A customer arrives at the checkout counter to pay for her selected items. The cashier scans each item's bar code and records quantity if greater than one. The cash register displays the price of each item, its description, and quantity. When all the items are entered, the cashier indicates the end of sale. The cash register displays the total cost of the purchase, including tax. Occasionally, a customer may have tax exempt status, so the cashier must check the certificate and remove the sales tax. Very often, a customer may come with a special coupon that the cashier may need to scan or record in order to apply discounts. The customer may select one of three methods to pay for the transaction:

   a. **Cash:** the cashier takes the money from the customer and puts it into the cash register, the cash register indicates how much change is due to the customer.
   b. **Check:** the cashier verifies that the customer is in good standing by sending a request to an authorization center via the cash register.
   c. **Credit card:** the customer slides her credit card and the cash register sends a request for authorization to an authorization center.

   After the payment, the cash register records the sale and prints a receipt, which the cashier gives to the customer.

3. Develop a decision tree and a decision table to represent how the gate control system in Figure 3 will control the gate in response to external conditions.

4. Develop functional requirements for the use case "validate a user via user ID and password" using the template provided in this chapter.

5. Suppose your company wants to develop a database to store and manage all its business rules. Please help design the database structure using any logical data model.

6. Create a business use case diagram for BizbyOrder Books. Here are some of its business processes. BizbyOrder Books is specialized in ordering books for two types of customers: individuals and businesses in lower Manhattan. This is how these two customers are different. When an individual customer orders books, he or she has to pay a 20% down payment. A business customer can establish a credit line

with BizbyOrder and pays 50% down payment if and only if the order amount exceeds its credit limit. The bookstore orders its books through five national distributors. Because of various special agreements in the book industry, each publisher sells its books exclusively to one distributor. This is how the bookstore runs its daily business. Each time a customer comes in to buy a book, the bookstore uses its database system to find the title and locate the distributor that sells the book. Then the customer will leave contact information and make a down payment (if needed) for BizbyOrder to send the order to a distributor. When an ordered book comes in, the customer will be contacted to pick up and pay the rest of the balance.

7. Develop a list of business rules based on the following text. Insure-A-Person Inc. provides health insurance services to employees and their family members across America. Due to the need to promote its customer relations, the company has decided to open up a web-based system for clinics and individual customers to be able to file claims on the Internet 24 hours a day and 7 days a week. The company has approached you to design a relational database for this purpose. According to the company, this is how the web-based system is supposed to work. Within 60 days of seeking treatments for himself or any of his family members, a customer needs to log on to the system and file a claim. First, you specify the name of a patient, the date and the place the service was provided, and the primary doctor providing the service. Then, you detail the procedures performed by the doctor. In the medical industry, all procedures have been standardized with fixed identification numbers and short descriptions. The insurance company will pay for the service based on all the procedures performed by the service.

Chapter 13

Requirements Elicitation and Validation

Introduction

In this chapter, we consider the initial and final steps of requirement development, i.e., requirements elicitation and validation. In particular, we will learn the techniques of requirements elicitation, including where to look for information and how to discover requirements from the sources, and approaches to requirements validation, with emphasis on requirements inspection and requirements-based tests.

Elicitation means to identify and discover unknown requirements, and validation means to check and test requirements for errors, omissions, and ambiguities. These two steps are pooled together because activities and participants involved in the steps are almost identical. Between these two steps is requirements documentation, which was discussed in the previous chapter. Validation occurs only after requirements are identified and documented; one cannot validate an unknown or implicit requirement.

Requirements Elicitation

To elicit requirements, a business analyst should possess the knowledge of: (1) what requirements are to be discovered, (2) where the requirements are discovered, and (3) through which channels or techniques the requirements may be discovered. These three elements — abbreviated as WWW that stands for what, where, and which — are the key components in a requirements elicitation plan or strategy.

In Chapter 12, we classified requirements into three levels — business (blue sky), user (sea level), and software (deep ocean) — and into two categories — functional and non-functional. For example, as functional business requirements, vision, scope, and major features are the key artifacts, and as non-functional business requirements, business rules are the key component. At the sea level, system use cases, information or entity objects, and user interfaces or system prototypes are the key functional and non-functional requirements. Table 1 of Chapter 12 provides an overview of the classification.

Chapters 3–11 essentially provide us with the knowledge for acquiring the "what" element of requirements elicitation. Chapters 5 and 8 cover concepts and techniques of how to identify business objects and how to represent business rules. Chapters 9–11 cover the techniques of how to identify use cases, how to describe use cases, and how to optimize use cases. Chapter 12 provides a general overview of how to develop requirements documents. All these chapters contribute to our understanding of what requirements are to be captured and how they are represented and documented.

Where are the requirements discovered? Some are inside people's mind, some are embodied in business practice, some are coded in computer programs, and some are written in business documents. The sources may vary from organization to organization and from project to project. Thus, the first step to form an elicitation strategy is to create a road map that pinpoints each type of requirements to its potential information sources.

Table 1 provides a general road map that suggests the most likely sources of information for each type of business requirements, including project rationale and vision, business problems, opportunities, threats, business risks, scope, main features, and business use cases. For example, the table suggests that business analysts should look for corporate strategic and marketing surveys or ideas from a few key visionaries for project rationale and vision.

Business rules are cross-board assets and may be treated separately from other business requirements. Thus, a separate table is included here to suggest possible sources for eliciting business rules (see Table 2). The table follows the classification system proposed in the previous chapter and treats the three types of business rules differently. The likely sources are different for different types of rules. For example, structured rules are typically coded into a legacy system's data structures and user interfaces, whereas algorithmic and behavioral rules are coded

Table 1.   The sources of information for business requirements.

Requirements	Source	Sub-class	Format
Vision	People	Visionaries	Ideas
	Documents	Mission Strategic plan Marketing surveys	Publications
Business objectives Business problems Opportunities Threats Business risks	Documents	Strategic plan Departmental objectives Marketing surveys Operational analysis reports Decision analysis reports Risk analysis reports Industrial reviews Accreditation criteria Interoffice memoranda Complaints Suggestion box notes Meeting minutes	Publications Written or digital recordings
	People	Business stakeholders	Attitudes Beliefs
Business use cases	Processes	Human interactions Organizational interactions	Behavior
	People	Processor engineers Operations managers	Cognitive knowledge of workflows
	Documents	Workflow charts Employee handbooks Professional books Bill of responsibilities Operating procedures Job outlines Task instructions	Publications
Scope Main features	People	Business stakeholders	Ideas Attitudes Beliefs
	Documents	Meeting minutes Strategic plan Decisions and referendums Process reengineering requests Problem diagnosis Decision analysis reports Marketing surveys	Written or digital recordings Publications

Table 2.   The sources of information for business rules.

Requirements	Source	Sub-class	Format
Structural rules	Systems	Data(base) structures User interfaces (screens/reports)	Codes Views
	Documents	Business forms Business statements Accounting records Performance reviews	Publications
	People	Employees	Cognitive knowledge of entities and relationships
Algorithmic rules	Systems	Algorithms	Codes
	Documents	Professional books Policy manuals Contracts and agreements Laws and regulations Notices and announcements	Publications
	People	Administrative assistants Professional workers Operations managers	Cognitive knowledge of constraints and procedures
Behavioral rules	Systems	Algorithms	Codes
	Documents	Workflow charts Employee handbooks Professional codes Bill of responsibilities Operating procedures Action scripts Employee contracts Customer agreements Vendor agreements	Publications
	People	Business stakeholders	Cognitive knowledge and psychomotor knowledge of event–response patterns
	Processes	Human interactions Organizational interactions	Behavior

into invisible algorithms. Their documentary sources are also different. Structural rules can often be discovered from business forms and statements. Algorithmic rules are found in professional books, legal documents, and corporate announcements. Behavioral rules can be found in workflow charts, employee handbooks, professional codes, and business agreements.

User requirements include user goals, resources, and tasks as represented as system use cases, information models, and interface prototypes. User requirements are derived from business requirements and business rules. Thus, the sources for discovering user requirements should include the business requirement document, business rules repository, and related artifacts. Additional user goals and resources must be identified from users, their interactions, or legacy systems. For example, business forms, statements, and workflow charts are important for identifying system use cases. In fact, each business form or statement, if computerized, probably suggests a system use case. For another example, users' knowledge of functional requirements provides an additional source of information for identifying user tasks, user's knowledge of usability and usefulness is an important source of information for identifying interface requirements, and business stakeholders' knowledge of data requirements adds extra value to the discovery of information models. Table 3 suggests the likely sources for eliciting use cases, information models, and interface prototypes.

The third element of an elicitation strategy is the channels or techniques through which requirements may be discovered from the respective sources. As a rule of thumb, requirements coded in legacy systems may be recovered through reverse engineering, ideas and cognitive knowledge may be elicited trough individual interviews and workshops, attitudes and beliefs may be elicited through questionnaire surveys and joint applications development, business documents and user interfaces may be sampled, and psychomotor knowledge and organizational behaviors may be discovered through observations and structured walkthroughs. Table 4 documents the various techniques along with their definitions, applicable source formats, benefits, and disadvantages.

Requirement Validation

"Software development is like sex; if you make a mistake, you have to support it for life." This is a joke, but it carries enough truth. Software errors are introduced in all stages of the development process. Among all, the most harmful ones are due to incomplete, ambiguous, and erroneous requirements. Erroneous requirements can lead developers to create a donkey although a customer wants a horse. In the face of incomplete

Table 3.   The sources of information for user requirements.

Requirements	Source	Sub-class	Format
System use cases	Documents	Vision statements Major features Scope Business use cases Workflow charts Action scripts Business forms Business statements Algorithmic rules Behavioral rules Operating procedures Job outlines Task instructions	Publications

	People	Users	Ideas
			Beliefs
			Cognitive knowledge
			  of functional
			  requirements

	Systems	Algorithms User interfaces (screens/reports)	Codes Views
	Processes	Human interactions	Behavior
Information Models	Documents	Structural rules Business forms Business statements Accounting records Performance reviews Work measure reviews Business use cases System use cases Algorithmic rules Behavioral rules	Publications
	Systems	Data(base) structures User interfaces (screens/reports)	Codes Views
	People	All business stakeholders	Cognitive knowledge of data requirements
Interface Prototypes	Documents	Systems use cases Business forms Business statements Business communications	Publications

Table 3.   (*Continued*)

Requirements	Source	Sub-class	Format

	People	Users	Ideas
			Beliefs
			Cognitive knowledge
			  of usability and
			  usefulness

	Systems	User interface (screens/reports) Systems interfaces (API/OS)	Views Codes

or ambiguous requirements, developers, under time pressure, often make their own interpretations, which may be incorrect.

Requirement defects are better found and corrected in the early stages of the development process. Studies (Booch *et al.*, 1999) have documented that, compared to the cost of fixing a defect during requirements discovery, it takes ten times more to correct it in the design stage and a hundred times more in the implementation stage.

Requirements validation is a process to discover and correct requirements defects. The major activities are conducted to achieve the three following objectives:

1. User requirements are in alignment with business requirements, and software requirements can be traced to user requirements.
2. All software requirements are complete, correct, feasible, necessary, unambiguous, and verifiable.
3. All user and business requirements are complete, correct, prioritized, consistent, unambiguous, and traceable.

The techniques for requirements validation include peer reviews, inspections, and requirement-based tests. Peer review is the easiest among the three approaches. It can be as simple as cross-checking by a colleague, which can happen whenever a use case is described or when a portion of the requirements document is developed. It can be a slightly more formal request for comments, where the requirements document is passed around to a few team members for comments and suggestions. A very formal review process is the structured walkthrough in which the author presents the requirements document to a group of team members to solicit their comments and suggestions.

Table 4.   Elicitation techniques.

Techniques	Definition	Applicable Sources	Advantages	Disadvantages
Document sampling	Randomly or systematically collect related documents	Publications, written/digital recordings	Accurate information, less demand on customers, flexible elicitation scheduling	Time consuming for information filtering and comprehension
Questionnaires	Mass-produce and distribute questions to many respondents	Attitudes, beliefs	Less demand on customers, uniform response formats, allow response anonymity	Low response rate, inflexible format, allows false or ambiguous responses
Interviews	Solicit responses from direct, face-to-face interactions	Cognitive knowledge	Flexible and open-ended questions to probe for in-depth knowledge, opportunity to observe sign language	Time and resource consuming, inflexible in elicitation planning
Day-in-the-life	Analysts observe the users or be an intern at work	Psychomotor knowledge, behavior	More accurate and reliable than verbal responses on workflows and behavioral responses, be able to identify problems with the current process, inexpensive and flexible to plan	May not expose all alternate or exceptional scenarios, users may not feel comfortable on being observed, errors may be recorded with no correction
Reverse engineering	Recover a data or procedural model from working applications	Codes	Fast and accurate information, directly useful for requirements modeling, not involving customers	Requires expensive CASE tools, may be illegal or infeasible

| Prototyping | Build a small working model of the user's requirements | Implicit knowledge, attitudes, beliefs, ideas | Allow users to recognize implicit knowledge through visual feel-and-look and expose requirements that are not well understood or difficult to articulate, risk reduction, serves as a training mechanism, allow the development of test cases | Analysts may need to be trained in developing prototypes, users may get unrealistic expectations, increases development costs |
| Joint requirements planning (e.g., brainstorming) | Structured group meetings involving users and managers, organized by a sponsor in top management, chaired by a facilitator, recorded by scribes (BAs), and attended by IT staff, to generate ideas, identify problems, and define systems requirements | Ideas, attitudes, beliefs, cognitive knowledge | Obtain group consensus on problems, objectives, and requirements, good for solving unstructured problems, identifying unconventional and multiperspective responses, generate new ideas through brainstorming | Time and resource consuming, difficult to schedule, opinionists may dominate, minor details may be sidetracked |

Requirements inspection is probably the most formal approach to ensure the quality of requirements. It was originally developed at IBM (Fagan, 1976) and has now been adopted as the best business practice. It is a multistage process involving various participants (Wiegers, 2003), including the author of the requirements document, the author of any predecessor work product for the item being inspected, people who will do work based on the item being inspected (such as developers), and people responsible for the systems that interface with the item being inspected. To ensure efficiency, the rule of seven is recommended to be followed, i.e., the inspection team should not exceed more than seven participants. All participants, including the author, look for defects and improvement opportunities. Some play special roles, as follows:

- *Author:* The author plays the passive role of listening to comments and responding to questions.
- *Moderator:* The moderator coordinates the inspection with the author, facilitates the inspection meeting, follows up on the corrections with the author, and reports inspection results to the management.
- *Reader:* One participant who is less familiar with the item being inspected can be assigned to the role of reader, who paraphrases one requirement at a time.
- *Scribe:* The scribe uses standard forms to document the issues raised and defects found during the inspection meeting. He or she should read what is written to confirm its accuracy.

The inspection process starts with the requirements document that has been well developed in the sense that the document conforms to the standard template, is formatted neatly, has been checked for spelling and grammar errors, and all reference materials are available. As a rule of thumb, *if the moderator cannot find more than three major errors in a ten-minute examination, the document may be ready for inspection.*

After the author and the moderator agree that it is ready for inspection, they jointly plan for inspection. This includes the selection of inspectors and the schedule of the inspection meeting. The length of the meeting shall consider the size of the item being inspected. A rate of 2~4 pages per hour is reasonable for deciding the length of the meeting.

Before the inspection meeting, there should be a period for all inspectors to read and examine the document for possible defects and issues. Studies have (Humphrey, 1989) found that 75% of errors were actually

Table 5.   Defect checklist for use case documents.

☐	Is the use case uniquely labeled and correctly named?	
☐	Do the pre- and postconditions properly frame the use case?	
☐	Does the summary state succinctly what the use case is and the value it brings to the primary actor?	
☐	Does the document contain the version number, creation and modification dates, and person in charge?	
☐	Are all supporting actors sufficient and necessary for the use case?	
☐	Are all alternate flows documented?	
☐	Are all exceptional flows documented?	
☐	Are steps to execute an inclusion use case, if any, documented?	
☐	Are extension points documented if extension use cases exist?	
☐	Are all referenced prompts, messages, and prototypes included?	
☐	Is each interaction statement clear, complete, and unambiguous?	
☐	Is each interaction uniquely labeled?	
☐	Is each interaction free from design and implementation details?	
☐	Is the condition that leads to an alternate or exceptional flow possible and verifiable?	
☐	All repeated interactions clearly indicated?	
☐	Are all other related artifacts, if referenced, accessible for review?	
☐	Is there any sub-sequence of interactions that can be split into a separate use case?	
☐	Can this use case extend to another one?	
☐	Is this use case kind of similar to another use case?	

discovered during this period. To improve the efficiency, a standard check-list may be used. Table 5 shows a simple checklist for inspecting system use cases. Organizations may develop their own custom checklists for use cases and other business, user, and software requirements based on their desired quality attributes.

During the inspection meeting, the reader reads the requirements one at a time in his or her own words. Other inspectors identify possible errors and raise issues. The scribe will capture and confirm the errors and issues. The inspection meeting may be held in multiple sessions. In the end, the group must collectively decide whether to accept the document as is, with minor revisions, or with major revisions. If a major revision is needed, the group may also decide how to correct the problems and whether another inspection meeting is needed after rework. Otherwise, the author shall correct all minor errors and issues and follow-up with the moderator to report how he or she has addressed the problems.

Process-Oriented Requirements Validation

Most validation techniques focus on the final requirements document. This section proposes a process-oriented approach to validation and quality measurement and a delta approach to analyzing process errors. In the process-oriented approach, defects are the manifestation of a defective process, and quality is the responsibility of all stakeholders involved in the requirements development process. Improving quality asks for improving the process. In contrast, in the document-based approach, defects are considered to be the source of other problems such as customer complaints, developers' frustrations, etc. Quality is considered to be the responsibility of the author who creates the requirements document. Improving quality simply means information scrap and rework.

Requirements development consists of two essential steps: elicitation (or collection) and documentation (or presentation). After requirements are discovered, requirements must be developed or documented to be validated. This is called presentation. The process-oriented approach shall focus on each of these steps.

We use fishbone diagrams, an analysis tool invented by Japanese quality control statistician Kaoru Ishikawa, to systematically examine the causes that contribute to good or bad quality. Note that the design of a fishbone diagram looks much like the skeleton of a fish. The head of fish shows the problem to be studied. Each bone of the fish labels a cause that leads to the problem. In addition, the tool suggests that one looks for causes from typical categories signified as the 4 M's — Methods, Machines, Materials, Manpower — the 4 P's — Place, Procedure, People, Policies — and the 4 S's — Surroundings, Suppliers, Systems, Skills.

Requirements elicitation

During requirements collection or elicitation, the most frequent quality problems come from instruments, people, and procedures. The problems include observation biases, missing observations, measurement errors,

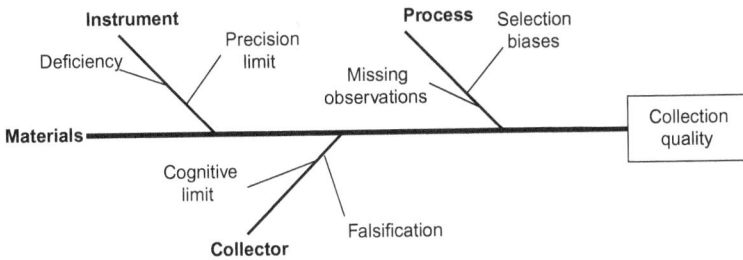

Figure 1.   A fishbone diagram for collection quality.

instrument deficiencies, and intentional falsification. Figure 1 shows the fishbone diagram for elicitation quality.

When selecting a sample process to observe or example user to interview, it is imperative that the sample be representative of a population under study. Otherwise, the collected data will lack objectivity. A typical example of observation biases is to interview a most technically capable user; the usability requirement for this user is biased because it does not reflect the skills of majority regular users.

Missing observations occur when there are objects that are supposed to be observed but not observed or when there are properties of an object that are supposed to be measured but not measured. Examples include cases when a user group is not represented or an important class of documents is not sampled.

Measurement errors are not avoidable since instruments have limited capacity (Deming, 1986). Measurement errors can also occur due to unintentional human mistakes. In either case, they reduce the accuracy of collected requirements.

Instrument deficiencies refer to the problem that an instrument is defective, i.e., it has no capacity and can produce statistically unstable observations.

Falsification refers to the behavior of making up false data. It includes the creation of data that does not correspond to genuine requirements and intentional distortion of observations. Both instrument deficiencies and falsification reduce the reliability of observed data.

According the common sources of errors during requirements collection, we propose four attributes to measure collection quality: objectivity, completeness, accuracy, and reliability, and Table 6 lists their definitions. Obviously, objectivity and completeness are semantically distinct from

Table 6.   Quality dimensions for requirements elicitation.

Attribute	Associated Errors	Definition
Accuracy	Limited capability of instruments, limited capability of data collectors	The extent to which collected data are free of measurement errors
Objectivity	Observation biases	The extent to which the sample selected for observation is representative of a population
Reliability	Defective instruments, falsification	The extent to which collected data are free of falsifications and defective readings
Completeness	Missing observations	All values that are supposed to be collected are collected

each other and from accuracy and reliability. In a sense, both accuracy and reliability measure the extent to which captured requirements are free from measurement errors. However, accuracy assesses the errors in terms of how capable the collection process is if the collection process is stable and has a statistical capability. It may be measured by the six-sigma of the variation of the collection process. In contrast, reliability assesses the extent to which the collection process is statistically stable and has a capability. It may be measured in terms of the probability that instruments are not defective and data falsification does not occur. Therefore, accuracy and reliability are conceptually distinct. In addition, the four attributes cover the content domain of collection quality: if data collection is unbiased and complete and collected data are reliable and accurate, then it is sufficient to infer that data collection process is free of errors and has high quality.

Requirements documentation

During requirements documentation or presentation, the most frequent causes of quality problems come from people, presentation designs (frameworks or templates), and presentation devices. The problems include typical human errors (e.g., typographical errors, grammatical errors, and computational errors), interpretation errors, information selection biases, sequence errors, lack of flow controls, layout design deficiency, device deficiency, and approximation errors. Figure 2 shows a fishbone diagram for identifying documentation quality issues.

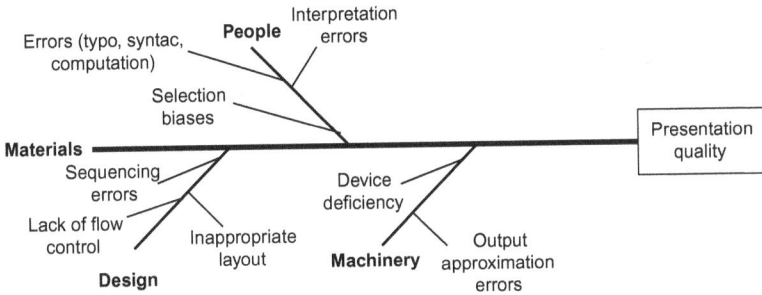

Figure 2.   A fishbone diagram for presentation quality.

Human errors will present incorrect information. Selection biases prevent one from presenting data objectively and neutrally. They include hiding data that have conflicts of interest and highlighting data that favor certain opinions. Interpretation errors occur when there are language or tool deficiencies, when data are difficult to understand, or when original data have ambiguity in meaning. For example, the same term may have different definitions, formats, and measurement units. In order to present related requirements in a unified view, interpretations and reinterpretations of certain data are needed, and errors may not be avoidable. Interpretation errors influence the faithfulness of a presentation to its original source. Sequence errors, a lack of flow controls, and inappropriate layout affect not only the appearance but also the comprehension of a requirements document. Finally, device deficiencies and approximation errors will distort the original information and affect the precision of presented data.

According to these sources of errors for documentation, we propose three dimensions of documentation quality: Preciseness, Faithfulness, and Formality. Table 7 summarizes their definitions and associated quality problems. Preciseness measures how much presented requirements are free from mechanical errors such as typographical errors, syntactical errors, computational errors, approximation errors, and problems due to defective presentation devices. Conceptually, it covers correctness, i.e., no typographical and grammatical errors, and precision, i.e., little approximation errors. Although correctness and precision are frequently cited in existing studies, they are semantically overlapping. In addition, correctness is more applicable to texts whereas precision is more applicable to multimedia data. Therefore, we choose preciseness to cover both correctness and precision. Faithfulness measures how much presented data are free of

Table 7.   Quality dimensions for requirements documentation.

Attribute	Associated Errors	Definition
Preciseness	Typographical errors, grammatical errors, computation errors, approximation errors, device deficiencies	The extent to which presented data are free of mechanical errors
Faithfulness	Interpretation errors, selection biases	The extent to which presented data are free of interpretation and presentation biases
Formality	Sequencing errors, inappropriate layout, lack of flow controls	The extent to which data are presented concisely, consistently, and attractively

subjective opinions and views. Conceptually, it dictates that requirements are faithful to the truth and semantically identical to its origin. Formality measures how much data are presented concisely, consistently, and attractively. It is an attribute that calls for concise content, consistent layout, and attractive visual and audio appeals. From these definitions, it is easy to see that the three dimensions are semantically distinct. Preciseness and faithfulness measure the content aspect of presentation quality, respectively, from the perspectives of objective and subjective errors, whereas formality measures the appearance aspect of presentation quality. Furthermore, the three dimensions semantically cover the domain of presentation quality. Requirements are well presented if and only if they are free of mechanical errors, free of subjective biases, and are laid out nicely.

Requirements-Based Tests

Requirements-based tests aim to check whether the software behaves as the requirements specify. Unlike the white-box tests that check internal program codes, requirements-based tests are "black-box" in nature. Requirements-based tests can be done only after the system is developed. However, test cases can be developed in the early stage upon the requirements are developed.

Developing test cases is a very important technique for requirements validation. In a sense, developing test cases is a process of "coding" the high-level system behavior using plain English instead of a programming language. Thus, it can help pinpoint problematic requirements. In fact, if the requirements are complete, accurate, and clear, the process of deriving the test cases is straightforward. If the requirements are not in good shape, the attempt to derive test cases will expose problems. For example, suppose we want to design a test case to see how the system responds to a request to withdraw cash if the withdrawal limit is not exceeded. We need to specify a test withdrawal limit, a test credit line, and a test withdrawal amount. Theoretically, any positive number can be a test number. What about $0.1 as a withdraw limit? That cannot be tested because an ATM does not handle coin transactions. However, according to a use case description like the one we saw earlier, such a withdrawal limit still allows the basic flow of the use case to be executed. The exposed problem will help clarify the specification of when the withdrawal limit is said to be exceeded.

Test cases may be described as a list of data inputs and user actions, system conditions, and expected results. They may be formally documented using templates. The following are a few test case examples:

- Test Case 1: The cashier scans an item, and the item information exists. Expected result: The screen displays item title, unit price, sub-total, and grand total.
- Test Case 2: The cashier scans an item, item information is found, the screen displays the item title, unit price, sub-total, and grand total, and the cashier chooses to remove the item. Expected result: The line item is removed from the screen and the grand total resets to zero.
- Test Case 3: The cashier scans an item, and item information does not exist. Expected result: Display message "Item is not found. Please ask a supervisor for assistance."

Test cases are derived from use cases. Yet they are by no means equivalent. A use case typically has basic flow of events as well as alternate and exceptional flows. In theory, each test case should be developed for each combination of conditions that lead to a unique path of interactions or system states. For example, for the "withdraw cash" use case, we need at least one test case to cover all possible combinations of values for the following event variables: card validity, pin validity, number of invalid pin entries, whether an account is on hold, whether a withdrawal limit is exceeded, whether a receipt is requested, whether a card is removed on time, and whether money is removed on time. If each event variable takes on two values, one would need $2^8 = 256$ test cases to cover distinct scenarios or flows of events. Clearly, the number of test cases exponentially explodes with the number of events/conditions.

In practice, it is not necessary to develop test cases that follow an entire use case. When events/conditions are not interrelated, they can be split into different test cases. This is how the explosive number of test cases may be reduced. For example, if we separate card validity condition out into its own independent test cases, the number of test cases will be reduced to $128 + 2$.

- Test Case 4: User inserts a card, and the card is valid. Expected result: Display message "Enter Your Pin Number."
- Test Case 5: User inserts a card, and the card is invalid. Expected result: the card is ejected and display message "Invalid Card" for two seconds.

Of course, conditions of whether the card holder requests a receipt, whether one removes cash, and whether one removes a card can be also tested independently of other conditions. With two test cases to cover each condition, we will need 6 test cases. The following are a few examples:

- Test Case 6: User chooses to have a receipt, ATM dispenses and holds cash for 5 seconds, but user does not remove cash. Expected result: ATM takes cash back and the Display message is "Your transaction is voided."
- Test Case 7: User chooses to have a receipt, ATM dispenses and holds cash for 5 seconds, and user removes cash. Expected result: ATM prints a receipt, ejects the card, and displays the message "Thanks for using ABC Bank. Please remove your card."
- Test Case 8: ATM ejects and holds a card for 5 seconds, but user does not remove the card within 5 seconds. Expected result: ATM swallows the card and displays the message "Sorry your card is taken by ABC Bank."
- Test Case 9: ATM rejects and holds a card for 5 seconds, and user takes the card within 4 seconds. Expected result: Display message "Thanks for using ABC Bank."

Table 8.   Invalid pin entries and ATM responses.

		1	2	3	4
C	Invalid pin entry	Y	Y	N	N
	Number of pin entries	1~2	3	1~2	3
D	Ask for pin reentry	Y	N	N	N
	Ask for withdrawal amount	N	N	Y	Y

Assume the remaining four conditions are tested jointly. We will need $16 + 2 + 6 = 24$ cases, a dramatic reduction from 256 test cases.

When two or more conditions are interrelated, test cases may be developed with the aid of a decision table that describes the expected result or system response under each of the possible combinations of the conditions. For example, Table 8 shows a decision table for deriving test cases considering jointly the conditions of pin validity and number of invalid pin entries. Since each event variable takes on two values, there are in total four combinations.

If a decision table exists, a rule of thumb is that one test case is developed for each column in the table. Of course, the rule does not have to be followed blindly. For example, in Table 8, one test case is needed for the last two columns because, when the pin is valid, the number of pin entries is irrelevant to system response; the ATM will then not ask for a re-entry. For this type of asymmetric events, a decision tree may be a better device for designing test cases. For example, using the decision tree given in Figure 7 in the previous chapter, four test cases may be developed to cover all possible combinations of three event variables: time, presence of an object, and light condition (assume work hours are 8:00 AM–5:00 PM on M–F):

- Test Case 10: Time is 5:01 PM on Friday. Expected result: Light is off.
- Test case 11: Time is 8:01 AM on Monday, an object is present, light is bright. Expected result: Light is off.
- Test Case 12: Time is 4:59 PM on Friday, an object is not present. Expected result: Light is off.
- Test Case 13: Time is 4:49 PM on Friday, an object is present, light is dim. Expected result: Light is on.

Note that a decision table or tree may prescribe multiple valid input data for a test but does not tell exactly which one is to be used for the test. For example, in Table 8, the first and third columns allow 1 or 2 as a valid number of pin entries. Similarly, for the decision tree in Figure 7 shown in Chapter 12, the valid values for work hours include 8:00 AM–5:00 PM on Monday through Friday by assumption.

Which value should be used for a test? In fact, multiple tests should be designed. The tests include both positive ones, where inputs are within the valid range of values, and negative ones, where inputs are outside the valid range. *Boundary value analysis* provides general principles for determining test data. Depending on whether values are ordered or not, it suggests different strategies, as described in the following.

If values are ordered and valid values are in non-consecutive intervals (like [500, 800], [801, 1,000], etc.) or ranges (like 1~3, 4, 5~8, etc.), then for each interval or range, create two positive tests, one at either end of the interval or range, and create two negative tests, one just beyond the interval or range at the low end and the other just beyond the high end. For example, for the decision tree shown in Figure 7 of Chapter 12, the two positive test values for working hours are 8:00 AM and 5:00 PM and the two negative test values are 7:59 AM and 5:01 PM.

If values are ordered and valid values are in consecutive intervals or ranges, then for each interval range, create two positive tests for each end of the interval or range. Then for all intervals or ranges create two negative tests, one is below the smallest acceptable value and one is above the largest acceptable value. For example, the positive numbers of pin entries are 1, 2, and 3, whereas the negative test values are 0 and 4 (see Table 8). For another example, suppose the discount rate is determined by order amount as follows: 5% for order of $500 or more, 8% for orders of $1,000 or more, and 15% for orders of $2,000 or more (see Table 2 of Chapter 12). The positive test values are, respectively, 500, 999.99, 1,000, 1,999.99, and 2,000. Because there is no maximum for invalid values, the only negative test value is 499.99.

If values are not ordered and valid values are in sets of one or more elements, then for each set, create one positive test using any value in the set and one negative test with any value outside the set. For example, suppose you offer free shipping to customers in OH, MI, PA, NJ, and NY. Then create a positive test using any of these states and one negative test using any of the other states.

Review Questions

1. What are the three elements in a requirement elicitation strategy?
2. Which sources will you be after to discover business use cases?
3. What requirements may be elicited by using a day-in-the-life technique?
4. What is structural walkthrough? How is executed?
5. Explain how affective knowledge such as attitudes and beliefs may be elicited differently from cognitive knowledge.
6. What techniques are effective if there are multiple conflicting objectives to be reconciled?
7. Describe the process of requirements inspections. What role does a moderator play in the process?
8. What is a test case? Is a test case the same as a use case?
9. What is boundary value analysis?
10. What is the philosophy of the process-oriented requirements validation?
11. What is a fishbone diagram? What are the typical sources that we look for causes to a problem?
12. What is Joint Requirements Planning? What role does the author play in the process?

Exercises

1. Develop a checklist for evaluating the quality of vision statements.
2. Develop test cases for the use case "validate a user via user ID and password."
3. Develop a fishbone diagram to document causes that affect the quality of the requirements inspection process.
4. If group discount is determined by the number of passengers in the group and valid ranges are, respectively, 1–3, 4–10, and 11–20. What test values are to be used?
5. If shipping costs is determined by the US regions, each of which consists of specific states, how would you determine test values?
6. Use decision tables to represent algorithmic or behavioral rules contained in the following text. Then develop test cases based on the tables. In the purchasing department, each purchase request is assigned to a caseworker within the department. This caseworker follows the

purchase request through the entire purchasing process and acts as the sole contact person with the person or unit buying the goods or services. The department refers to its fellow employees buying goods and services as "customers." The purchasing process is such that purchase requests over 1,500 must be out for bid to vendors, and the associated request for bids for these large requests must be approved by the department. If the purchase is under 1,500, the product or service can simply be bought from any approved vendor, but the purchase request must still be approved by the department and they must issue a purchase order.

7. Identify all business rules embodied in the following description. Develop decision tables on how to handle down payments for BizbyOrder Books. Then develop test cases accordingly. Here are some of its business processes. BizbyOrder Books is specialized in ordering books for two types of customers: individuals and businesses in lower Manhattan. This is how these two different customers are different. When an individual customer orders books, he or she must pay 20% down payment. A business customer can establish a credit line with BizbyOrder and pays 50% down if only if the order amount exceeds its credit limit. The bookstore orders its books through five national distributors. Because of various special agreements in the book industry, each publisher sells its books exclusively to one distributor. This is how the bookstore runs its daily business. Each time a customer comes in to buy a book, the bookstore uses its database system to find the title and locate the distributor that sells the book. Then the customer will leave contact information and make a down payment (if needed) for BizbyOrder to send the order to a distributor. When an ordered book comes in, the customer will be contacted to pick it up and pay the rest of the balance.

Chapter 14

Collaboration

Introduction

In Chapters 9–11, we learned use case diagramming to capture functional requirements and use case storyboarding to describe each use case as a sequence of interactions between the user and the system. Remember that the system is a collection of classes in the static view or a collection of running objects in the dynamic view, and thus each action or activity performed by the system will have to be performed by one or more of the objects in collaboration.

In Chapters 5–8, we learned how to allocate operations into objects based on the data flow reduction principle. An operation allocated to a object does not have to be performed entirely by the object; the whole or a part of it can be delegated to other more capable objects. This is the essence of collaboration. In this chapter, we will first introduce a few heuristics or principles on how to achieve collaboration. Then we will carry out a few examples from earlier chapters further to illustrate the concept of collaboration.

Heuristics for Achieving Collaboration

When allocating tasks to objects, just think about a team of co-workers and who should do what in a collaborative endeavor. For example, assume you are a student pursuing a higher learning degree. Of course, in this endeavor, you are a hero. Yet you cannot do it alone; you need help from others. You may need parents to finance your pursuit, a college to offer a program of courses, and professors to teach the courses, etc. In this collaborative endeavor, you as a student have capability to study and pursue a degree. Your parents have capabilities to provide financial resources to assist you. Your school has capabilities to offer programs and courses and award degrees. Your professors have capabilities to teach the courses. Thus, it is clear who should do what in the collaboration. If we are to allocate the actions (operations) to the stakeholders, you should be assigned the tasks to study and to get a degree; your parents to offer financial assistance; your college to offer programs, offer courses, and award degrees; and your professor to teach courses.

Heuristics are the summary of experience in terms of simple guidelines and principles. They are often followed by experts and may be used to distinguish bad designs from good ones.

Heuristics 1: Operations symbolize object capabilities

In the real world, each object has its own unique capabilities. What does an airplane do? It flies. Thus "fly" is the unique operation that an airplane object should have that distinguishes an airplane from other types of objects. Similarly, in computer programming, objects are created or abstracted for their unique capabilities. These capabilities should be operations of the objects. For example, what does a card reader in an ATM do? It reads, verifies, and ejects a card, and thus it should have three operations for doing those. For another example, what does a transaction object do? It performs transactions such as obtaining and/or adjusting transaction details.

Sometimes it may not be immediately clear which object should house a capability or an operation that needs to be captured. For example, to model the fact that books are put on shelves, we have Book and Shelf classes and need to allocate an operation so that books can be put on shelves. Which objects should carry out the responsibility? Do books locate themselves on shelves or do shelves shelf books? The common sense is that a book does not have the capability to locate itself; a book's responsibility is to house written texts or pictures, and it will be too much to do for a book to know how to locate itself. By the way, a book can be placed on desks too. Does it mean that the books should also have the knowledge of a desk to locate themselves? In contrast, a shelf has a capability of housing books; it is exactly what a bookshelf does. Thus, a shelf shall fulfill the responsibility (see Figure 1).

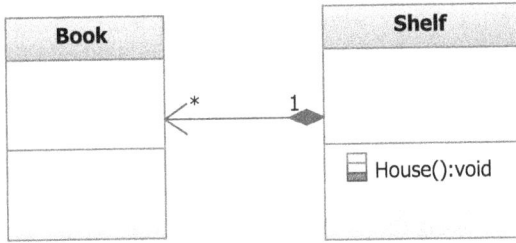

Figure 1.   Books and shelves.

A similar conundrum is seen in other examples. For example, in an object-oriented dairy farm, should an object-oriented cow possess the operation to milk itself, or should an object-oriented milk to un-cow itself?

Heuristics 2: Operations fulfill responsibilities

Objects live in the community of other objects, and each supports the community by contributing its own service, i.e., performing its operations.

As we stated before, objects are assumed to be *small, encapsulated, and smart objects*. They are small because each has limited functionalities (operations) and limited knowledge (attributes) that have one focus. They are encapsulated because each has its own body with a clear boundary from its surroundings. They are smart because each owns capabilities of using their own knowledge to process their own data or to provide services based on their own data. Now we are adding the fourth adjective to the objects: *collaborative*. Objects are collaborative because each has the responsibilities to perform actions that it is capable of.

This heuristic is employed by a well-known technique called *CRC cards* for identifying and allocating operations. Class responsibility collaborator (CRC) cards are standard index cards, which record and play the role of objects and engage in phantom communications. They can be used for identifying object collaborations via their responsibilities (Beck and Cunningham, 1989; Wilkinson 1995; Ambler 1995). Each index card records one class. Vertically divide a card into two sections. The top 4/5 of the card is used to write the name and the attributes. The bottom 1/5 of the card is to list the role of the class. Each role is described by one verb, two numbers, and the name of another class. For example, "take (0, 6) courses" may be a good description of the role of a student. It indicates that a student may take at least 0 and at most 6 courses.

The CRC card method is also very useful for effective communication among a group of analysts and users. Suppose one person holds a card and reads aloud to others as follows, "I am a student. I need to take courses." Another person holding the card for the course realizes a role "enroll (6, 35) students." A third person holding the dorm card reads aloud, "I am a dorm. I house students." Then the first person holding the student card realizes that he missed the role "stays in (0, 1) dorm." A fourth person reads aloud, "I am a student club. I need to recruit students." The first person then adds another role "join in (0, 1) club" to his card. This communication continues until all the roles are identified.

Keep everything about one object in one card. If there are too many attributes that you do not have enough space to write them all, it may indicate that the objects have too many attributes. Some of them may be unnecessary. Some of them may be redundant. And still some may be better re-allocated into another object. Similarly, if an object has too many roles to play, it may be too general, playing the role of an amalgamation of two or more objects that should be separated. Ask yourself whether each role is played by the entire set of objects or by only a sub-group of objects. If it is the latter case, splitting the class into sub-classes can reduce the number of roles each class plays.

Heuristics 3: A hero delegates but does not relay

After we identify a hero to house an operation based on capabilities or responsibilities, the hero object can delegate some portions of the job to other objects. This heuristic encourages delegations but discourages simple relays when it comes to decide where to house an operation. It means that each object shall do what it is capable of and delegates what it is not to other objects that are more capable. Of course, when an object can do nothing but relay the whole task, the object should not be made a hero to house the operation.

Let us consider a simple example. Suppose employees go to work by driving. Here, there are two types of objects involved: `Employee` and `Car`. Their relationships are easy to perceive: Each employee has one or more cars while each car belongs to one employee, and so we have associations between `Employee` and `Car` classes as in Figure 2. The question is what operations should be allocated to each object. Apparently, the major functionality is `GoToWork()`, and it is the responsibility or capability of the `Employee` class and so should be housed in there. However, to implement

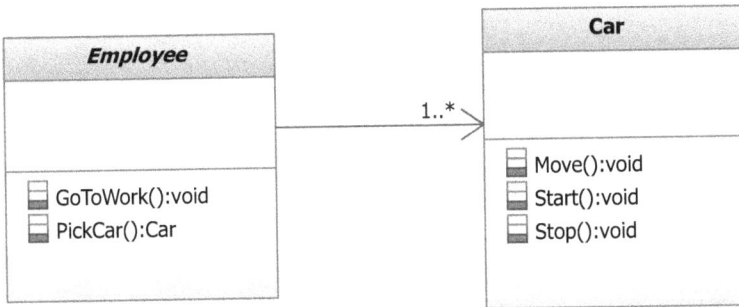

Figure 2.   Who should carry the operation of Drive?

the functionality, employees would need to pick a car and drive the car to the workplace. Does an employee have the full knowledge to move a car, including how to push gas into cylinder, how to burn the gas, and how to turn the wheels? As an alternative, the `Employee` class can delegate the task to an expert. Who is it? Using CRC cards, the person holding the `Employee` class can announce "I am an employee and need to go to work, for which I need a car to take me there." The person holding the `Car` class shall realizes the responsibilities: start car, move car, stop car, etc. The `Car` class has full knowledge of doing so. Since a car already has the capability to move, why do not we just send a message to a car after it is picked and have the car to carry an employee to work? Thus, the `Employee` class should delegate the sub-tasks to the `Car` class, and `GoToWork()` operation may be described without operation `DriveCar()`. The following is a code segment to implement the idea:

```
GotoWork()
{
        Car myCar = PickCar();
        myCar.Start();
        myCar.Move();
        myCar.Stop();
}
```

Collaboration via Examples: Compute Order Amount

For complex operations described in a use case description, *the three heuristics are often applied jointly*. We should first apply heuristics of

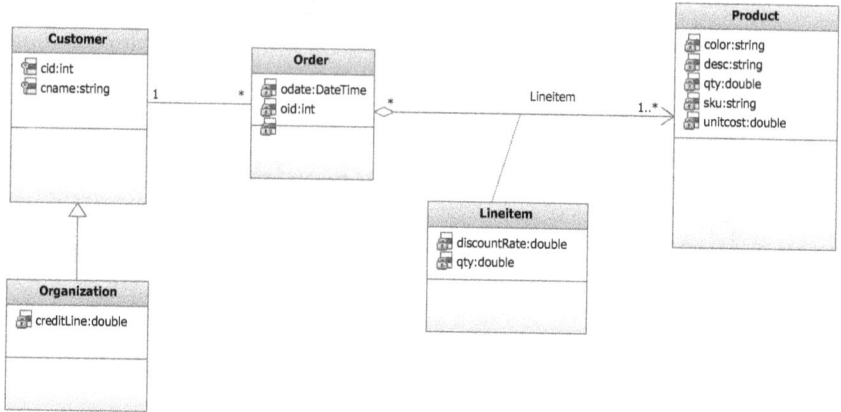

Figure 3.   A class diagram for customer orders.

capabilities or the principle of data flow reduction to determine where the operation should be allocated. Then we apply the delegation heuristics to delegate some or all parts of the job to capable objects.

In this section, let us use a familiar example to learn how to apply the heuristics. Figure 3 shows a class diagram of domain objects: Organization, Customer, Order, LineItem, and Product. Which one of this classes should house the operation to compute the total amount ordered by a customer? It is not immediately clear because no objects have the information readily available. First let us see which object is most capable or has all or most of the information to compute the total amount. To answer this question, let us implement the class diagram using the following C# code:

```csharp
public class Customer
{
    protected int cid;
    protected string cname;
    protected List<Order> orders;
}

public class Organization: Customer
{
    private double creditLine;
}
```

```csharp
public class Order
{
    private int oid;
    private DateTime odate;
    private Customer madeBy;
    private List<Product> orderedItems;
    private Dictionary<Product, LineItem> orderLines;
}

public class Product
{
    private string sku, color, desc;
    private double qty, unitcost;
}

public class LineItem
{
    private double qty, discountRate;
    private Order;
    private Product item;
}
```

Which object has all the information it needs to compute the total amount ordered by a customer? A `Product` object only knows its own unit cost, and each `LineItem` object knows the quantity and discount rate for each item ordered. The following C# code implements these capabilities:

```csharp
public class Product
{
    private string sku, color, desc;
    private double qty, unitcost;

    public double GetPrice()
    {
        return unitcost;
    }
}

public class LineItem
{
    private double qty, discountRate;
    private Order;
    private Product item;
```

```
public double GetQty()
{
    return qty;
}

public double GetDiscountRate()
{
    return discountRate;
}
}
```

However, to compute the total amount, we would need to know all the items and their quantities ordered. What about Order objects? According to the code, each Order object knows what items and order lines it contains, and so it should be able to compute its own total amount, as implemented by the following C# code:

```
public class Order
{
    private int oid;
    private DateTime odate;
    private Customer madeBy;
    private List<Product> orderedItems;
    private Dictionary<Product, LineItem> orderLines;

    public double GetOderAmount()
    {
        double amount = 0d;
        double price, qty, rate;
        foreach (Product p in orderedItems)
        {
            price = p.GetPrice();
            qty = orderLines[p].GetQty();
            rate = orderLines[p].GetDiscountRate();
            amount += price * qty * (1 - rate);
        }
        return amount;
    }
}
```

Note that in the above code, the Order class delegates the jobs of getting price to Product and getting quantity and discount rate to LineItem class. Now what does an Order object, myOrder, as shown on page 335, do?

```
Order myOrder = new Order();
```

Well, it can compute the total amount of its own by calling the function

```
myOrder.GetOderAmount()
```

as programmed inside the `Order` class. However, an `Order` object does not know the items and order lines contained by other orders. To house the operation inside the `Order` class, we would have to tell the operations what other orders, beside the order object in question, were also placed by a customer.

Who knows all the orders placed by a customer? Of course, it is the `Customer` object, as we can see from the C# code above. Thus, it is the best choice to place the operation inside the `Customer` class. Indeed, as shown in the following C# code, we can implement the computation inside the `Customer` class easily:

```
public class Customer
{
    protected int cid;
    protected string cname;
    protected List<Order> orders;

    public double GetOrderTotal()
    {
        double total = 0d;
        foreach (Order o in orders)
        {
            total += o.GetOderAmount();
        }
        return total;
    }
}
```

Note that the `Customer` class delegates the job of finding the order amount by each order to the `Order` class.

In sum, we made the `Customer` a hero to house the operation `GetOrderTotal()`, but the customer object does not and cannot do it by itself. It delegated the job of computing an order amount for each order to the `Order` class, which in turn delegated the job of getting prices, quantities, and discount rates to `Product` and `LineItem` classes.

It is correct to allocate `GetOrderTotal()` into the `Customer` class according to information needs or the data flow reduction principle. However, we should be aware that knowledge or information is necessary but not sufficient for deciding a hero, and we may still need to make a judgmental call on what an object is supposed to be capable of and, based on the judgment, to allocate operations. For example, in the current example, does an `Order` object have enough knowledge to compute the total? Yes, it does. It has necessary information because each order knows who made the order, and so it can delegate the job to the customer. For example, we could implement the operation as follows:

```
public class Order
{
    private int oid;
    private DateTime odate;
    private Customer madeBy;
    private List<Product> orderedItems;
    private Dictionary<Product, LineItem> orderLines;

    public double GetTotalbyEachCustomer()
    {
        return madeBy.GetOrderTotal();
    }
}
```

However, there are at least two problems in doing so. First, an `Order` object makes no contribution by its own. It simply relays the whole task to another object. *An object that does nothing but relay a task should not be made a hero to host the task.* Second, it is problematic to send a message to an order object to find the total amount made by a customer. Which object can we send the message to? Each order knows its owner, but the owner may not be the customer that we intend to send the message to. Thus, we will have to send the message to all order objects and ask each one if its owner matches the customer we intend to send to.

Of course, the function `GetTotalbyEachCustomer()` can be still useful to an `Order` object. For example, when making a new order and deciding whether we should discount the order amount, we may need to find out the total amount made by the owner. However, in this case, the order object delegates the job to the customer object and shall not claim to be the hero of the capability.

Collaboration via Examples: Compute Grade Point Average

Now let us carry the student registration example from earlier chapters further and try to model and implement additional operations by following the heuristics.

Figure 4 shows a familiar class diagram that we have seen earlier in Chapter 7. Now we need to code one more operation: Compute grade point average (GPA). We programmed the operation as an isolated function in Chapter 4, but now we will program it via collaboration of related objects.

First, let us implement the Course class along with recursive association for prerequisites. The two operations are merely data getters:

```
public class Course
{
    private string cno;
    private double credits;
    private List<Course> prerequisites;

    public double GetCreditHours()
    {
        return credits;
    }

    public List<Course> GetPrerequisites()
    {
        return prerequisites;
    }
}
```

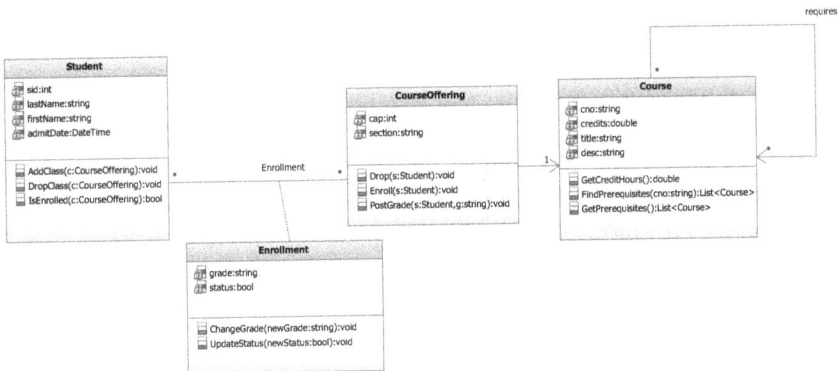

Figure 4.   Class diagram for student registration system.

The following is the code snippet from Chapter 7 with one extra instance variable `crsUnder` in the `CourseOffering` class to reflect its unidirectional association with `Course` (see the boldfaced code).

```
public class CourseOffering
{
    private string section;
    private int cap;
    private Course crsUnder;
    private List<Student> enrollees;
    private Dictionary<int, Enrollment> roster;

}

public class Enrollment
{
    private bool status;
    private string grade;
    private Student student;
    private CourseOffering section;

}

public class Student
{
    private DateTime admitDate;
    private int sid;
    private string firstName, lastName;
    private List<CourseOffering> enrolledSections;
    Dictionary<CourseOffering, Enrollment> transcript;
}
```

To compute GPA, we will need to know all the courses that a student has finished. Since the `Student` class has a list of past enrollment records or transcripts, it should be enough to do the calculation. Thus, the `GetGPA` operation should be located inside the `Student` class. Does a `Student` object do everything? Let us look at the operational logic (may be modeled by an activity diagram here): GPA = total points / total credits. Each course grade is translated into points as follows: A → 4, B → 3, C → 2, D → 1, and F → 0. Multiplying the course point by the course credit leads one to arrive at the grade points for each course. Adding the grade points of all courses results in the total points. According to this analysis, we identified some sub-tasks: get credit hours for each course, get grade for each finished course, and convert grades to points.

Let us contemplate a CRC session. The person holding the Student class announces, "I am a student and needs to compute GPA, for which I need to know credit hours of each course I took and the grade I got." "I have grades for each of your course," the person holding the Enrollment class answers. The person holding the Course class also answers, "I know credit hours."

Because Enrollment objects know grades and Course objects know credits, Student objects should delegate GetCreditHours to Course and GetGrade to Enrollment. What about converting grades to points? We can put it anywhere, or even as a utility class. However, it would be the best to put it in the Enrollment class; doing so will remove the need for the input parameter "grade" and we do not even need the GetGrade() method. In the next section, we will find that there is another added benefit of doing so; we will be able to tell if grade A is better than B, which in turn is better than C, etc., simply by comparing the points in checking prerequisites. Otherwise, we will have to find a way to tell if a student has a C or better grade. GetCreditHours() is already implemented in the code above, and GetPoints() method is implemented below in the Enrollment class (see the boldfaced code):

```
public class Enrollment
{
    private bool status;
    private string grade;
    private Student student;
    private CourseOffering section;

    public double GetPoints()
    {
        double points = 0;
        switch(grade)
        {
            case "A":
                points = 4;
                break;
            case "B":
```

```
                points = 3;
                break;
            case "C":
                points = 2;
                break;
            case "D":
                points = 1;
                break;
            case "F":
                points = 0;
                break;
        }
        return points;
    }


}
```

The `Student` class has a list of enrolled course offerings but does not know courses directly according to the class diagram in Figure 4. So, in order for the `Student` class to call `GetCreditHours()` method, it has to do it indirectly through the `CourseOffering` class, which of course can delegate (or relay) the actual job to the `Course` class, which is the real hero for housing `GetCreditHours()` option (see the boldfaced code). Besides getting credit hours for a course, we also add a helper method to get the course object from `CourseOffering`, which may be an alternate method in lieu of `GetCreditHours()`.

```
public class CourseOffering
{
    private string section;
    private int cap;
    private Course crsUnder;
    private List<Student> enrollees;
    private Dictionary<int, Enrollment> roster;

    public double GetCreditHours()
    {
        return crsUnder.GetCreditHours();
    }

    public Course GetCourse()
    {
        return crsUnder;
```

```
    }


}
```

With the above two helper functions, we can now implement the GetGPA() method in the Student class as follows (see the boldfaced code). The code simply goes through each enrolled section for a credit hour and the corresponding transcript entry, or enrollment object, for a point. Along with GetGPA(), we also include a similar method to compute total credit hours.

```
public class Student
{
    private DateTime admitDate;
    private int sid;
    private string firstName, lastName;
    private List<CourseOffering> enrolledSections;
    Dictionary<CourseOffering, Enrollment> transcript;

    public double ComputeGPA()
    {
        double totalCredits = 0;
        double totalPoints = 0;
        foreach (CourseOffering s in enrolledSections)
        {
            totalCredits = totalCredits +
                s.GetCreditHours();
            totalPoints = totalPoints + s.GetCreditHours()
                * transcript[s].GetPoints();
        }
        return totalPoints / totalCredits;
    }

    public double GetTotalCredits()
    {
        double result = 0;
        foreach (CourseOffering s in enrolledSections)
        {
            if (transcript[s].GetPoints() >= 1)
                result += s.GetCreditHours();
        }
        return result;
    }

}
```

Collaboration via Examples: Check Prerequisites

In this section, we continue the student registration example (see Figure 4) and add a function to check if a student has fulfilled the prerequisites. The function can improve the existing operations: AddClass() in the Student class and Enroll() in the CourseOffering class, by adding a precondition.

To check prerequisites, we need a list of courses that are designated as prerequisites and a list of courses that a student has finished. No object has data on both lists, so where should the operation be allocated? The Course class has a list of designated prerequisites. However, it cannot obtain a list of finished courses from a Student object because it cannot send messages to the Student object according to unidirectional navigability. The CourseOffering class is a possible choice because it can ask the Course class for the list of prerequisites and ask a Student object for a list of finished classes. The only parameter needed is a Student object or a student ID. The reader may consider this choice as a homework exercise. The second choice is the Student class because it has a list of finished courses, and it can also ask CourseOffering for help to get a list of prerequisites. Now let us implement the second choice.

First, the Student class needs the CourseOffering class to help get a list of prerequisites. This can be done easily because CourseOffering can delegate the actual job to the Course class:

```
public class CourseOffering
{
    private string section;
    private int cap;
    private Course crsUnder;
    private List<Student> enrollees;
    private Dictionary<int, Enrollment> roster;

    public List<Course> GetPrerequisites()
    {
        return crsUnder.GetPrerequisites();
    }

    public double GetCreditHours()
    {
        return crsUnder.GetCreditHours();
    }

}
```

Next, let us create a helper method in the `Student` class to check if a student has successfully finished a given course, assuming the passing grade is C or better. Note that we are able to use `GetPoints()` to tell if a student has a passing score.

```
public bool IsFinished (Course c)
{
    foreach (CourseOffering s in enrolledSections)
    {
        if (transcript[s].GetPoints()>=2 &&
            s.GetCourse() == c)
        {
            return true;
        }
    }
    return false;
}
```

With these helper methods, we can now implement `CheckPrerequ isites(CourseOffering c)` as follows by checking each course in the list of the prerequisites to see if a student has passed the prerequisites (see the boldfaced code). The code tests each prerequisite, and if anyone has not finished the course successfully, the student is deemed as not meeting the prerequisites.

```
public class Student
{
    private DateTime admitDate;
    private int sid;
    private string firstName, lastName;
    private List<CourseOffering> enrolledSections;
    Dictionary<CourseOffering, Enrollment> transcript;

    public bool IsFinished (Course c)
    {
        foreach (CourseOffering s in enrolledSections)
        {
            if (transcript[s].GetPoints()>=2 &&
                s.GetCourse() == c)
            {
                return true;
            }
        }
        return false;
    }
```

```
public bool CheckPrerequisites(Course c)
{
    List<Course> prereqs = c.GetPrerequisites();
    foreach (Course crs in prereqs)
    {
        if (this.IsFinished(crs) == false)
            return false;
    }
    return true;
}

public double ComputeGPA()
{
    double totalCredits = 0;
    double totalPoints = 0;
    foreach (CourseOffering s in enrolledSections)
    {
        totalCredits = totalCredits +
            s.GetCreditHours();
        totalPoints = totalPoints + s.GetCreditHours()
            * transcript[s].GetPoints();
    }
    return totalPoints / totalCredits;
}

public double GetTotalCredits()
{
    double result = 0;
    foreach (CourseOffering s in enrolledSections)
    {
        if (transcript[s].GetPoints() >= 1)
            result += s.GetCreditHours();
    }
    return result;
}

}
```

Collaboration via Examples: Check Time Conflicts

Besides checking prerequisites, we must also make sure the time for the course in which one is to be enrolled is not conflicting with ones that a student has already registered. To check for time conflicts, we need to know what courses a student has registered for as well as the meeting times for each individual one. Thus, the best class to house the operation is the `Student` class.

Now let us analyze the logic of the operation. First, we need each course offering to tell its meeting times. Then we need to compare the meeting times of the course in which one wishes to be enrolled with that of each of registered courses to see if there is any overlap.

This sounds easy but is actually a complex job, partially because the time for course offerings are typically stated like Mondays and Wednesdays between 1:30 PM and 2:55 PM during a period between January 15, 2020 and May 5, 2020. Currently, we use a text field to record the class time; it is difficult to be quantified for the implementation of the `CheckTimeConflicts` operation. Built-in types such as `Date` in Java or `DateTime` in C# will not serve the purpose; `Date` and `DateTime` objects are absolute points in time, and since the time for a class meeting is like 1:30 PM–2:55 PM every Monday between January 15, 2020 and May 10, 2020, for example, this will not suit.

To better model meeting times, we first need to have the `Period` class with attributes `beginDate` and `endDate` to represent values like "between January 18, 2020 and May 10, 2020." Then we need a data type for a point in time during a day. It is a relative time, i.e., time without year, month, and day values. Thus, the new `Time` class has data members `hour`, `minute`, and `second` only. Then we can use the `Time` class to create `TimeSlot` class with data members: `day` (for week day 1, 2, 3, etc.), `beginTime`, and `endTime`. The latter two are of the `Time` type. Then the class time for each course will be specified by a time period like between January 15, 2012 and May 10, 2020, and one or more time slots like Monday 1:30 PM–2:55 PM, Wednesday 1:30 PM–2:55 PM, etc. With these helper classes, we now update the class diagram accordingly (Figure 5).

In order to check if two class times are overlapping, we can delegate the task to `CourseOffering`, and so we create a method `Overlap(CourseOffering c)` in there. But in order to check if one course offering has time conflicts with another, we need to check if two periods are overlapping and if two times slots are overlapping. So, the `CourseOffering` class is going to delegate the sub-tasks to `Period` and `TimeSlot` classes; thus, we create helper operations accordingly in there.

It is easy to compare if two periods are overlapping by simply comparing the `beginDate` of one period with the `endDate` of the second period.

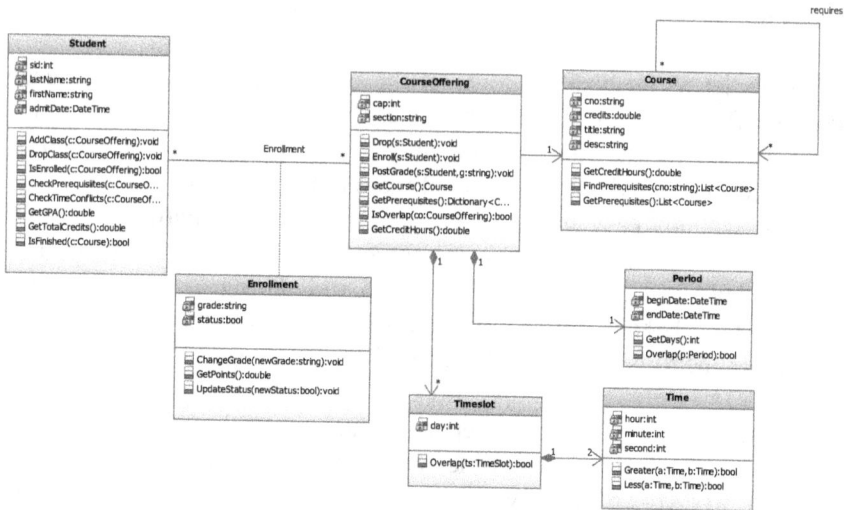

Figure 5.   Revised class diagram for student registration system.

```
public class Period
{
    private DateTime beginDate;
    private DateTime endDate;

    public bool NotOverlap(Period p)
    {
        if (this.endDate < p.beginDate || p.endDate <
            this.beginDate)
            return true;
        else
            return false;
    }

    public bool IsOverlap(Period p)
    {
        if(this.endDate >= p.beginDate && p.endDate >=
            this.beginDate)
            return true;
        else
            return false;
    }
}
```

In order to tell if one time slot is overlapping with another time slot, we need to have a way to compare two times to see which is earlier and which is later. For that purpose, we delegate the job to the Time class and create an operator < there so that we can compare if time *a* is less than time *b* by using the operator *a* < *b*. The following code shows the implementation of the Time class with the operator <. The basic idea is to count the total number of seconds from 00:00:00 of a day to see which time is smaller.

```
public class Time
{
    int hour, minute, second;
    public static bool operator <(Time t1, Time t2)
    {
        if (t1.hour * 3600 + t1.minute * 60 + t1.second
                < t2.hour * 3600 + t2.minute * 60 +
                    t2.second)
            return true;
        else
            return false;
    }

    public static bool operator >(Time t1, Time t2)
    {
        if (t1.hour * 3600 + t1.minute * 60 + t1.second
                > t2.hour * 3600 + t2.minute * 60 +
                    t2.second)
            return true;
        else
            return false;
    }
}
```

To implement the TimeSlot class, we need a data type for weekdays. Note that since a weekday takes values 0, 1, 2, ..., 6, so we can create an enum type Day for it.

```
public enum Day
{
    Sunday,
    Monday,
    Tuesday,
    Wednesday,
    Thursday,
    Friday,
    Saturday
}
```

If two time slots are on different weekdays, they will not be overlapping. If they are on the same weekday, we check if the end time of one slot is smaller than the begin time of another time slot.

```
public class TimeSlot
{
    Day d;
    Time beginTime;
    Time endTime;

    public bool NotOverlap(TimeSlot ts)
    {
        if (this.d != ts.d)
            return true;
        else
        {
            if (this.endTime < ts.beginTime ||
                ts.endTime < this.beginTime)
                return true;
            else
        return false;
        }
    }
}
```

Now we can check if one course offering has overlapping time with another. First check if they have overlapping periods. If they do, check for all possible overlapping time slots; two courses are deemed to have time conflicts if any two time slots overlap (see the boldfaced code).

```
public class CourseOffering
{
    private string section;
    private int cap;
    private Course crsUnder;
    private List<Student> enrollees;
    private Dictionary<int, Enrollment> roster;
    private Period term;
    private List<TimeSlot> meetingTimes;

    public bool HasConflict(CourseOffering s)
    {
        if (this.term.NotOverlap(s.term))
```

```
          return false;
      else
      {
          foreach (TimeSlot ts1 in this.meetingTimes)
          {
              foreach (TimeSlot ts2 in s.meetingTimes)
              {
                  if (ts1.NotOverlap(ts2) == false)
                      return true;
              }
          }
          return false;
      }
  }

  public Course GetCourse()
  {
  return crsUnder;
  }

  public double GetCreditHours()
  {
  return crsUnder.GetCreditHours();
  }

}
```

Finally, let us implement `CheckTimeConflicts(CourseOffering c)` in the `Student` class. The `Student` class is a hero here; it does the most important jobs in the registration system. Yet, see how little it does. It delegates most of the job to other classes. The hero becomes a hero with a lot of helpers.

```
public class Student
{
    private DateTime admitDate;
    private int sid;
    private string firstName, lastName;
    private List<CourseOffering> enrolledSections;
    Dictionary<CourseOffering, Enrollment> transcript;

    public bool CheckTimeConflict(CourseOffering s)
    {
        foreach (CourseOffering enS in enrolledSections)
```

```
    {
        if (enS.HasConflict(s))
            return true;
    }
    return false;
}

public bool HasFinished (Course c)
{
    foreach (CourseOffering s in enrolledSections)
    {
        if (transcript[s].GetPoints()>=2 &&
            s.GetCourse() == c)
        {
            return true;
        }
    }
    return false;
}

public bool CheckPrerequisites(Course c)
{
    List<Course> prereqs = c.GetPrerequisites();
    foreach (Course crs in prereqs)
    {
        if (this.HasFinished(crs) == false)
            return false;
    }
    return true;
}

public double ComputeGPA()
{
    double totalCredits = 0;
    double totalPoints = 0;
    foreach (CourseOffering s in enrolledSections)
    {
        totalCredits = totalCredits +
            s.GetCreditHours();
        totalPoints = totalPoints + s.GetCreditHours()
            * transcript[s].GetPoints();
    }
    return totalPoints / totalCredits;
}

}
```

We must not forget to call the newly added `CheckPrerquisistes` and `CheckTimeConflicts` methods before we enroll students. So, we change `AddClass(CourseOffering c)` in the `Student` class and `Enroll(Student s)` in the `CourseOffering` class coded in Chapter 7 (see the boldfaced code). To be perfect, we also add a method in the `CourseOffering` class to check if the offering is currently available.

```
public class CourseOffering
{
    private int sectionNo;
    private bool status;
    private Course crsUnder;
    private List<Student> enrollees;
    private Dictionary<int, Enrollment> roster;
    private Period term;
    private List<TimeSlot> meetingTimes;

    public bool HasConflict(CourseOffering s)
    {
        if (this.term.NotOverlap(s.term))
            return false;
        else
        {
            foreach (TimeSlot ts1 in this.meetingTimes)
            {
                foreach (TimeSlot ts2 in s.meetingTimes)
                {
                    if (ts1.NotOverlap(ts2) == false)
                        return true;
                }
            }
            return false;
        }
    }

    public bool IsOpen()
    {
        return status;
    }

    public Course GetCourse()
    {
        return crsUnder;
    }
```

```
public double GetCreditHours()
{
    return crsUnder.GetCreditHours();
}

public void Enroll(Student stu)
{
    if (stu.CheckPrerequisites(crsUnder)
            && !stu.CheckTimeConflict(this) &&
            this.IsOpen())
    {
        enrollees.Add(stu);
        Enrollment e = new Enrollment(stu, this);
        roster.Add(stu.SID, e);
        stu.Sync("add", this, e);
    }
    else
    {
        throw new Exception("Can't enroll");
    }
}

public void Drop(Student stu)
{
    enrollees.Remove(stu);
    roster.Remove(stu.SID);
    stu.Sync("drop", this);
}

public void Sync(string type, Student s,
    Enrollment e = null)
{
    if (type == "add")
    {
        enrollees.Add(s);
        roster.Add(s.SID, e);
    }
    else if (type == "drop")
    {
        enrollees.Remove(s);
        roster.Remove(s.SID);
    }
}

public void PostGrade(Student stu, string g)
{
```

```
      roster[stu.SID].ChangeGrade(g);
}

public List<Course> GetPrerequisites()
{
    return crsUnder.GetPrerequisites();
}


}

public class Student
{
    private DateTime admitDate;
    private int sid;
    private string firstName, lastName;
    private List<CourseOffering> enrolledSections;
    Dictionary<CourseOffering, Enrollment>
        transcript;

    public bool CheckTimeConflict(CourseOffering s)
    {
        foreach (CourseOffering enS in
           enrolledSections)
        {
            if (enS.HasConflict(s))
               return true;
        }
        return false;
    }

    public bool HasFinished (Course c)
    {
        foreach (CourseOffering s in
           enrolledSections)
        {
          if (transcript[s].GetPoints()>=2 &&
             s.GetCourse() == c)
          {
             return true;
          }
        }
        return false;
    }

    public bool CheckPrerequisites(Course c)
    {
```

```
         List<Course> prereqs = c.GetPrerequisites();
         foreach (Course crs in prereqs)
         {
            if (this.HasFinished(crs) == false)
               return false;
         }
         return true;
      }

      public double ComputeGPA()
      {
         double totalCredits = 0;
         double totalPoints = 0;
         foreach (CourseOffering s in enrolledSections)
         {
            totalCredits = totalCredits +
               s.GetCreditHours();
            totalPoints = totalPoints +
               s.GetCreditHours() * transcript[s].
               GetPoints();
         }
         return totalPoints / totalCredits;
      }

      public double GetTotalCredits()
      {
         double result = 0;
         foreach (CourseOffering s in enrolledSections)
         {
            if (transcript[s].GetPoints() >= 1)
               result += s.GetCreditHours();
         }
         return result;
      }

      public int SID
      {
         get {return sid;}
         set {sid = value;}
      }

      public bool IsEnrolled(CourseOffering sec)
      {
         return enrolledSections.Contains(sec);
      }
```

```
public void AddClass(CourseOffering sec)
{
   if (this.CheckPrerequisites(sec.GetCourse())
      && !this.CheckTimeConflict(sec) && sec.
      IsOpen())
   {
      enrolledSections.Add(sec);
      Enrollment e = new Enrollment(this, sec);
      transcript.Add(sec, e);
      sec.Sync("add", this, e);
   }
   else
   {
      throw new Exception("Can't enroll");
   }
}

public void DropClass(CourseOffering sec)
{
   enrolledSections.Remove(sec);
   transcript.Remove(sec);
   sec.Sync("drop", this);
}

public void Sync(string type, CourseOffering sec,
   Enrollment e = null)
{
   if (type == "add")
   {
      enrolledSections.Add(sec);
      transcript.Add(sec, e);
   }
   else if (type == "drop")
   {
      enrolledSections.Remove(sec);
      transcript.Remove(sec);
   }
}
}
```

Exercises

1. Think of a strategy of applying all guidelines or heuristics for modeling objects and then write an essay to discuss your strategy.

2. Implement the operations "bool CheckPrerequisites (Student s)" in the CourseOffering class to check if student s has passed all the courses designated as prerequisites. Implement the operation "bool CheckTimeConflicts (Student s)" in the CourseOffering class to check if a student has time conflicts if taking the current course offering. Update AddClass (Course Offering c) in the Student class and AddStudent (Student s) in the CourseOffering class accordingly.

3. To create a new order, the necessary sub-tasks include add items, update inventory, compute totals, and compute tax. The classes involved include Customer, Order, Product, and OrderLine. Program the MakeOrder() operation along with all the helper methods in collaborating classes. Then draw the class diagram accordingly.

4. To receive a shipment, an inventory manager has to check each received item, quantity, and quality against the original order and package list. She has to note the differences, update the order lines and orders, and update the inventory level. Create a graphical user interface for the manager to do the job. Then create the class diagram with appropriate data members and operations.

5. A professor makes various assignments throughout the course for students. Each assignment will carry a certain maximum score. Students' submissions are graded, and each is awarded with a certain point at or below of the maximum score. Program the method to compute the total points earned by each student throughout the course. Program another method to find the average student grade for each assignment. Draw the class diagram accordingly.

6. Airlines often announce flights for a fixed period of one or two years in their catalogs. For example, Flight 1023 flies from New York to Beijing every Sunday at 12:00 AM for Year 2013 and every Wednesday and Sunday at 11:30 PM for Year 2014. Some flights have multiple stopover airports on the way to the final destination. Airlines then schedule flights according to the above catalogs. For example, a flight a passenger actually flies is Flight 1023 from New York to Beijing from July 30, 11:30 PM to July 31, 2:00 PM. When a customer books flights, he or she can do it for one or more passengers, such as family members or company employees. For each booking, the airlines require each passenger name and birth date to be recorded. Create a class diagram for the airline booking system and make sure you have appropriate date types for flight schedules.

7. Hospitals offer various types of tests such as X-rays, autopsy, cholesterol counts, etc. When performing actual tests for a patient, they need to track the physician who ordered the test, and the physician who read the test results, and the physician or employee who performed the test. The test results can be varied. Sometimes, it is a simple value for an attribute such as blood count. Sometimes it has multiple values as the result. Sometimes it contains one or more image or sound files along with experts' reading comments and diagnosis. Create a class diagram for a medical record system to handle tests and test results.

8. Implement the following class diagram (Figure 6) for a clinic appointment system. The key functionality is to make or take appointments for patients with doctors. Of course, each new appointment must be during the working hours for the doctor the patient wants to see and must not conflict with other existing appointments. Assume each new appointment will last for at least 15 minutes each. Each appointment has a specific begin time and end time. But the trouble is the doctor's working hours are recorded text, and a typical example will read like this: between January and June of 2014, Dr. Johnson works at Boston Women's Hospital on Tuesdays and Fridays from 8:00 AM to 4:00 PM, and Cambridge College Office on Mondays and Thursdays from 9:00 AM to 2:00 PM. Determine where to house the operations MakeAppointment and CheckTimeConflict. Then implement the operations and create a helper classes and methods if needed. Submit the final class diagram and class code.

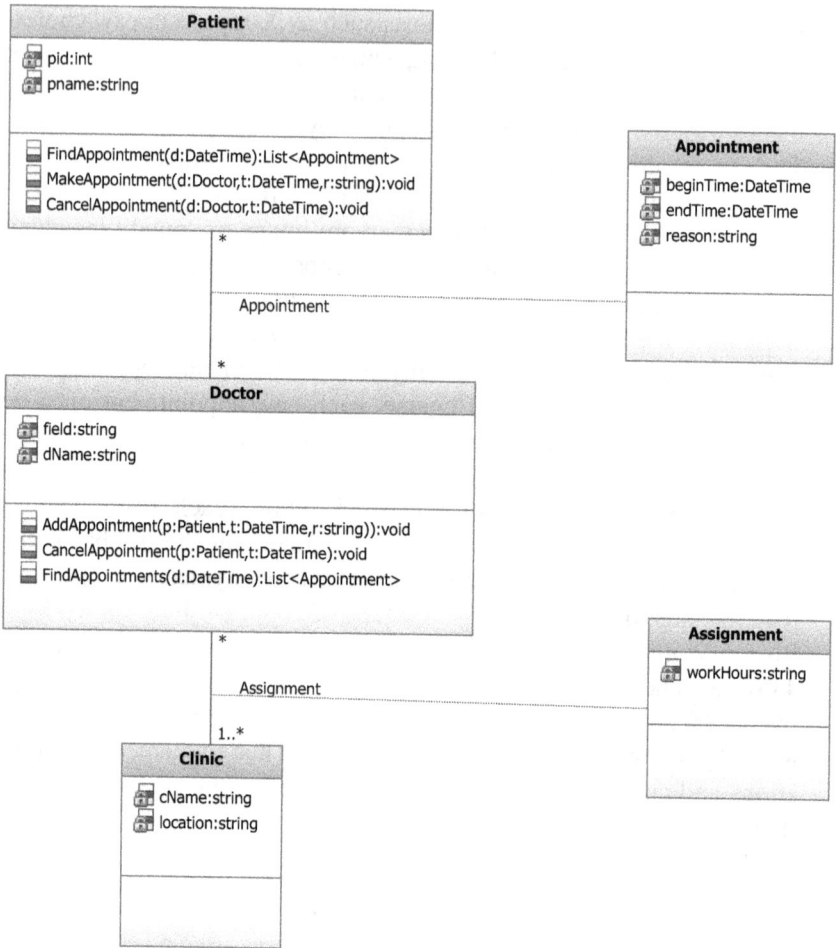

Figure 6.　Class diagram for clinic appointment system.

Chapter 15

Collaboration Modeling

Introduction

A use case is a sequence of interactions, and each interaction is in the form of either user requesting the system to perform a service, or the system performing the service, providing feedback, or asking the user to provide additional inputs.

What is the system? In earlier chapters, we understood the system as a collection of objects. Thus, after all, *all the services and requests by the system must be performed by one or more objects that constitute the system.*

Collaboration modeling graphically depicts which objects perform the services or display the information on behalf of the system and how these objects collaborate with each other to perform a use case. In a sense, a collaboration model is a graphical representation of the use case story with one important distinction: Each interaction involving the system must be allocated to one or more constituent objects. In this way, *collaboration modeling bridges classes and use cases* and becomes the formal method for identifying and capturing operations in class diagramming.

Collaborating objects, including actors and the constituent objects of the system, interact with each other by sending messages. A collaboration model represents the messages either spatially using *communication diagrams* or chronologically using *sequence diagrams*. These two models are almost equivalent and can be converted to each other mechanically. The difference is that communication diagrams emphasize communication links, while sequence diagrams emphasize the lifelines of objects.

Communication Diagrams

Objects collaborate by sending messages to each other. *A message is a function call*; that object A sends a message to object B simply means that A calls or invokes a function owned by object B. We discussed functional calls in Chapter 4. The exception here is that objects are the owners/containers of functions, and to call a function, we need to send a message to its container object. For example, an order system contains `Order` and `Item` objects. When computing the order amount, an order object may ask an `Item` object to get item information. This may be expressed as the `GetItemInfo()` message (see Figure 1).

In collaboration models, the objects are named in the following format: "`object name:class name`". For example, we may name a pilot object as `james:Pilot` and an airplane object as `spitfire:Airplane`, etc. Objects can be anonymous, i.e., with names omitted, such as `:Order` and `:Item` in Figure 1.

A communication diagram connects participating objects, and even actors, using *communication links* and depicts messages over the links. A link, drawn as an undirected line, is an instance of associations and corresponds to the association relationship in a class diagram. One or more messages may be communicated through a link, and each message has a name indicating the message content, a number indicating the message sequence order, and an arrow indicating the direction of the message (see Figure 1).

To create a communication diagram using Rhapsody, right click on a package and select Add New → Diagrams → Communication Diagram menu. A dialog will prompt for a diagram name. After dismissing the dialog, a blank canvas along with the diagram tools will be displayed (see Figure 2).

There are two ways to create object nodes using Rhapsody. We can drag an existing class from the model browser to a communication diagram. To create an object of an association class, we will need to make an object from the association by first right clicking on the association in

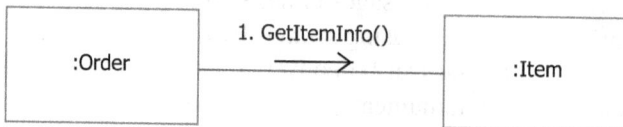

Figure 1.   Sending message between objects.

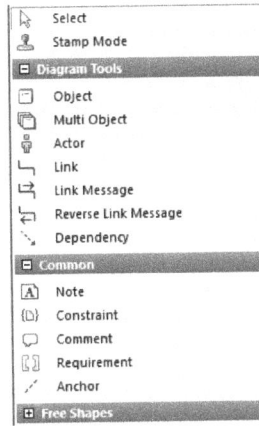

Figure 2.   Communication diagram tools.

the model browser and selecting the Make an Object menu item and then dragging the created object from the model browser to the diagram. We can also create a new object by using the Object tool from the Diagram Tools (see Figure 2 for a snapshot of the toolbox). In this second approach, a new class may be created for model synchronization: upon naming a new object, Rhapsody will ask whether we want to create the class if it does not exist yet.

To create a message, we will initially make a communication link and later attach the message to the link by first selecting the Link Message tool and then clicking on the link line. Each message will be automatically labeled with a sequence number, which may be changed if we need to.

Collaboration usually involves many objects sending many messages to each other to perform a whole use case. Putting all these messages in a spatial order, we have a communication diagram. Figure 3 shows a communication diagram to carry out the use case "Post Grades" performed by the Professor actor: When a course is finished, a professor will post grades to the registration system, and the system will update the object data accordingly, including grades, earned credits, grade point average (GPA), and enrollment status. Assume a professor uses a user interface, called `PostGradeForm`, to post grades. The system needs the collaboration of the user interface object and four domain objects to perform the use case: `Student`, `Enrollment`, `Course`, and `CourseOffering`.

As we learned earlier, updating enrollment status and changing grades are best done by an `Enrollment` object, retrieving credit hours can be

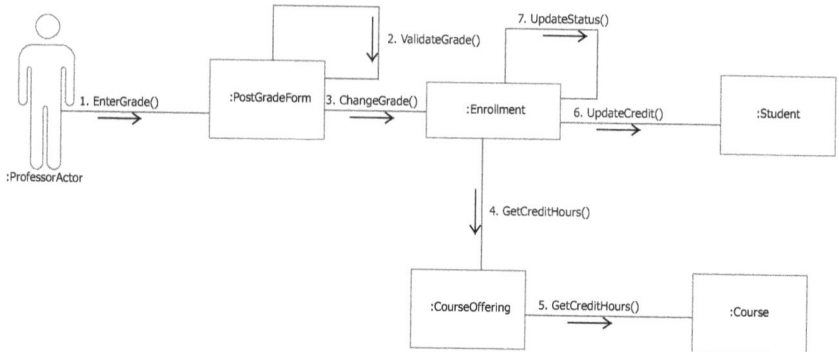

Figure 3.   Communication diagram for posting grades.

done by a `Course` object, and updating total credits and GPA can be done by a `Student` object. The collaboration can be designed as follows. First, a professor enters grades into the user interface `PostGradeForm`, which can then send the message `ChangeGrade` to an `Enrollment` object. The `Enrollment` object may act as a coordinator, delegating some parts of the job in updating credits and updating GPA to other objects. To retrieve credit hours, the `Enrollment` object sends the `GetCreditHours()` request to the `CourseOffering` object, which in turn sends the request to the `Course` object. Then the `Enrollment` object can send a message to the `Student` object to update the total credit and GPA.

Communication links

Note in Figure 3 that the `Enrollment` object does not send the message `GetCreditHours()` directly to the `Course` object, rather it does so indirectly through the `CourseOffering` object. This is because the `Enrollment` object is not associated with the `Course` object and does not know the `Course` object directly in the class diagram (see Figure 4). Two objects can send messages to each other directly only if there is a communication link between them, or equivalently if they have an association relationship in the class diagram. In addition, if navigability is unidirectional, messages are permitted to go in one direction. For example, an order is associated with one or more items, and the association is one-way from `Order` to `Item`, and so an order can send `GetItemInfo()` or other messages to an item. However, due to unidirectional navigability (see Figure 5), an item object cannot send any message to an order object.

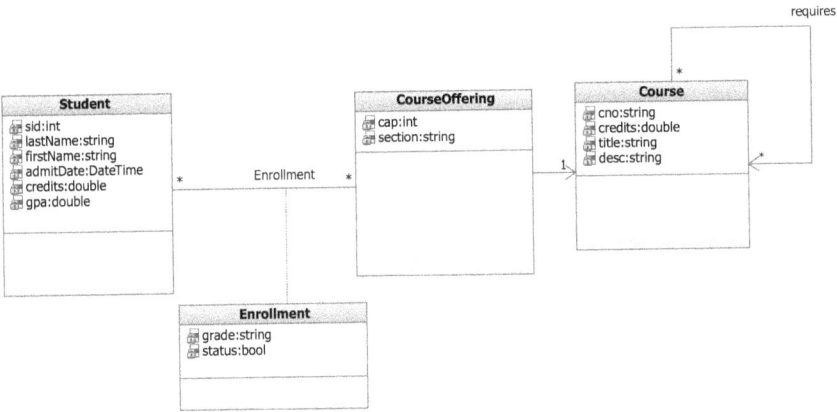

Figure 4.   Business objects in collaboration to perform the "Post Grades" use case.

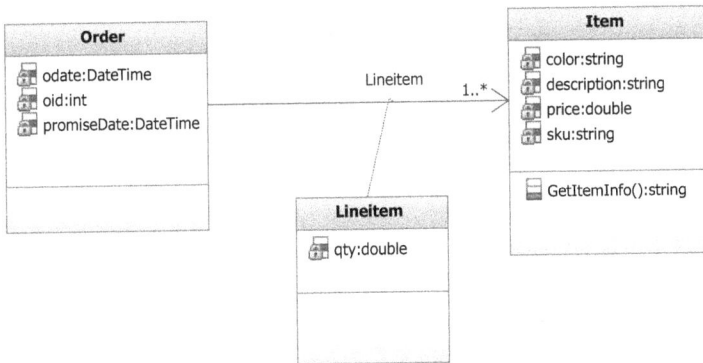

Figure 5.   One-way navigable associations.

A collaboration model links to class diagramming in two important ways. First, *messages suggest object operations*. For example, according to Figure 3, the message UpdateCredit() implies that its receiver, the Student class, must have an operation to update credits. Incorporating all the messages in Figure 3, the class diagram in Figure 4 will be enriched into the one in Figure 6 with the added operations.

Second, a communication link implies the existence of an association. If a collaboration diagram shows the need for collaboration between two objects but they are not associated with each other, it may indicate that the class diagram has deficiency and requires a structural change to reflect the need for collaboration.

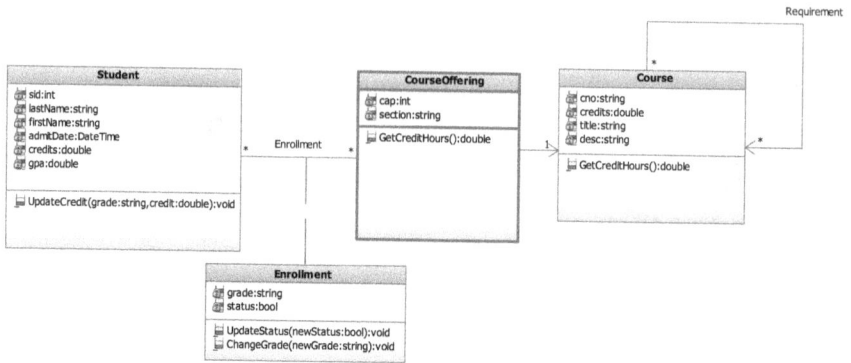

Figure 6.   Enriched version of class diagram in Figure 4.

Therefore, after drawing a collaboration model, we need not only enrich a class diagram by capturing required operations but also must make structural changes to the class diagram. After all, the most important deliverable is the class diagram. It is the classes that constitute the business application to be developed.

It is also worth noting in Figure 3 that collaboration to perform a use case involves not only business or domain objects such as Student, Course, CourseOffering, etc., but also non-domain objects such as user interfaces and external agents such as actors.

With regard to class diagramming, thus far in this book, we have focused on domain objects and their relationships. When modeling object collaboration to perform a use case, we may have to include non-domain objects since a typical system consists of not only domain objects but also interface objects such as forms and reports as well as control objects that are responsible for transaction coordination. As a rule of thumb, *whenever there is a message from an actor to the system or a message from the system to the actor, there will be a need of a user interface.* Similarly, *whenever there is a complex task that requires multiple objects to collaborate, a control object may be needed.*

In Figure 3, the Enrollment object plays the role of a coordinator in interacting with and delegating jobs to two or more domain objects and coordinating their operations. This role is usually played by a *control object.* Unlike user interfaces, control objects embody business logics. Unlike domain objects, control objects carry no business data members. We may imagine a control object as a command center that can send messages to domain objects, sequence or coordinate their operations, integrate their results, or orchestra collaboration of the objects.

Due to a lack of object-oriented database systems, object data are often persisted into and populated by relational databases. As such, there may be a need for special *data access objects* to which other objects may delegate data persistence-related operations, including building connections to a database and retrieving data from or saving data to a database.

A final type of non-domain class consists of universal helpers, called *utility classes*. Utility classes help all other classes; they hold useful functions that can be called for help in tasks like converting dates into different formats, looking up tax rates, checking input errors, doing mathematical calculations, etc. Since their functions are universal, utility classes do not really belong to a system, and they can be packed as a library to be imported to any system. Therefore, we do not need to model messages to utility classes.

In sum, collaboration modeling involves communications among four different types of objects: domain objects, user interfaces, control objects, and data access objects. Then, how do we determine their communication links in communication diagrams (or equivalently the associations of these objects in class diagrams)? A*n overarching criterion is reusability: A class is less reusable if it has more associations with other classes.* Just imagine a scenario in which a domain object can send messages to a user interface, or equivalently a domain class has an association with a user interface class. Such a domain class will not be reusable because it has the user interface as a data member, and so it is always reused along with the user interface together. This situation is rare. A system has many different use cases, and an organization may have different systems for different areas or at different times. Different use cases or systems usually have dramatically different user interfaces and control logics, but they are all based on the same domain objects. Therefore, a simple heuristic is to make domain classes more reusable than control classes, which in turn are more reusable than user interfaces. A *step-and-skip model* reflects this heuristic (see Figure 7): rank four categories of classes in order, an object at a higher rank can send messages to any object at a lower rank. These four categories from high to low ranks are, respectively, user interface objects, control objects, domain objects, and data access objects. Thus, a user interface object can send messages to control objects, domain objects, and data access objects, but not vice versa. Similarly, a control object may send messages to domain and data access objects. Data access objects are of the lowest rank; all other objects can execute the functionalities of data access objects to persist data into a database.

The step-and-skip model is a general guideline, and its implementation is dependent on the context. For example, the model suggests that

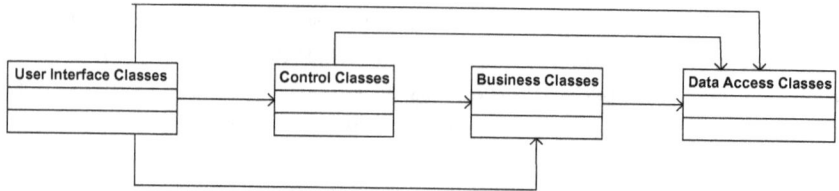

Figure 7.   The step-and-skip model for communication links.

domain objects can send messages to data access objects. This would imply that domain classes are reusable with data access classes together as a bundle. This makes sense sometimes, and the practice has been followed by some formal technologies such as the entity framework. After all, domain objects are data containers, and binding them with data access classes will free us from plumbing jobs related to the persistence of the object data and the CRUD activities on the object data.

However, reusing domain and data access classes together may not be an optimal solution in situations with potential future changes in data sources and/or data access techniques. For example, an organization may switch its database system down the road from Oracle to MySQL or from a relational DBMS to a non-relational one. A better solution in these situations is perhaps to make domain objects independent of data sources and techniques of accessing the data sources. In doing so, the resulting domain classes will be more reusable over time.

By the same token, user interface classes can achieve a higher level of independence if restricted them from sending messages to other objects except for control ones. Interface classes should focus on how to present data and how to respond to user actions. All data to be presented should be fed by control objects, and all user actions should be delegated to control objects as well.

Therefore, a less questionable alternative to the step-and-skip model is the VCM (View–Control–Model) model considering the above arguments. The VCM model suggests that control objects can send messages to both domain objects and data access objects, while user interfaces can send messages to control objects only (see Figure 8). The VCM model reduces coding flexibilities because of reduced communication links, but it improves code reusability because a change in data access classes would not affect domain classes and user interfaces, and a change in domain classes would not affect user interfaces.

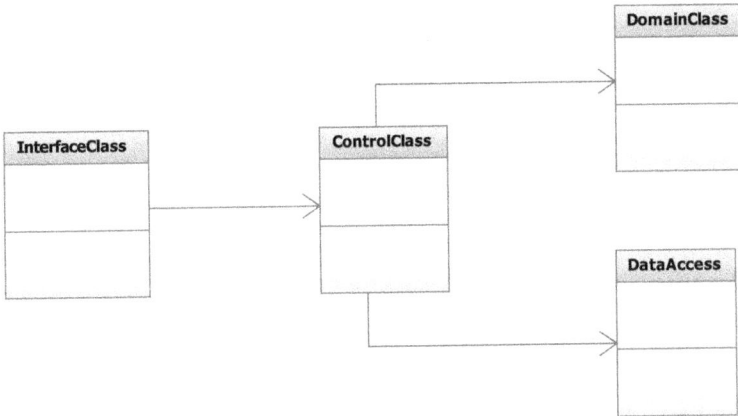

Figure 8. The VCM model for communication links.

Communication diagramming via examples: Enroll classes

Let us model the collaborations to perform the "enroll classes" use case. Assume Student is the primary actor, and Student uses a login form for authentication and a registration form (see Figure 9) to search for and sign up to courses. Assume the login form merely collects user credentials and passes account data to UserControl, a control class, for account validation. RegistrationForm displays a list of course offerings that meet the search criteria and whether each one is available for signing up and whether the student meets the prerequisites or has time conflicts. The Student class performs the tasks of checking prerequisites and time conflicts (see Chapter 14 for more details). Per the VCM model, we create a control class, called RegistrationManager, to act as a middleman between RegistrationForm and the domain classes. RegistrationForm delegates the job of searching for course offerings as well as registering classes to the control object, which will, in turn, notify Student, CourseOffering, and Enrollment objects to execute the actions, e.g., add a student to a course offering's roster, add a class to a student's transcripts, etc.

Figure 10 depicts an extended class diagram of the student registration system with non-domain classes, LoginForm, UserControl, RegistrationForm, and RegistrationManager included. Note that in the diagram ---> is a dependence relationship between classes, and here it means that RegistrationManager depends on

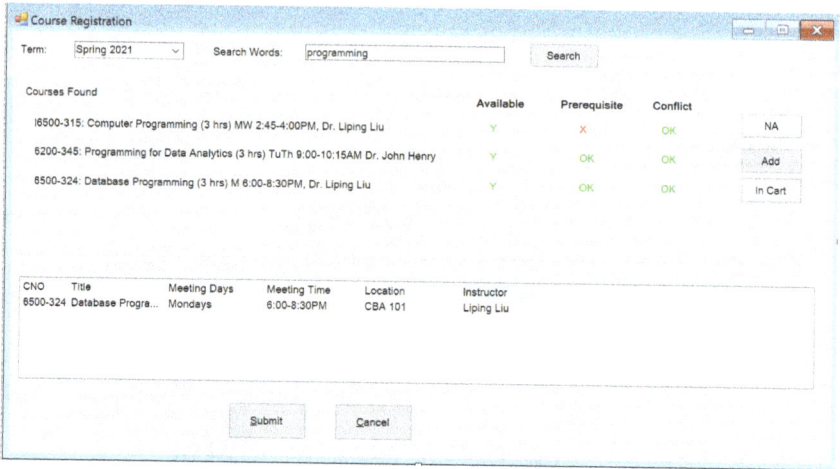

Figure 9.   Registration form.

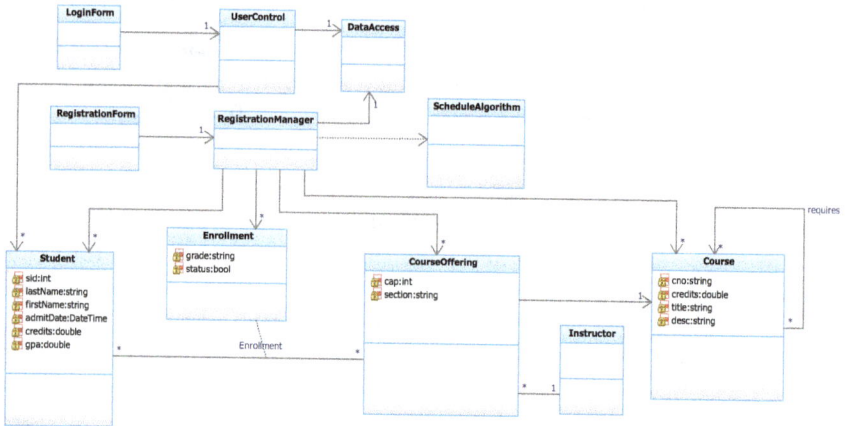

Figure 10.   An extended class diagram of a student registration system.

ScheduleAlgorithm; if the latter changes, then Registration Manager changes accordingly.

To create a communication diagram for a use case, we need to follow the sequence of events in the use case's description. Each step in the description is represented by one or more messages in the communication diagram. A request by an actor to the system or by the system to an actor is done through a user interface object, which merely displays information and/or collects user inputs but delegates the main functions such as obtaining or processing data to a control object. An activity performed by the system is done through collaboration between one or more domain, control, and/or data access objects.

Figure 11 shows a communication diagram for performing the "enroll classes" use case, whose basic flow is listed in the following. Here, UI1, which is LoginForm, is omitted, and UI2 (RegistrationForm) is shown in Figure 9.

Basic Flow:

1. Student enters user name and password (UI1)
2. The system verifies the account
   [do steps 3–14 for each class]
3. Student searches for a class (UI2)
4. The system checks for the availability of the class
5. The system requests for the prerequisites of the class
6. The system retrieves the student's finished courses
7. The system checks for satisfaction of prerequisites
8. The system requests active enrolled courses
9. The system checks for time conflicts
10. The system confirms no time conflicts
11. The system displays the search results (UI2)
12. Student requests to add a class to the shopping basket (UI2)
13. The system adds the class into the shopping basket
14. The student clicks Submit button (UI2)
15. The system creates new enrollment records
16. The system notifies the accounting system of updated credit hours
17. The system displays the confirmation message (PM1)

The messages are roughly grouped into three categories: messages 1–4 validate login, messages 5–13 search for course offerings, and the remaining messages register for a course offering.

First, to validate login, the user enters login credentials into the login form, which delegates the actual job to UserControl, which will in turn load all user data and send a message to each Student object to see if the credentials match.

Second, to search for course offerings, the user enters search criteria into RegistrationForm, which then delegates the actual query

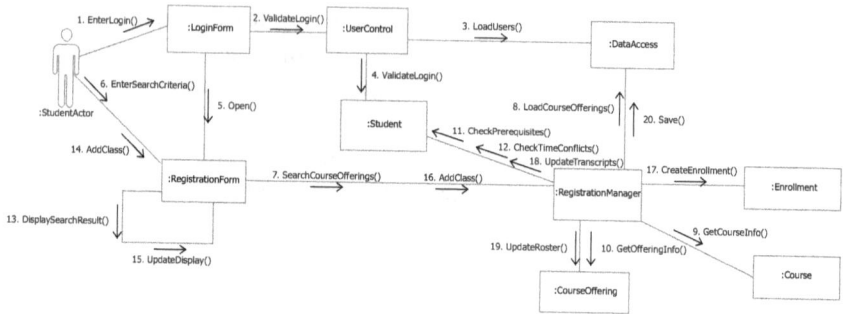

Figure 11.   Collaboration diagram for "Register Courses" use case.

to RegistrationManager. The control object will in turn ask the data access class to load and initiate Course and CourseOffering objects that meet the search criteria. It will ask each Course and CourseOffering object to get detailed information on each retrieved offering. It will also ask the Student object to check for prerequisites and possible time conflicts so that RegistrationForm can inform the user of feasible choices.

Finally, when the user selects a course to enroll in, Registration Form will first update its display to show the added course in the basket and then send a message to RegisgrationManager to actually perform the registration, which is translated into four messages: create a new enrollment record, add the record to the student's transcripts, add the record to the course offering's roster, and save the data to the database.

Sequence Diagrams

The basic elements of a sequence diagram include participants, lifelines, and messages (see Figure 12). The participants, including actors and constituent objects, are shown on the top of the diagram. Each participant has a corresponding lifeline running down the page. A lifeline simply indicates that the participant exists at that point in the sequence considering that an object can be created and/or deleted during a sequence.

Interactions are in the form of sending and/or receiving messages. Messages on a sequence diagram are specified using an arrow from the sender to the receiver. Messages can flow in whatever direction makes

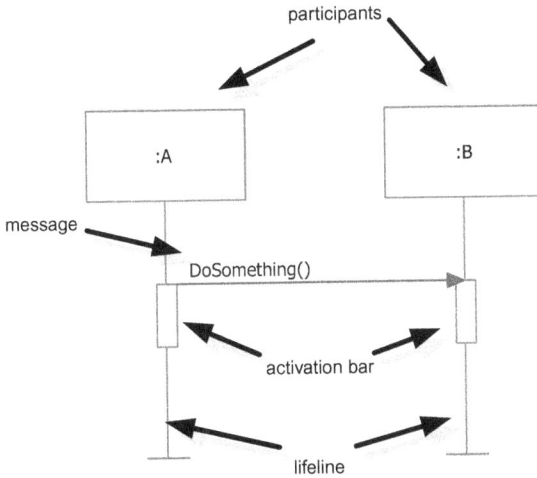

Figure 12.   Elements of a sequence diagram.

sense for the required interaction — from left to right, right to left, or even back to the sender itself.

When sending or receiving a message, an activation bar can be shown on the sending and receiving ends of a message (see Figure 12). An activation bar indicates that the sending participant is busy while it sends the message and that the receiving participant is busy while the message is received.

Messages are sent or received in chronological order from the top of the lifelines to the bottom. All the messages are labeled with a sequence number indicating the order of messages in the overall sequence. Note that the time on a sequence diagram is all about ordering, not duration.

Sequence diagramming in Rhapsody

To create a sequence diagram using Rhapsody, right click with the mouse on a package and select Add New → Diagrams → Sequence Diagram menu. A dialog will prompt for a diagram name and choice of Analysis or Design as the operation mode. In the design mode, messages will be automatically captured as operations and class diagrams are synchronized. In the analysis mode, the synchronization is not automatic, but the user can choose to capture a message as an operation later by right clicking on

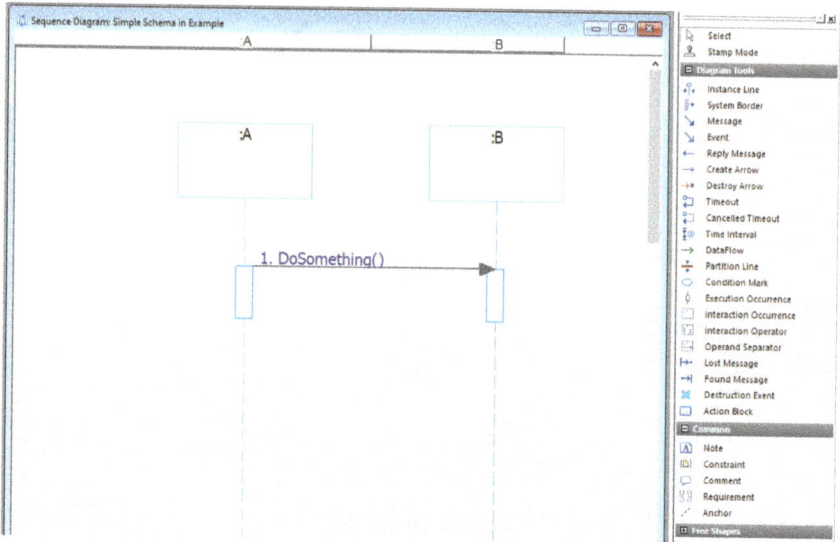

Figure 13.   Sequence diagramming tools in Rhapsody.

the message and selecting the Auto Realize menu item. After dismissing the dialog, a blank canvas along with the diagram tools will be displayed (see Figure 13).

To create participants and their lifelines, we can drag Instance Line from the diagram tools. Upon naming the participant, if the diagram is in the design mode and the name is not that of an existing object, a dialog will appear asking whether we want to create a class, of which the participant is an instance. We can also simply drag an actor, a class, or an object from the model browser to the canvas, and a lifeline will be created automatically. Note that if the participant is an association object, we will first need to create the object by right clicking on the association in the model browser and selecting the Make an Object menu item.

To create a message, select Message from the diagram toolbox, and first click on the sender's lifeline and then on the receiver's lifeline. We can add activation bars by right clicking on the message and selecting Add Execution Occurrences menu item.

By default, a diagram may not show the sequence numbers of the messages. We can turn on sequence numbers by resetting the diagram property: right click anywhere in the diagram, select Features menu, and go to the Properties tab. We can also set the Auto Create Execution Occurrences property in the same place.

Representing a use case story

Sequence diagramming starts with a use case story from which we first identify participants, including actors and user interfaces. We can use class diagrams to identify additional participants such as domain or control objects. The system may also include hardware components as participants. For example, an ATM consists of a card reader, a cash dispenser, a printer, a keypad, a screen, a modem, a cash holder, a paper holder, and a controller that coordinates other devices.

Then we follow the use case story to model each interaction between the user and the system as one or more messages between participants, from the user to an object, from one object to another object, or from an object to the user. The following are a list of examples of how interactions are translated into messages for the "withdraw cash" use case:

- "Card Holder inserts a card" will be modeled as a message from the card holder to the card reader.
- "ATM validates the card" will be modeled as a message that the card reader sends to itself to validate the card.
- "ATM asks for a pin" will be turned into a message on the screen to the card holder.
- "Card Holder enters a pin" will be a message from the card holder to the keypad.
- "ATM validates the pin" will be modeled as several messages from the controller to the modem and then to Card Network.
- "ATM asks for a withdrawal amount" will be a message on the screen to the user, and "Card Holder enters a withdrawal amount" will be a message from the user to the keypad.
- "ATM dispenses cash" is a message from the controller to the cash dispenser.
- "ATM prints a receipt" is a message from the controller to the printer.

In sum, if an interaction is that the user uses the system, it must be translated into a message that the user sends to a user interface of the system. If the system performs a function, the function must be performed by an object or a group of objects through collaboration, i.e., sending messages among the participating objects.

Note that sequence and communication diagrams are not expressive in describing alternate and exceptional flows or showing advanced optimization concepts such as extension points. If desirable, one may attach separate notes to the message calls.

Sequence diagramming via examples: Food order system

Figure 14 shows the use cases for the food order system of a typical sit-in restaurant. One of the most important use cases for the system is to take customer orders. By describing the use case, it will become clear that this use case requires collaboration among the following objects — user interfaces OrderScreen, LoginScreen, and KitchenOrder — along with domain classes Employee, Order, Table, Food, and OrderLine.

Figure 15 demonstrates the preliminary class diagram that shows how the objects are structurally related. Here, the OrderItem object is

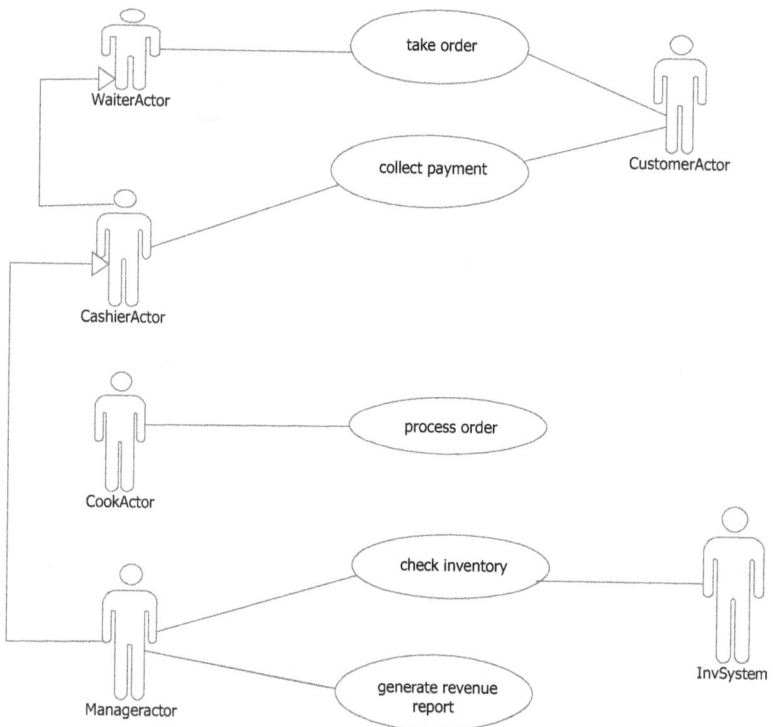

Figure 14.   Use case diagram for a food order system.

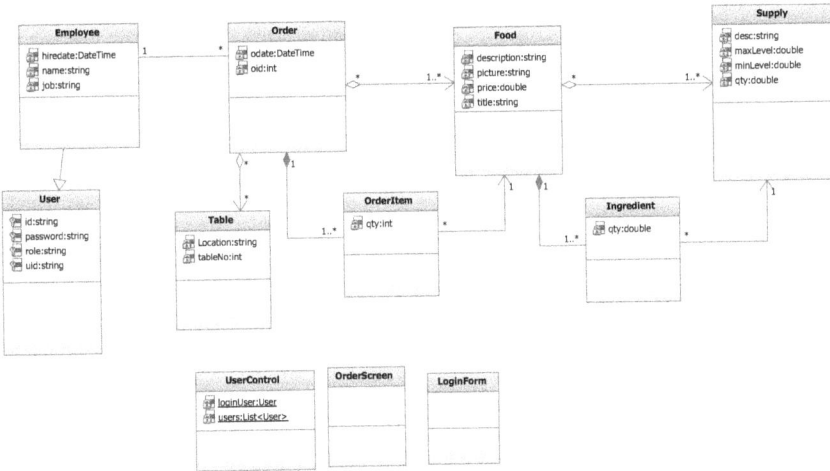

Figure 15.   The class diagram for a food order system.

used to represent the need that a customer may order the same food in multiple quantities and/or add a specification to a food. OrderItem is an association class, but it is represented here as an equivalent ordinary class. Each food is made of one or more supplies, and the Ingredient object models the quantity of each supply in making the food. Ingredient is an association object but equivalently represented as an ordinary object.

The order screen (see Figure 16) uses a combo box to show a list of available tables and a graphical panel or combo box to show all the foods and drinks. The basic flow of the "take order" use case is listed here. In the following, instead of showing an entire sequence diagram, we show the messages segment by segment.

Basic Flow:

1. Waiter swipes Employee ID
2. The system validates the ID
3. The system loads Food Order Screen (UI1)
   [Steps 4–9 repeat for each order item]
4. Waiter selects a table (UI1)
5. The system updates table status
6. Waiter selects a seat (UI1)
7. Waiter selects a food and adds cooking instructions (UI1)

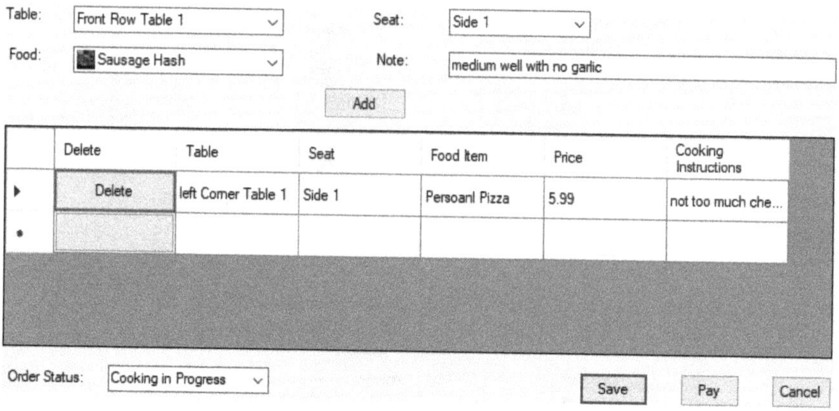

Figure 16.   Food order screen prototype.

8. Waiter presses Add button
9. The system displays the order items
10. Waiter presses Save button (UI1)
11. The system changes order status
12. The systems updates inventory
13. Waiter logs out
14. The system goes idle

Steps 1–2 deal with user authentication. There are many ways to do this. For example, in real life, you will probably send user account information to an authentication gateway to check if the account is valid. Here, a more object-oriented approach is taken, which is free from a relational database, and assumes that all data are held by objects. So, a control class, UserControl, is contemplated that holds a list of all user objects as coded in the following:

```
public class UserControl {
        private List<User> users;
        private User loginUser;
        //
}
```

To validate an account, UserControl will load data to initialize all the user objects, and then use a loop to ask each user object to see if the account information matches it. If it does match one of them (that is what the basic flow assumes anyway), then `UserControl` will set the login user object for the entire transaction session and ask the object to obtain user details such as employee name, job, and ID to be shown on the order screen. This segment of the sequence is shown in Figure 17, and the code is illustrated below.

```
public class LoginForm
{
        Public void Swipe()
        {
                //
        }
}
public class User
{
        private string uid, role;
        public bool VerifyUser(string id) { //}
        public string GetUserInfo() { //}

}
public class UserControl {
        private List<User> users;
        private User LoginUser;
```

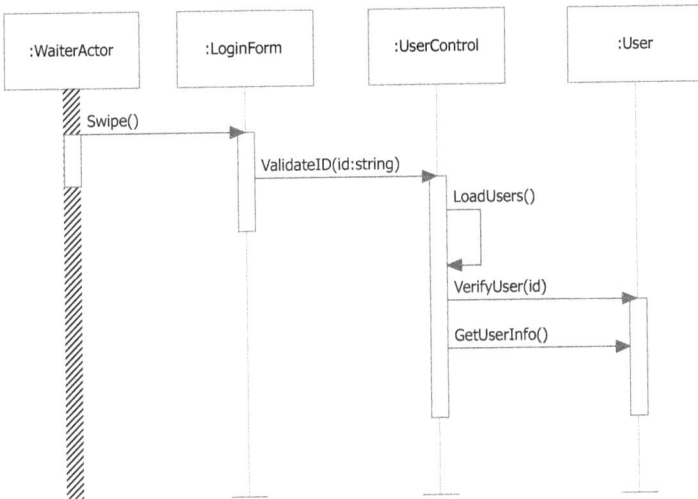

Figure 17.   Sequence diagram for user authentication.

```
public void loadUsers()
{
        users = new List<User>();
        //code to initialize users
}
public void ValidateUser(string id)
{
        loginUser = null;
        foreach (User u in users)
        {
                if (u.Verify(id))
                loginUser = u;
                break;
        }
    }
}
```

As we can see, a sequence diagram re-expresses a use case story as interactions between the user and the system, but the system does not appear in a sequence diagram as a participant. Rather it is expanded into and replaced with its constituent objects so that an interaction between the user and the system is replaced by one between the user and the objects that constitute the system. Operations performed by the system are transformed into one or more intra-system requests and responses.

Figure 18 shows the updated classes involved in the collaboration with the messages captured as operations. In order to help the reader understand the sequence diagram, the skeleton code for these classes is listed above.

Step 3 is to load the order screen if the user login is authenticated. This involves several collaborating objects to act in coordination: (1) `log-inFrom` is closed and `OrderScreen` is displayed, (2) a new order is

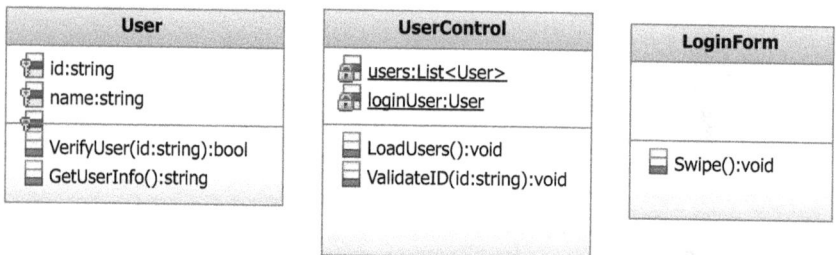

Figure 18.   Updated classes reflecting the sequence diagram in Figure 17.

created, (3) all tables, if available, are loaded into the combo box, (4) all food and drink items, if available, are loaded into an imaged combo box. The sequence diagram is shown in Figure 19.

Steps 4–5 deal with selecting a table for the order (see Figure 20). Of course, the system needs to check if the table is still available because the other waiters may have assigned it to another customer after the current order screen was loaded. If available, then the system needs to add the table to the current order and update the table status.

Steps 6–9 are mostly about adding order items, including seat selection, food selection, and cooking instruction; all are user actions performed using the order screen with exceptions to re-check for food availability and add an order item to the order object (see Figure 21).

Steps 10–12 are about saving the order. These include updating order status from Pending to Sent to Kitchen and updating the inventory of ingredients used. The latter must be done via the collaboration of many participating objects. First, each food item can get the quantity of each ingredient used to make the food from the Ingredient object. Using this quantity, a Food object can send a message to each Supply object to change the inventory quantity. Second, an Order object can get the quantity of each food ordered from each OrderItem object and can then

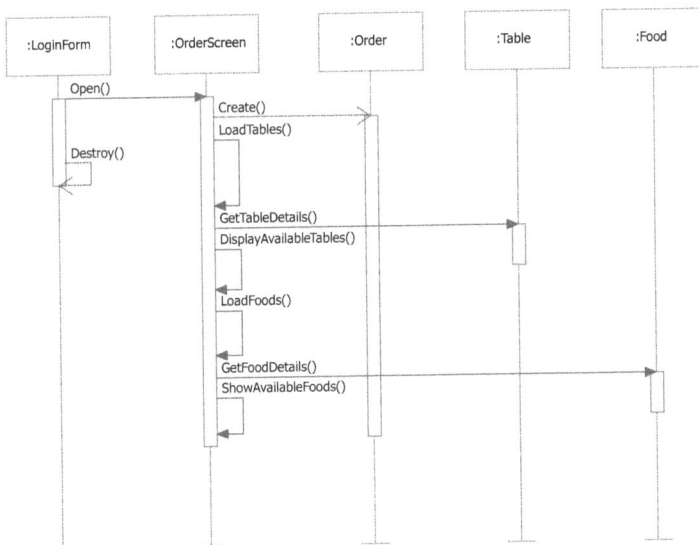

Figure 19. Sequence diagram for loading order screen.

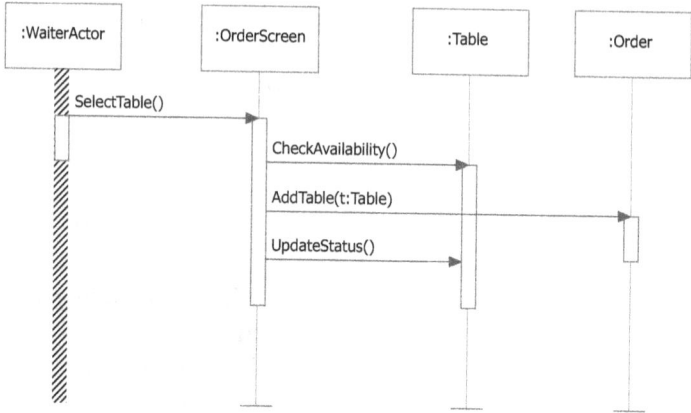

Figure 20.   Sequence diagram for adding table to food order.

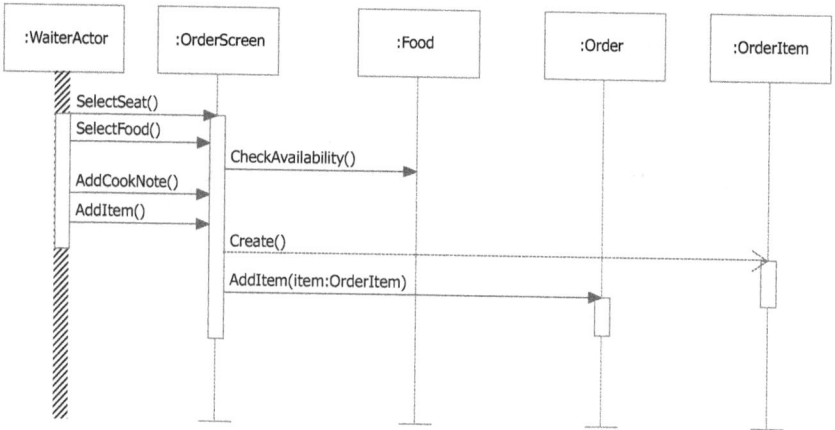

Figure 21.   Sequence diagram for adding food to order.

send this quantity to each `Food` object to update the inventory, which in turn calls each `Supply` object to update the inventory used. The following implementation shows the collaboration of the related objects as depicted in Figure 22:

```
Public class Supply {
        private double qty;
        public void UpdateInventory(double qtyChange)
        { qty += qtyChange; }
```

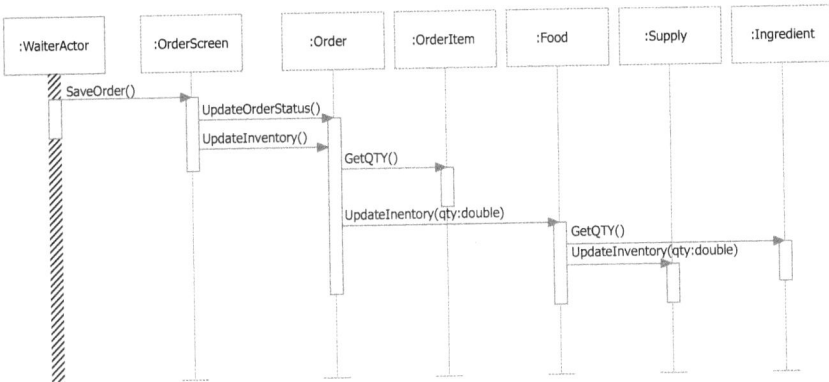

Figure 22. Sequence diagram for saving orders.

```
}
public class Ingredient {
        private Supply supply;
        private Food food;
        Private double qty;
        Public double GetQTY() { return qty;}
        Public Supply GetSupply() { return supply;}
}

public class Food {
        Private List<Supply> itemsUsed;
        Private Dictionary<Supply, Ingredient>
            ingredient;
        Public void UpdateInventory(double q) {
                foreach (Supply s in itemsUsed) {
                        double qty = ingredient[s].
                            GetQTY();
                        s.UpdateInventory(-q*qty);
                }
        }
}

public class OrderItem {
        private Order order;
        private Food food;
        Private double qty;
        Public double GetQTY() { return qty;}
}
```

```
public class Order {
        private List<Food> foods;
        private Dictionary<Food, OrderItem> orderLiness;
        public void UpdateInventory(){
                double q;
                foreach (Food f in foods) {
                        q = orderliness[f].GetQTY();
                        f.UpdateInventory(q);
                }
        }
}
```

The last step is that the user logs out of the system. This can be performed by the UserControl object with the operation Logout(), which can be implemented by one command:

```
public class UserControl {
private List<User> users;
private User LoginUser;
...

Public void Logout()
{ loginUser = null; }
}
```

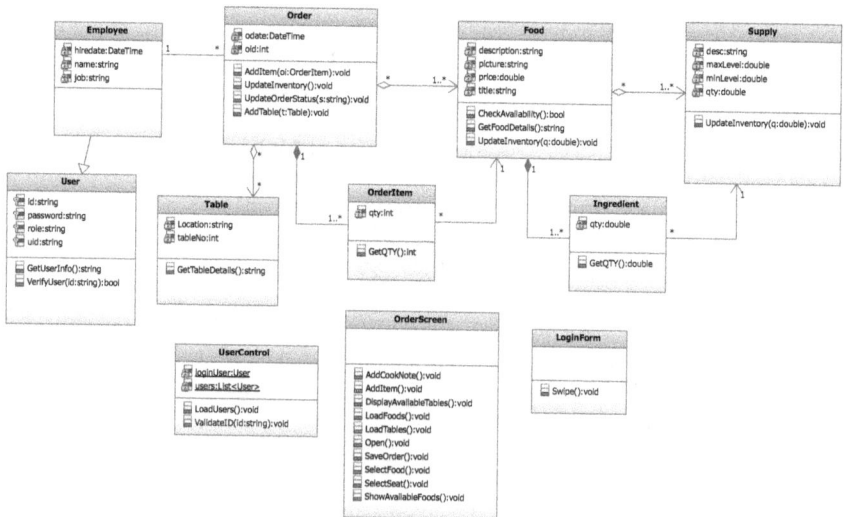

Figure 23. The expanded class diagram for a food ordering system.

Summarizing the interactions in all the sequence diagrams, the `Employee` object must be capable of authenticating a user. `Order` objects must be able to create a new order, assign a table to the order, add food item(s) to the order, update order status, and update inventory used for making the foods in the order. `Food` objects shall be able to get item information and update inventory for the ingredients used to cook the food. Finally, `OrderItem` shall be able to get the item quantity of each order. By capturing these operations into the class diagram in Figure 15, we obtain the expanded class diagram shown in Figure 23.

An afterthought

Collaboration modeling is indispensable in object-oriented development. Beginners in object modeling tend to find it easy to come up with a list of attributes for objects. Capturing functions or operations, on the other hand, is always a difficult task. However, thanks to use case storyboarding and collaboration modeling, the object-oriented method has made it possible for there to be no lack of operations to be captured. Everything comes together with ease because the skill for storyboarding seems to be easy to grasp, and the idea of collaboration is not hard to comprehend.

Like use case stories and activity diagrams, collaboration models are procedural ones. Each of these models has a niche in software development. A collaboration model may not be as expressive as other alternatives. For example, a collaboration model cannot express alternate and exceptional flows as well as a use case story can and cannot represent loop and decision controls as well as an activity diagram can. However, collaboration modeling offers benefits that use case storyboarding and activity diagramming cannot. A collaboration diagram visually shows the chronological or spatial order of interactions as a sequence of messages. It is more concise than a use case story. Collaboration modeling bridges objects with use cases; if the objects are designed in such a way that they are capable of performing interactions on behalf of the system, the use case can be realized. Thus, a collaboration diagram elaborates the responsibilities of the objects for realizing a use case. A collaboration model is useful for the model-driven software engineering approach to requirements specification, i.e., code generation from a model. For example, the messages are realized into the operations of the receiving object. Thus, one may trace back to a class diagram to find missing operations.

Exercises

1. Consider the ATM as a collection of collaborating objects: Card Reader, Cash Dispenser, Printer, Screen, Keyboard, etc. Then draw a sequence diagram such that all the services done by the ATM must be reallocated to the correct object.

2. Each of following is a subsequence of interactions involving a few objects. Please draw the class diagram of the involved objects and a sequence diagram showing the interaction. Then add the appropriate operation to the related classes.

   a. In the POS, the cashier scans an item, then the system retrieves and displays the item information on screen.

   b. In the student registration system, Student enters login information, and the system validates the account.

   c. In the inventory system, Manager checks the discrepancy box for an ordered item, and the system displays another popup screen to allow the manager to enter details on the discrepancy such as defects, incorrect quantity, etc.

   d. In the food order system, waiter selects a table from the list, and the system validates the availability status of the table.

3. In a library circulation system, two of the most important use cases are checkout and return. The business objects important to the system are borrowers, checkouts, returns, books, copies, etc.

   a. Create a class diagram to represent the business objects.

   b. Design graphical user interfaces for checkout and return.

   c. Create sequence diagrams for the following segments of use case descriptions:

      i. Return: The circulation desk employee scans a book, the system retrieves the rental record, the employee marks the return condition, the system updates the rental record, and the system update the inventory.

      ii. Checkout: The circulation desk employee scans the borrower ID, the system retrieves and displays the borrower record, the system checks overdue records, the system displays checkout screen, the employee scans a book copy, the system retrieves the book copy record, the employee confirms the book copy condition, the employee verifies the due date with the borrower, the employee affirms checkout, the system updates the inventory, the system creates a rental record, and the system prints a due date notice.

4. (Student Registration System): For the use case "schedule course offerings", write a use case story and then create a sequence diagram. Finally, expand the relevant classes by including the required functions shown in the sequence diagram.

5. (Restaurant): Taking food orders is the most important use case that involves many computer screens such as Food Order Form and business objects Food, Customer, Order, etc. List and define these screen and business objects and then create a sequence diagram that shows their collaboration in performing the use case. Finally, expand the relevant classes by including the captured functions shown in the sequence diagram.

6. (Point of Sale System): Checking out is a use case for any point of sale system for retail stores. From its story, we know that many objects will be involved in performing each use case such as checkout screen, payment screen, receipt, etc., and business objects such as Item, Transaction, Employee, etc. Create a list of these screens and business objects and then create a sequence diagram to show the collaboration among them to achieve this use case. Finally, expand the relevant classes by including the captured functions shown in the sequence diagram.

7. (Inventory System): Receiving new shipments is a typical use case in the inventory management system. It will update inventory, notify the vendor of discrepancies, update order status, and record unshipped items, etc.

    a. Create a class diagram to model the business objects involved.
    b. Create flow of events for this use case.
    c. List participating screen, report, and business objects that will perform the use case.
    d. Create a sequence diagram to show the collaboration in performing the use case.
    e. Expand the relevant classes by including the captured functions shown in the sequence diagram.

Chapter 16

A Complete Use Case Implementation

Introduction

This chapter will put class diagrams, use case diagrams, use case story-boarding, and collaboration modeling and programming together into a workable mini system. In order to limit the project size and at the same time illustrate all aspects of modeling and coding, we will continue the registration system project we developed in prior chapters, in particular, Chapter 7 on coding association classes and Chapters 14 and 15 on collaboration modeling, where we have developed a class diagram and its implementation with many useful functions. In this chapter, we will base on the class diagram and the code and create additional artifacts to realize a simple use case, "Post Grade", performed by professors. This chapter involves a lot of nitty-gritty details, and so the reader is advised to read and follow it as a hands-on project.

Use Case and Storyboarding

The following (Figure 1) is a simple use diagram that shows the function-alities that we will develop. It assumes that we factor out the interactions for user login into a separate inclusion use case.

The use cases are described as follows using the template provided in Chapter 10 as follows:

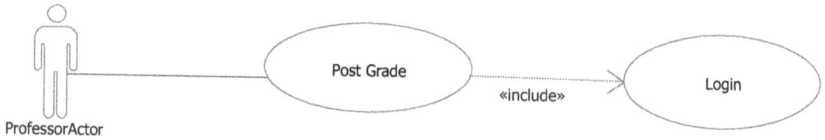

Figure 1.   "Post Grade" use case.

Use Case: Login
Type: Inclusion use case

Flow of Events:

Basic Flow:
1.   The system display login screen (UI1)
2.   User enters username, password, and role (UI1)
3.   The system validates login

Alternate Flows:
3a. incorrect login:
.1 The system counts the number of errors
.2 The system checks login error limit
.3 Go to step 1

Exceptional Flows:
3a.2a. too many login errors:
.1 The system displays "too many login errors" message (PM1)
.2 The system freezes for 10 minutes

Use Case: Post Grade
Type: Base use case

Flow of Events:

Basic Flow:
1. Include (login)
2. The system displays Post Grade Form (UI2)
3. The system retrieves all classes
4. The system displays the classes in Post Grade Form (UI2)
5. The professor selects a class (UI2)
6. The system displays the grade sheet of the selected class (UI2)
7. The professor enters grades (UI2)
8. The professor saves the grades (UI2)
9. The system saves the grades
10. The system displays "Grade Saved" confirmation
11. The professor logs out

Alternate Flows:

9a. invalid grade:

.1 The system displays "Invalid Grade" message (PM2)

.2 Go to 7

Exception Flows:

3a. no class retrieved:

.1 The system displays "no class available" in class list (UI2)

.2 The system disables Load button (UI2)

.3 The professor logs out

Prompts and Messages:

1. Too many logon errors: You have tried too many times. Your account is locked, please contact us at 800-908-1234 for assistance.

2. Invalid grade: An invalid grade is entered.

User Interfaces:

1. Login Screen

User Name:	rsmith
Password:	********

Login   Cancel

2. Post Grade Form

Section: 6500-324-001: Database Programming (M 6:00-8:30PM)

Student ID	Student Name	Grade	Comment
1234567	John Smith	A	
1023876	John Doe	A-	
9876543	Lisa Johnson	B+	
2222222	Bill Henry	B	
3333333	Laura Smith	B-	

Submit   Reset

Collaboration Modeling

The mini registration system consists of domain classes, user interfaces, control classes, and data access classes. We design one data access class, `FileAccess`, to be responsible for retrieving data from a data source such as data files and writing data to the sources. Domain classes such as `Student, Instructor,` and `CourseOffering` are data holders and processors. The "Post Grade" use case is mostly about retrieving and saving data. It does not have much need for domain objects to process data, but it heavily relies on object persistence for initializing and saving the domain objects.

User interfaces present data to the user and take in user inputs. Two graphical user interfaces involved in the "Post Grade" use case are `LoginForm` and `PostGradeForm`, as shown in the use case story above.

Control classes are responsible for coordinating tasks among multiple domain objects and data access objects. Typically, each use case requires at least one control class. We create two control classes: `UserControl` class for loading, finding, and verifying users; and `InstructorControl` class to load and manage all course offerings (or sections) taught by an instructor.

In collaboration modeling, we usually ignore operations related to object persistence. Thus, we will just model collaboration among user interface objects and control objects for the "Post Grade" use case. The sequence diagram for the use case is shown in Figure 2.

From this sequence diagram, we captured eight functions to be required by the two control classes and two user interface classes as shown the class diagram in Figure 3. These classes will be explained in detail and implemented in later sections of this chapter. The two domain classes involved in the collaboration acquired one simple operation each, which will be implemented in the following section.

Domain Classes

Since we need data on instructors and the course offerings (or sections) they teach, the class diagram completed by Chapter 14 is slightly extended to include the `Instructor` class. Also, we want both students and instructors to login to the system, and so we generalize `Instructor` and `Student` into `User` class with common attributes `userid, password,`

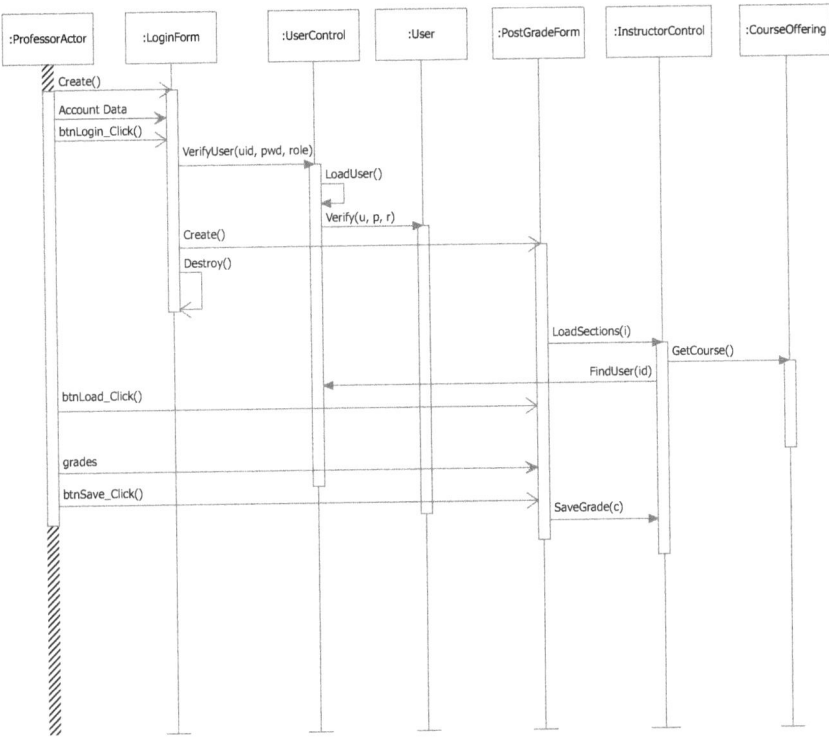

Figure 2.   The sequence diagram for "Post Grade" use case.

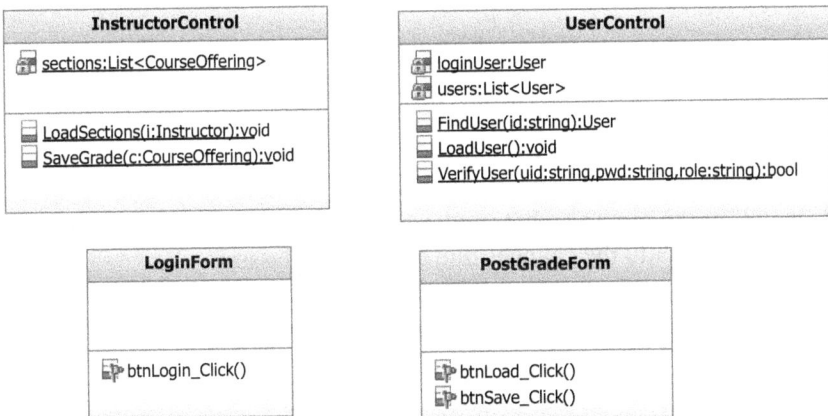

Figure 3.   User interfaces and control classes for the "Post Grade" use case.

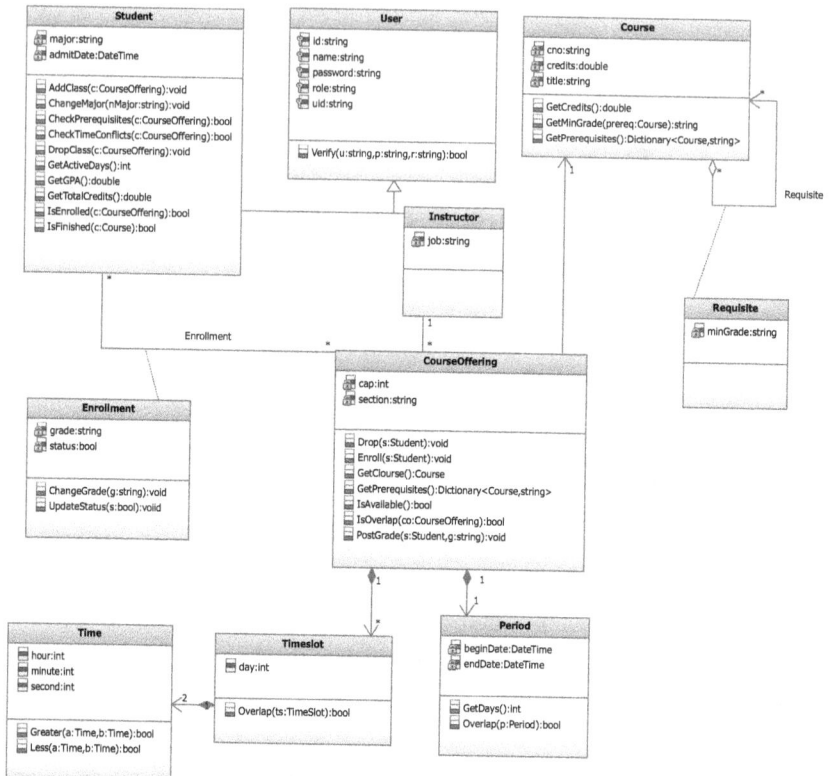

Figure 4.   Domain classes.

role, id, and name. Here, id is either student ID or employee ID issued by a university. Also, we now allow different minimum scores for different prerequisites, and so we use the association class Requisite to capture the new requirement. The extended class diagram is shown in Figure 4.

Except for the new classes and resulting changes in functions, all the code will be the same as what we developed in Chapters 7 and 14. The following shows only the changes due to the new classes.

User Class

```
public class User
{
        protected string u;
        protected string p;
        protected string r;
```

```csharp
protected string id;
protected string name;

public string ID
{
    get {return id;}
    set {id = value;}
}

public string Name
{
    get {return name;}
    set {name = value;}
}

public string Role
{
    get {return id;}
    set {id = value;}
}

public bool Verify(string uid, string
pwd, string role)
{
    if (uid == u && p == pwd && r ==
    role)
        return true;
     else
        return false;
}

public User()
{
    u = "";
    p = "";
    r = "";
    id = "";
    name = "";
}

public User(string userid, string
password,
      string workid, string workrole,
      string fullName)
{
    u = userid;
```

```
                        p = password;
                        id = workid;
                        r = workrole;
                        name = fullName;
                }
        }
```

Requisite Class

```
public class Requisite
{
    private string minGrade;
    private Course crs;
    private Course prerequisite;

    public string GetMinimumGrade()
    {
        return minGrade;
    }

    public Requisite()
    {
        crs = new Course();
        prerequisite = new Course();
        minGrade = "C";
    }

    public Requisite(string minG, Course c, Course
        preq)
    {
        crs = c;
        prerequisite = prerequisite;
        minGrade = minG;
    }
}
```

The following shows the classes and operations that have been changed from Chapters 7 and 14. Note that the snippet only shows the attributes and operations that have been changed.

Course Class

```
public class Course
{
```

```
private string cno;
private string title;
private double credits;
private Dictionary<Course, string>
   prerequisites;

public string CNO
{
    get {return cno;}
    set {cno = value;}
}

public string Title
{
    get {return title;}
    set {title = value;}
}

public double Credits
{
    get {return credits;}
    set {credits = value;}
}

public Course()
{
    //
}
public Course(string cnumber)
{
    cno = cnumber;
    title = "";
    credits = 0;
}

public Dictionary<Course, string>
   GetPrerequisites()
{
    return prerequisites;
}

public string GetMinGrade(Course prereq)
{
    return prerequisites[prereq];
}
}
```

CourseOffering Class

```
public class CourseOffering
{
    private string section;
    private Course courseUnder;
    private int cap;
    private Dictionary<string, Student> roster;
    private Dictionary<Student, Enrollment>
        gradeBook;
    private TimeSlot[] timeslots;
    private Period period;

    public int Cap
    {
        get {return cap;}
        set {cap = value;}
    }

    public string Section
    {
        get {return section;}
        set {section = value;}
    }

    public Course CourseUnder
    {
        get {return courseUnder;}
        set {courseUnder = value;}
    }

    public Dictionary<string, Student> Roster
    {
        get {return roster;}
        set {roster = value;}
    }

    public Dictionary<Student, Enrollment>
        Gradebook
    {
        get {return gradeBook;}
        set {gradeBook = value;}
    }
```

```
public CourseOffering()
{
    //
}

public CourseOffering(string cnumber, string
    snumber)
{
    courseUnder = new Course(cnumber);
    section = snumber;
    cap = 0;
}

public Dictionary<Course,string>
    GetPrerequisites()
{
    return courseUnder.GetPrerequisites();
}

public bool IsAvailable()
{
    return (cap > roster.Count);
}
}
```

Instructor Class

```
public class Instructor : User
{
    private string job;
    private List<CourseOffering> sections;

    public string Job
    {
        get {return job;}
        set {job = value;}
    }

    public List<CourseOffering> Sections
    {
        get {return sections;}
        set {sections = value;}
    }
}
```

Student Class

```
public class Student:User
{
    private string major;
    private DateTime admitDate;
    private List<CourseOffering> classes;
    private Dictionary<CourseOffering,
        Enrollment> transcript;

    public Student():base()
    {
        major = "";
        admitDate = DateTime.Now;
    }

    public Student(string id)
    {
        base.id = id;
    }

    public double GetTotalCredits()
    {
        double result = 0;
        foreach (CourseOffering co in classes)
        {
            if (RegisterTool.
                GetPoint(transcript[co].Grade) >= 1)
                    result += co.GetCourse().
                        Credits;
        }
        return result;
    }

    public bool CheckPrerequisite(CourseOffering co)
    {
        bool result = true;
        foreach (KeyValuePair<Course,string> p in
            co.GetPrerequisites())
        {
            if (IsFinished(p.Key) == false ||
                RegisterTool.
                GetPoint(transcript[co].Grade) <
                RegisterTool.GetPoint(p.Value))
            {
                result = false;
```

```
                    break;
                }
            }
            return result;
        }
    }
```

Note that `RegisterTool` is a utility class with one static function "`int GetPoint(string  grade)`", which converts a letter grade into numerical points as the one in Enrollment class in Chapter 14.

Note also that because of new minimum score requirements for pre-requisites, the function `CheckPrerequisite()` will now not only check if a student has finished a course but also if the student has earned a minimum score as specified by a `Requisite` object (see the boldfaced code in the function).

Object Persistence

Objects are created, modified, and destroyed in transient memory, and their data will be lost after the system is closed. Therefore, there is a need to save object data, or persist domain objects, for future runs of the system. There are a few frameworks such as Entity Framework and Hibernate that are widely used for object persistence using relational databases. This book will not explore these frameworks due to scope limitation. In this chapter, we assume the reader is not familiar with database concepts, models, and languages, and so we choose to use a plain data file to save object data. For readers who are familiar with XML, they may slightly modify the `FileAccess` class to persist objects using XML files. Both techniques for object persistence are simple but help to understand the issues of object persistence.

To make domain objects more reusable and maintainable, we follow a persistence strategy that makes domain objects free from the concerns of how and where their data are saved. In other words, domain objects are made such that they do not know objects that do inputs and outputs with files or databases. This strategy may be achieved in many ways. For example, the entity framework uses the technique of partial classes to separate each domain class into two or more partial classes: one holds domain data and the other handles inputs and outputs. The strategy here is to allow domain objects to pass out and take in a generic data structure such as a table row, or an XML node, or a comma-separated text, and then, based on the data structure, create two generic functions `Serialize()` and `Deserialize()` in each domain class, where `Serialize()` packs object data into the data structure, and `Deserialize()` extracts values from the data structure to recover objects.

In this chapter, I will use comma-separated strings to pass in and out object data. In the following, I will show each domain class's `Serialize()` and `Deserialize()` functions along with the corresponding data files.

Since both `Instructor` and `Student` objects are kinds of `User` objects, we will just create the "users.txt" file as follows to persist both types of objects:

```
Student, lisa, hello, 009911, LisaJohnson, Business,
    12/2/2015
Instructor,  liu,  hello,  120911,  Liping  Liu,
    Professor
...
```

In the `User` class, we have five attributes common to both `Student` and `Instructor` objects. So, its `Serialize()` function will use the common data to create strings made of the first five values of each row in the "users.txt" file. We make `Serialize()` virtual so that each child class can modify it or, in this case, add additional attribute values specific to each child. Since each row in the "users.txt" file can create either one `Student` or one `Instructor` object and each user is either a student or an instructor, there is no need for a function to create separate `User` objects. However, we want both `Instructor` and `Student` to have the same contract for deserialization to take advantage of polymorphism. Therefore, we create an abstract `Deserialize()` function in the `User` class.

```
public abstract class User
{
    protected string u;
    protected string p;
    protected string r;
    protected string id;
    protected string name;
    public abstract void Deserialize(string data);
    public virtual string Serialize()
    {
        return Role + "," + u + "," + p + "," + id +
            "," + name;
    }
}
```

Now we override these two functions inside Student and Instructor classes. For Serialize(), we merely attach the additional attributes to the result of the Serialize() function in the parent class using base.Serialize(). For Deserialize(string data), we split the comma-separated parameter "data" into individual values and use the values to initialize object attributes.

```
public class Student:User
{
    private string major;
    private DateTime admitDate;
    private List<CourseOffering> classes;
    private Dictionary<CourseOffering, Enrollment>
        transcript;

    public override void Deserialize(string data)
    {
        string[] values = data.Split(',');
        base.r = values[0].Trim();
        base.u = values[1].Trim();
        base.p = values[2].Trim();
        base.id = values[3].Trim();
        base.name = values[4].Trim();
        major = values[5].Trim();
        admitDate = Convert.ToDateTime(values[6].
            Trim());
    }

    public override string Serialize()
    {
        return base.Serialize() + "," + name + ","
            + admitDate.ToShortDateString();
    }
}
```

We can serialize and deserialize Instructor object in the same way as Student objects. Since there is no separate association class handling the persistence of the relationship data between Instructor and CourseOffering, we create two more functions, SerializeTeaching() and DeserializeTeaching(string[] data), to persist the relationship data. Note that each instructor may teach multiple sections, so the DeserializeTeaching() function will need a list of comma-separated strings, one from each row of the "teaching.txt" file as follows:

```
325, 001, 120911
324, 801, 120911
...
```

The updated code for Instructor class is as follows:

```
public class Instructor:User
{
    private string job;
    private List<CourseOffering> sections;

    public override void Deserialize(string data)
    {
        string[] values = data.Split(',');
        base.r = values[0].Trim();
        base.u = values[1].Trim();
        base.p = values[2].Trim();
        base.id = values[3].Trim();
        base.name = values[4].Trim();
        job = values[5].Trim();
    }

    public override string Serialize()
    {
        return base.Serialize() + "," + job;
    }

    public string[] SerializeTeaching()
    {
        int count = sections.Count;
        string[] results = new string[count];
        for (int i=0; i<= count-1; i++)
        {
            results[i] = sections[i].CourseUnder.CNO
                + "," + sections[i].Section + "," +
                    base.id;
        }
        return results;
    }

    public void DeserializeTeaching(string[] data)
    {
```

```
    sections = new List<CourseOffering>();
    string[] values;
    CourseOffering co;
    foreach (string s in data)
    {
        values = s.Split(',');
        co = new CourseOffering(values[0].Trim(),
            values[1].Trim());
        sections.Add(co);
    }
}
}
```

The `Serialize()` and `Deserialize()` functions for other objects are straightforward, and so I will simply list them below without further elaboration:

Course Class

Data File "courses.txt":

```
325, Systems Analysis and Design, 3
324, Database Management, 3
  ...
```

Function Code:

```
public class Course
{
    private string cno;
    private string title;
    private double credits;
    //private List<Course> prerequisites;
    private Dictionary<Course, string> prerequisites;

    public string Serialize()
    {
        return cno + "," + title + "," + credits;
    }
    public void Deserialize(string data)
    {
        string[] values = data.Split(',');
        cno = values[0].Trim();
        title = values[1].Trim();
        credits = Convert.ToDouble(values[2].Trim());
    }
}
```

CourseOffering Class

Data File "sections.txt":
```
325,  001,  30
324,  801,  30
325,  801,  30
   ...
```

Function Code:

```
public class CourseOffering
{
    private string section;
    private Course courseUnder;
    private int cap;
    private Dictionary<string, Student> roster;
    private Dictionary<Student, Enrollment> gradeBook;
    private TimeSlot[] timeslots;
    private Period period;

    public string Serialize()
    {
        return courseUnder.CNO + "," + section + ","
        + cap;
    }

    public void Deserialize(string data)
    {
        string[] values = data.Split(',');
        courseUnder = new Course(values[0].Trim());
        section = values[1].Trim();
        cap = Convert.ToInt32(values[2].Trim());
    }
}
```

Enrollment Class

Data File "enrollment.txt":
```
325,001,009911,True,A
325,001,101911,False,N
   ...
```

Function Code:

```
public class Enrollment
{
    private string grade;
    private bool status;
    private Student student;
    private CourseOffering courseOffering;

    public string Serialize()
    {
        return courseOffering.CourseUnder.CNO + "," +
            courseOffering.Section + "," +
            student.ID + "," +
            status + "," + grade;
    }

    public void Deserialize(string data)
    {
        string[] values = data.Split(',');
        courseOffering = new
            CourseOffering(values[0].Trim(),
            values[1].Trim());
        student = new Student();
        student.ID = values[2].Trim();
        status = Convert.ToBoolean(values[3].
            Trim());
        grade = values[4].Trim();

    }
}
```

Requisite Class

Data File "prerequisites.txt":

```
325, 324, C
643, 641, B
...
```

Function Code:

```
public class Requisite
{
    private string minGrade;
    private Course crs;
```

```
        private Course prerequisite;

        public string Serialize()
        {
            return crs.CNO + "," + prerequisite.CNO +
                "," + minGrade;
        }

        public void Deserialize(string data)
        {
            string[] values = data.Split(',');
            crs = new Course(values[0].Trim());
            prerequisite = new Course(values[1].Trim());
            minGrade = values[2].Trim();
        }
    }
```

Data Access and Control Objects

Note that we show data files to help understand `Serialize()` and `Deserialize()` functions. In fact, all domain classes are free from how and where data are saved. For example, we may store all data in a relational database or XML file if we have objects that can format data from those sources into a comma-separated text value or save a comma-separated text value into those sources. Therefore, all our domain classes are still reusable after adding functions dealing with object persistence.

Then what objects do we need to connect data sources and domain objects? Here, we design two different objects to fill in the void.

The first is data access objects, which are responsible for reading data from and writing data to data sources. In our case, since we are dealing with plain text files, we will create `FileAccess` class and utilize `StreamReader` and `StreamWriter` objects inside the `System.IO` package to read and write data. The basic functionalities for `FileAccess` objects are to read data from a specific file, return the data as an array of comma-separated strings, and write an array of strings into a specific file. We can add other more advanced read functions, such as the following, to read a subset of lines from a text file by using a specific column value:

```
    public static string[] ReadData(string fileName, int
        filterColumn, string filterValue)
```

This function can be further extended to include a list of columns, a list of filter values, and a list of comparison operations so that we can filter the text lines by using multiple criteria such as, for example, the first column value is equal to a certain filter value and the third column value is less than a certain filter value. We may also add advanced write functions, for example, ChangeLine(), to substitute one line of text with another.

```
using System.IO;
namespace Registration
{
    public class FileAccess
    {
        public static string[] ReadData(string fileName)
        {
            List<string> results = new List<string>();
            StreamReader myReader =
                File.OpenText(fileName);
            string line = myReader.ReadLine();
            while (line != null)
            {
                results.Add(line);
                line = myReader.ReadLine();
            }
            myReader.Close();
            return results.ToArray() ;
        }

        public static void WriteData(string fileName,
            string[] data)
        {
            StreamWriter myWriter = new
                StreamWriter(fileName);
            foreach (string s in data)
            {
                myWriter.WriteLine(s);
            }
            myWriter.Close();
        }

        public static string[] ReadData(string fileName,
            int filterColumn, string filterValue)
        {
            List<string> results = new List<string>();
            StreamReader myReader = File.
                OpenText(fileName);
```

```
        string line = myReader.ReadLine();
        string[] values;
        while (line != null)
        {
            values = line.Split(',');
            if (values[filterColumn].Trim() ==
                filterValue)
                  results.Add(line);
            line = myReader.ReadLine();
        }
        myReader.Close();
        return results.ToArray();
    }

    public static void ChangeLine(string fileName,
        string oldLineText, string newLineText)
    {
        string[] data = ReadData(fileName);
        for (int i = 0; i< data.Length; i++)
            if (data[i] == oldLineText)
                  data[i] = newLineText;
        WriteData(fileName, data);
    }

    public static void ChangeLine(string fileName,
        int lineNumber, string newLineText)
    {
        string[] data = ReadData(fileName);
        for (int i = 0; i < data.Length; i++)
            if (i == lineNumber-1)
                  data[i] = newLineText;
        WriteData(fileName, data);
    }
  }
}
```

The second is control objects, which in general play roles of coordinating and sequencing operations of multiple objects and, by the step-and-step model (see Chapter 15), have access to both data access objects and domain objects. Thus, controls objects can be rightly used to persist object data and initialize objects for the system.

To perform the "Post Grade" use case, professors first need to login. To do so, we need to get a list of users, including all students and instructors. For this purpose, it makes sense to create a UserControl class that can load all users, find users, and verify users. Remember that users are either students or instructors. So, after we load the "users.txt" file, we can use each row to initialize either a Student object or an Instructor object, depending on whether the first value, i.e., role, value, in the line. Actual initialization is done by using the Deserialize() function. Besides initializing User objects, we want the UserControl object to verify a user login and keep the login User object for future reference. So, the UserControl class has two static data members: a list of users and a logon user. Since the "Post Grade" use case needs to access Student objects later, instead of recreating those student objects, we want UserControl objects to maintain the user list and, when necessary, find a user object using the id attribute.

```
public class UserControl
{
    private static List<User> users;
    private static User loginUser;

    public static User LoginUser
    {
        get { return loginUser; }
        set { loginUser = value; }
    }

    public static bool VerifyUser(string uid, string
        pwd, string role)
    {
        bool result = false;
        foreach (User u in users)
        {
            if (u.Verify(uid, pwd, role))
            {
                loginUser = u;
                result = true;
                break;
            }
        }
        return result;
    }

    public static void LoadUsers()
    {
        users = new List<User>();
```

```
            string[] values;
            User user;
            string[] userData = FileAccess.
               ReadData("users.txt");
            foreach (string s in userData)
            {
                values = s.Split(',');
                if (values[0].Trim()=="Student")
                {
                    user = new Student();
                    user.Deserialize(s);
                    users.Add(user);
                }
                else if (values[0].Trim()=="Instructor")
                {
                    user = new Instructor();
                    user.Deserialize(s);
                    users.Add(user);
                }
            }
        }

        public static User FindUser(string id)
        {
            User user = null;

            foreach (User u in users)
            {
                if (u.ID == id)
                {
                    return u;
                }
            }
            return user;
        }
    }
```

To perform the "Post Grade" use case, a professor needs to have access to all his or her sections, each of which includes a list of students and a grade book. Therefore, we create an InstructorControl class to load or initialize all those objects that belong to the professor. InstructorControl will maintain all the CourseOffering objects that belong to the login user.

The `LoadInstructor()` function will first load the "teaching.txt" file to find all the sections that belong to the instructor and call the `DeserializeTeaching()` function to create the list of sections for the `Instructor` object.

```
string[] values;
string[] data = FileAccess.ReadData("teachings.txt",
    2, theInstructor.ID);
theInstructor.DeserializeTeaching(data);
```

Note that all `CourseOffering` objects in the list have not been initialized with actual data at this point. We need to continue to use the "sections.txt" file to initialize each section object. For each `CourseOffering` object `co` in the sections list, we load its data by searching for the line in the "sections.txt" file and use the line to deserialize `co`:

```
data = FileAccess.ReadData("sections.txt");
foreach (string s in data)
{
    values = s.Split(',');
    if (values[0].Trim() == co.CourseUnder.CNO &&
        values[1].Trim() == co.Section)
        co.Deserialize(s);
}
```

In order to access course titles and credits, we also need to use the "courses.txt" file to initialize each associated course object inside a `CourseOffering` object:

```
data = FileAccess.ReadData("courses.txt");
foreach (string s in data)
{
    values = s.Split(',');
    if (values[0].Trim() == co.CourseUnder.CNO)
        co.CourseUnder.Deserialize(s);
}
```

Note that since we do not have functions to read one line at a time, we will have to read the entire file and then use a loop to find the right line from the "sections.txt" and "courses.txt" files. If we were going to get data from a relational database, the search may be unnecessary.

Finally, we need to initialize a roster, a list of students in each section, a grade book, and a list of enrollment records. Both lists are dictionaries, and we will load the "enrollment.txt" file.

```
co.Roster = new Dictionary<string, Student>();
co.Gradebook = new Dictionary<Student, Enrollment>();
data = FileAccess.ReadData("enrollment.txt");
```

Then, using and splitting each line of the "enrollment.txt" file, we can find a student ID, a course number, and a section number. Using the student ID, we can ask UserControl class to find the corresponding Student object in its users list and add the object to the roster list:

```
Student stu;
if (values[0].Trim() == co.CourseUnder.CNO &&
        values[1].Trim() == co.Section)
{
    stu = (Student) UserControl.FindUser(values[2].
      Trim());
    co.Roster.Add(values[2].Trim(), stu);
}
```

Using the rest of the line, we can call the Deserialize() function to initialize an Enrollment object and add it to the gradebook list.

```
Enrollment e;
Student stu;
if (values[0].Trim() == co.CourseUnder.CNO &&
        values[1].Trim() == co.Section)
{
    stu = (Student) UserControl.FindUser(values[2].
      Trim());
    co.Roster.Add(values[2].Trim(), stu);
    e = new Enrollment(stu, co);
    e.Deserialize(s);
    co.Gradebook.Add(stu, e);
}
```

The other function is to save Enrollment object data to the "enrollment.txt" file. When a professor changes grades for a section, Enrollment objects will be changed. Since we cannot write a specific row into a file, we will have to load all the current "enrollment.txt" files into a list of strings; match each row using course number, section number, and student id; and substitute a matching row by the corresponding object data. After finishing the substitution of all matching rows, we write the list of strings back to the "enrollment.txt" file. The complete Instructor class is reproduced in what follows:

```
public class InstructorControl
{
    private static List<CourseOffering> sections;
    public static List<CourseOffering> Sections
    {
            get { return sections; }
            set { sections = value; }
    }

    public void SaveGrade(CourseOffering section)
    {
        string[] data = FileAccess.
           ReadData("enrollment.txt");
        string[] values;

        for (int i = 0; i < data.Length; i++)
        {
             values = data[i].Split(',');
             foreach (KeyValuePair<Student,
                Enrollment> p in section.Gradebook)
             {
                 if (values[0].Trim() == section.
                    CourseUnder.CNO
                        && values[1].Trim() == section.
                            Section
                        && values[2].Trim() == p.Key.ID)
                     data[i] = p.Value.Serialize();
             }
        }
    FileAccess.WriteData("enrollment.txt", data);
    }

    public static void LoadInstructor(Instructor
        theInstructor)
    {
        string[] values;
        string[] data = FileAccess.ReadData("teachings.
            txt", 2, theInstructor.ID);
```

```
theInstructor.DeserializeTeaching(data);
sections = theInstructor.Sections;
foreach (CourseOffering co in theInstructor.
    Sections)
{
    data = FileAccess.ReadData("sections.txt");
    foreach (string s in data)
    {
        values = s.Split(',');
        if (values[0].Trim() == co.CourseUnder.CNO
            && values[1].Trim() == co.Section)
            co.Deserialize(s);
    }
    data = FileAccess.ReadData("courses.txt");
    foreach (string s in data)
    {
        values = s.Split(',');
        if (values[0].Trim() == co.CourseUnder.CNO)
            co.CourseUnder.Deserialize(s);
    }
    co.Roster = new Dictionary<string, Student>();
    co.Gradebook = new Dictionary<Student,
        Enrollment>();
    data = FileAccess.ReadData("enrollment.txt");
    foreach (string s in data)
    {
        values = s.Split(',');
        Enrollment e;
        Student stu;
        if (values[0].Trim() == co.CourseUnder.CNO
            && values[1].Trim() == co.Section)
        {
            stu = (Student) UserControl.
                FindUser(values[2].Trim());
            co.Roster.Add(values[2].Trim(), stu);
            e = new Enrollment(stu, co);
            e.Deserialize(s);
            co.Gradebook.Add(stu, e);
        }
    }
}
}
}
```

Interface Classes

We follow the VCM model to implement the two user interfaces: `LoginForm` and `PostGradeForm`. The login form will use `UserControl` to verify user accounts.

```
public partial class LoginForm:Form
{
    private void btnLogin_Click(object sender,
        EventArgs e)
    {
        string txtRole;

        if (rbInstructor.Checked == true)
            txtRole = "Instructor";
        else
            txtRole = "Student";
        UserControl.LoadUsers();
        if (UserControl.VerifyUser(txtUID.Text,
            txtPWD.Text, txtRole))
        {
            PostGradeForm myPostGradeFrom =
                new PostGradeForm((Instructor)
                    UserControl.LoginUser);
            this.Hide();
            myPostGradeFrom.Show();
        }
        else
        {
            lblMessage.Text = "Incorrect Login. Try
                again!";
        }
    }
}
```

The `PostGrade` form needs to have a list of sections that belong to an instructor and, for each section, a list of students and a list of grades. All these objects can be delivered by the `InstructorControl` class. For example, all the sections, including course number, section number, and course title, to be displayed in the combo box come from `InstructorControl`.

```
InstructorControl.LoadInstructor(i);
foreach (CourseOffering co in InstructorControl.
   Sections)
{
    cboSections.Items.Add(co.CourseUnder.CNO + "-"
            + co.Section + ":" + co.CourseUnder.Title);
}
```

The main difficulty with the `PostGradeForm` is about how to present the data. Since each section may have a different number of students, we will need to programmatically create text boxes based on the number of the students in the section. Also, we want the form to be repainted if a different section is loaded. To this end, we will paint all the labels for student ID and name and text boxes for grades into a panel so that we can remove the panel and reload a new one when changing classes. The size of the panel is determined by the number of students in the section.

```
Dictionary<Student, Enrollment> d =
InstructorControl.Sections[cboSections.
    SelectedIndex].Gradebook;
int count = d.Count;
foreach (Control c in this.Controls)
    if (c.Name=="pan") { this.Controls.Remove(c);
Panel panel = new Panel();
panel.Name = "pan";
panel.Location = new Point(150, 100);
panel.Size = new Size(450, 100 + 25 * count);
panel.BorderStyle = BorderStyle.Fixed3D;
this.Controls.Add(panel);
```

Then we create three column heads using Label controls on the panel. Note the location is relative to the panel; the top left corner of the panel has (0,0) coordinate.

```
lbl = new Label();
lbl.Text = "Student ID";
lbl.Location = new Point(30, 20);
lbl.Font = new Font(lbl.Font, FontStyle.Bold);
panel.Controls.Add(lbl);

lbl = new Label();
lbl.Text = "Full Name";
lbl.Location = new Point(130, 20);
```

```
lbl.Font = new Font(lbl.Font, FontStyle.Bold);
panel.Controls.Add(lbl);

lbl = new Label();
lbl.Text = "Grade";
lbl.Location = new Point(280, 20);
lbl.Font = new Font(lbl.Font, FontStyle.Bold);
panel.Controls.Add(lbl);
```

Then we can use a loop to create a list of text boxes. We will need to refer to those textboxes by their indices, rather than names, and so we will keep them in a list called grades.

```
List<TextBox> grades = new List<TextBox>();
int i = 0;
foreach (KeyValuePair<Student, Enrollment> p in d)
{
    lbl = new Label();
    lbl.Text = p.Key.ID;
    lbl.Location = new Point(30, 50 + 25 * i);
    panel.Controls.Add(lbl);

    lbl = new Label();
    lbl.Text = p.Key.Name;
    lbl.Location = new Point(130, 50 + 25 * i);
    panel.Controls.Add(lbl);

    t = new TextBox();
    t.Text = p.Value.Grade;
    grades.Add(t);
    grades[i].Location = new System.Drawing.
      Point(280, 50 + (i * 25));
    panel.Controls.Add(grades[i]);
    i++;
}
```

Note that the textbox list must be a global variable for PostGradeForm because we need to refer to these boxes when pressing the save button, which we can also programmatically put on the panel.

```
Button btnSave = new Button();
btnSave.Text = "Save";
btnSave.Location = new Point(150, 60 + 25 * i);
panel.Controls.Add(btnSave);
```

To create an event handler for the button `btnSave`, we will add the following code and create a function called `btnSave_Click(object sender, EventArgs e)` to respond to the event `btnSave_Click`:

```
btnSave.Click += new EventHandler(btnSave_Click);
```

The following code is used to respond to `btnLoad_Click` event after a section is selected:

```
private void btnLoad_Click(object sender, EventArgs e)
{
    if (cboSections.SelectedIndex >= 0)
    {
        grades = new List<TextBox>();
        Label lbl;
        TextBox t;
        Dictionary<Student, Enrollment> d =
                InstructorControl.Sections[cboSections.
                    SelectedIndex].Gradebook;
        int count = d.Count;

        //remove the panel to redraw for a new section
        foreach (Control c in this.Controls)
            if (c.Name=="pan") { this.Controls.
                Remove(c); }

        Panel panel = new Panel();
        panel.Name = "pan";
        panel.Location = new Point(150, 100);
        panel.Size = new Size(450, 100 + 25 * count);
        panel.BorderStyle = BorderStyle.Fixed3D;
        //panel.SendToBack();
        this.Controls.Add(panel);

        panel.Controls.Clear();

        lbl = new Label();
        lbl.Text = "Student ID";
        lbl.Location = new Point(30, 20);
        lbl.Font = new Font(lbl.Font, FontStyle.Bold);
        panel.Controls.Add(lbl);

        lbl = new Label();
        lbl.Text = "Full Name";
```

```
lbl.Location = new Point(130, 20);
lbl.Font = new Font(lbl.Font, FontStyle.Bold);
panel.Controls.Add(lbl);

lbl = new Label();
lbl.Text = "Grade";
lbl.Location = new Point(280, 20);
lbl.Font = new Font(lbl.Font, FontStyle.Bold);
panel.Controls.Add(lbl);

int i = 0;
foreach (KeyValuePair<Student, Enrollment> p in d)
{
lbl = new Label();
lbl.Text = p.Key.ID;
lbl.Location = new Point(30, 50 + 25 * i);
panel.Controls.Add(lbl);

lbl = new Label();
lbl.Text = p.Key.Name;
lbl.Location = new Point(130, 50 + 25 * i);
panel.Controls.Add(lbl);

t = new TextBox();
t.Text = p.Value.Grade;
grades.Add(t);
grades[i].Location = new System.Drawing.
    Point(280, 50 + (i * 25));
panel.Controls.Add(grades[i]);
i++;
}

Button btnSave = new Button();
btnSave.Text = "Save";
btnSave.Location = new Point(150, 60 + 25 * i);
btnSave.Click += new EventHandler(btnSave_Click);
panel.Controls.Add(btnSave);
    }
}
```

Grades are saved into the "enrollment.txt" file, and data saving is handled by the InstructorControl class:

```
private void btnSave_Click(object sender, EventArgs e)
{
```

```
int i = 0;
foreach (KeyValuePair<Student, Enrollment> p in
            InstructorControl.
                Sections[cboSections.
                SelectedIndex].Gradebook)
{
    p.Value.Grade = grades[i].Text;
    p.Value.Status = true;
    i++;
}
InstructorControl.SaveGrade(InstructorControl.
    Sections[cboSections.SelectedIndex]);
MessageBox.Show("All Grades have been saved!")
}
```

Putting all the above classes together, we shall now have a working code for professors to post grades. Domain classes have many useful operations coded for enrolling classes by students. Thus, it will be a similar exercise to program the "Enroll Classes" use case, which will also involve reading and writing data to the "enrollment.txt" file. Also, additional use cases such "create offerings," "update course catalog," and "admit students" can be envisioned to add more functionalities to the registration system. Those use cases will involve reading and writing data to other data files.

Exercises

1. Compare the VCM model with the step-and-skip model. What advantages and disadvantages does each model have?
2. Give a scenario which shows clear benefits when domain classes do not know data access classes.
3. What is object persistence? Do a little research to compare the two most popular object persistence technologies.
4. Model and code "Enroll Classes" use case performed by the student actor.

Chapter 17

From Structured to Object-Oriented Development

Introduction

The first chapter charted two courses of systems development, respectively, via structured and object-oriented approaches and the models to be developed along the way. This last chapter provides a brief comparison of key models. It is meant for those readers who have prior knowledge in structured development to take advantage of the knowledge, or their old way of thinking may hinder their learning of a new approach to modeling and coding. The readers who do not have such knowledge may skip this chapter; they will not have disadvantages or confusions caused by the knowledge.

Structured development approaches were developed in the early 1970s. Early structured approaches were largely process-oriented (e.g., using data flow diagrams or program flow charts), while the latest are more data-oriented (e.g., using entity–relationship diagrams). In general, a comprehensive structured approach consists of both data modeling to represent business objects and process modeling to represent business processes. In the late 1980s, object-oriented techniques were proposed as an alternative approach. There have been more than 19 different object-oriented techniques proposed since 1988 (Wieringa 1998).

When compared with structured development, the proponents of object-oriented techniques claim that object-oriented development improves the communication between users and analysts, enhances reusability of code, increases productivity and reliability, and reduces the load of code maintenance (Booch 1991; Eaton and Gatian 1996; Garceau *et al.*, 1993;

Johnson 2000; Yourdon 1994). Additionally, object-oriented development supports abstraction at the object level and encourages good programming techniques. It allows a seamless transition among different phases of software development by using the same language for analysis, design, and programming (Graham 1994).

Due to the above-proclaimed benefits, in the last two decades, there was a phenomenal growth of interest in object-orientation. Many authors believed that it is critical to migrate to object-oriented development to develop bigger applications (Liberty 1997) in support of e-business and business process integration. Many organizations also concluded that such a migration is necessary, and its potential benefits can be realized (Levine and Rossmoore 1993).

However, the nature of the migration from structured to object-oriented development is seldom understood. On one hand, due to many conceptual as well as historic connections, one tends to believe that object-orientation represents an evolutionary advancement from structured development. On the other hand, the proponents of object-oriented development tend to emphasize the differences of the approaches and claim that object-orientation is a revolutionary new methodology. For example, Booch (1991) and Korson and McGregor (1990) stated that object-oriented approaches require a different way of thinking about decomposition. Lee and Pennington (1994) called object-orientation a new "paradigm" for software development. In the same vein, Fichman and Kemerer (1992) claim that the shift to object-oriented development represents a radical change from previous approaches to software development.

Morris *et al.* (1999) state that understanding the nature of the migration has important ramifications for systems analysts and the management of systems development projects. As they argued, if the migration is evolutionary, then system analysts currently trained in procedural methods (data flow and entity–relationship modeling) should be able to learn and effectively apply object-oriented development. On the other hand, if the migration is indeed revolutionary, prior experience in procedure methods might hinder an effective migration and a different mindset would be required to approach the object-oriented model.

Unfortunately, existing studies along this line have been not only limited but also inconclusive. For example, Vessey and Conger (1994) found that object-oriented methods were more difficult to learn and apply than process-oriented methods in a group of novices. In contrast, Lee and Pennington (1994) found that object-oriented design is easier and faster to learn than procedural design. Boehm-Davis and Ross (1992) had a balanced finding that object-oriented methods reduce the complexity and design time while increasing the completeness of the solution.

Similarly, very little empirical research has been conducted to determine the influence of previous experience in traditional or procedural approaches on the application of object-oriented development. Agarwal *et al.* (1996) studied the effects of prior experience in procedural modeling and task characteristics on performance in applying object-oriented development. Morris *et al.* (1999) examined whether prior experience in procedural modeling helps or hinders the performance of applying object-oriented development and compared procedural and object-oriented methods on the subjective mental workload (SMW). It is not a surprise that the results of these studies are also uncertain. For example, considering solution quality as a measure of performance, Morris *et al.* (1999) found that procedurally experienced individuals generate higher quality solutions using procedural methods than using object-oriented methods. Agarwal *et al.* (1996), however, did not support the same hypothesis. Moreover, Agarwal *et al.* (1996) found that the procedurally experienced group performed significantly better than the inexperienced group when solving an object-oriented problem, while Morris *et al.* (1999) found the opposite.

The purpose of this chapter is to understand the nature of migration from the perspective of the conceptual as well as cognitive connections between two approaches. We compare three typical systems analysis models, namely, entity–relationship diagrams, data flow diagrams, and class diagrams. We show their conceptual connections and disconnections and how they are different in terms of cognitive tasks. Finally, we present accumulated empirical evidence related to the debate.

In addition, this chapter attempts to shed some light on further empirical studies on the nature of the methodology migration. First, any reasonable model of performance in any domain ought to relate to accepted standards of good practice in that domain (Jeffries *et al.* 1980). This chapter will provide a basis on how to evaluate the performance of applying object-oriented development. Second, formalized approaches were written by experts in the area, trying to convey to others the procedures they use to perform the task. Most expert designers are familiar with them and may incorporate facets of them into their designs. Therefore, information-processing theories (Newell and Simon 1972) dictate that an analysis of the approaches is necessary to develop any research model that theorizes the impact of prior design experience on actual or perceived performance of applying the approaches. As a matter of fact, this chapter provides a theoretical foundation for two other empirical studies, respectively, on how prior design experience affects the performance of applying object-oriented development (Grandon and Liu 2001) and how prior experience affects the system analysts' perception of ease of use and self-efficacy of object-oriented development (Liu and Grandon 2002).

Here, we do not attempt to provide a survey of design approaches, like Wieringa (1998), or a tutorial on the approaches. Instead, we restrict our attention to three models that characterize both structured and object-oriented development and focus on the analysis of their conceptual as well as cognitive similarities and differences. However, besides revealing the nature of the migration from structured to object-oriented development, this chapter provides a template for both structured and object-oriented modeling. The nature of systems analysis is to break a complex business into simple units such as objects and processes, which allow detailed modeling, programmable specifications, and modular management. Following the same spirit, the chapter breaks the task of business modeling into manageable sub-tasks, such as discovering and representing entities and their inter-relationships, and processes and their collaborations. Each sub-task has a well-defined template, and certain sub-tasks have a well-defined modeling goal. Such an analytical view of system design models fills in the many voids in the existing methodology literature. For example, there have been some guidelines and tools (e.g., identifying nouns in use cases) to help with the identification of objects (Rosenberg 1999). However, there do not exist any defined processes or tools, for example, in the Unified Process (Jacobson *et al.*, 1999), to explicitly and systematically assist in eliciting relationships or to document them in class diagrams or entity–relationship diagrams. This chapter fills in the gap by specifying the nature and goal of relationships modeling that assist the systems analyst in discovering what relationships exist among objects and processes, thus greatly enhancing their modeling performance.

Requirement Models

In this section, we review data flow diagrams and entity–relationship diagrams as the representation of structured development and class diagrams as the representation of object-oriented development (Booch *et al.*, 1999).

Data flow diagrams

A data flow diagram depicts business processes and the flow of data among them. The elements of a data flow diagram include functions, data stores, and external entities. The elements are connected by data flows. A function represents a data activity to be performed. A data store represents data at rest. External entities are the sources and/or sinks of data, which are logically outside the system but communicate data with the system. There are two different kinds of relationships to connect diagramming elements together. First, data flows are used to represent informational collaboration among functions. To perform a function, as its name suggests, a function must have enough input flows provided by other functions, data stores, and/or external sources. It must also generate reasonable output flows to serve the information needs of other functions and/or external entities. Second, there is a whole-part relationship between functions and their sub-functions. The functional decomposition represents the delegation of responsibilities in the sense that the sum of sub-functions supports a high-level function, and the sum of all functions supports the mission of the entire system. Correspondingly, data flow diagrams are usually organized into a hierarchy of nested diagrams, where a function at a higher level maps to a decomposed diagram detailing sub-functions at the next lower level.

The goal of data flow diagramming is to ensure that all the responsibilities of a system are captured and allocated to functions. As stated by Fichman and Kemerer (1992), the goal of structured analysis and design is to develop a top-down decomposition of the functions to be performed by the system. There are some other additional criteria to be followed in process modeling. For example, functions are cohesive and loosely coupled to achieve reusability (see Gibson and Hughes 1994; Hoffer *et al.*, 1999; Whitten *et al.*, 2001 for further details).

Entity–relationship diagrams

An entity–relationship diagram graphically represents business objects and their relationships. The basic diagramming elements include entities (objects) and attributes (data). An entity is anything that can be distinctly identified. For instance, customers, events, or accounts are all entities. An attribute is a property or characteristic of an entity. Entities are grouped into entity types (or classes). Entities are connected by relationships like *associations* and *generalizations*. The eventual goal of entity–relationship modeling is to capture data requirements and organize the data efficiently.

The entity relationship model incorporates some of the important semantic information about the real world (Chen 1977). Entities are any business objects that have data, and conceptually they are the same objects in object-oriented analysis. The difference is that the objects have not only data but also capabilities of performing functions. The relationships in entity–relationship models have multiple semantics. First, they can be physical or logical associations between entities. For example, relationships between customers and accounts might represent who owns which account. As another example, relationships between dogs and animals represent the logical connection that dogs are a special kind of animal. Second, a relationship represents a data connection or an information navigation channel through which one can travel from the data about one entity to the data about another entity. For instance, a relationship between a customer and an account provides a navigational channel for one to look up the account given the customer details or vice versa. (For a more detailed description of entity–relationship modeling, see Chen 1977; Rob and Coronel 2000; Hoffer *et al.*, 1999.)

Class diagrams

Objects are the most important construct in object-oriented techniques. In object-oriented programming, an object is a self-contained program module that encapsulates both data and functions. A software system is then simply a collection of discrete classes that can be easily replaced, modified, or reused. In object-oriented analysis, an object often corresponds to a real-world object, like an airplane, an account, or a customer. It can be tangible or intangible. Different from an entity in entity–relationship models, an object has data as well as functionalities. As Rob and Coronel (2000) put it in a simple manner, an object is an abstract representation of a real-world entity that has a unique identity, embedded properties, and the ability to interact with other objects and itself.

A class diagram graphically depicts the static design view of a system: classes, collaborations, and their relationships. The basic diagramming elements include classes, attributes, and operations. A class is conceptually equivalent to an entity type or an entity set in entity–relationship models. It is a set of objects that share common attributes and functionalities. For example, we can group all customer objects to form the `Customer` class and all account objects to form the `Account` class. Objects are connected by relationships like association (including aggregation) and generalization (or inheritance). Like those in an entity–relationship model, association represents a physical or conceptual connection between two or more objects. Generalization represents the IS-A-KIND-OF relationships between related objects. Aggregation represents the IS-A-PART-OF relationships between objects. An entity–relationship model represents aggregation using associations between weak entities and the strong entities on which the weak entities depend, whereas a class diagram represents it using a special type of association, called *containment*.

There are two specific goals to be achieved in class diagramming. The first one is the same as that for an entity–relationship model. That is, a class diagram must capture both object data and their navigational relationships. The second goal is to capture object functions and their collaborative relationships. In addition, to ensure the highest level of code reusability and maintainability, data (knowledge) and functionalities (responsibilities) must be distributed to all objects evenly and coherently such that each object has the best knowledge to perform its own functions and no object has all the data or does everything (Coad and Nicola 1993; Liberty 1997).

Conceptual Connections

Even though both entity–relationship diagrams and class diagrams use similar concepts such as objects (entities), classes (entity type), and data members (attributes), and both represent a static structure of objects using relationships like associations and generalizations, they have some differences. For example, besides capturing data requirements, a class diagram also represents what functions the objects can perform and how the objects collaborate with each other to achieve overall system responsibilities. It is often tempting to say that an entity–relationship diagram is a class diagram without functionality specifications. In addition, an entity–relationship diagram has limitations in representing certain type of relationships such as aggregation and composition (Silberschatz *et al.*, 1999).

In structured development, a software system is viewed as a collection of programs (or functions) and a separate collection of data. As Wirth (1975) defined it in his book that is interestingly titled *Algorithms + Data Structures = Programs*, a software system is a set of mechanisms for performing certain action on certain data. Compared with object-oriented techniques, the structured ones have the following four overall features:

- Data and functions are separated. This dictates that we use two different models to capture data requirements and functional requirements. Specifically, the entity–relationship model and the data flow model are the two main vehicles used in structured development.
- Functions are the basic programming units that are callable by other functions and reusable by other software.
- Functions are participants in collaborations. The functionalities of a whole system are realized by the collaboration of functions, which is documented by data flow diagrams and structured charts as well.
- A record is the basic unit of data storage. Each record represents the attributes of one entity (object). Records are interlinked by navigational relationships, which are documented by entity–relationship diagrams.

In object-oriented techniques, on the other hand, a software system is viewed as a collection of objects that encapsulate both attributes (data) and methods (or functions). Compared with the structured techniques, the object-oriented ones have the following distinctive features:

- Data and functions are no longer isolated. Instead, they are both contained by high-level abstractions — objects. The integration of data and functions dictates that we use one model to capture data requirements and functional requirements. Specifically, the class diagram is the main vehicle used in object-oriented development.
- Classes are the basic programming units that are callable by other classes and reusable by other software.
- Objects are participants in collaborations. The functionalities of a whole system are realized by the collaboration of objects, which is documented by class diagrams and collaboration diagrams as well.
- Objects are the basic unit of data storage. Each object represents a real-world entity (object). Objects are interlinked by navigational relationships, which are documented by class diagrams.

The entity–relationship and the data flow models reasonably represent structured development while the class model represents object-oriented development. Table 1 compares the basic mechanics of the three

Table 1.   A comparison of requirement models.

	Elements	Relationships	Modeling Goals
Entity–relationship diagrams	• Entities • Attributes • Entity types	• Association • Generalization	• Data requirements are captured by entities and attributes • Data navigations are captured by relationships • Data are organized for efficient storage and processing
Data flow diagrams	• Functions • Data stores • External entities	• Data flow • Decomposition	• System responsibilities are captured by functions • The collaboration of functions is captured by data flows • Complex functions are decomposed into simple low-level sub-functions • Functions are cohesive and loosely coupled to achieve reusability
Class diagrams	• Objects • Attributes • Methods • Classes	• Association • Generalization • Aggregation	• Data requirements are captured by objects and attributes • System responsibilities are captured by objects and methods • The collaboration of objects is captured by relationships • Classes are cohesive and loosely coupled to achieve reusability

models as well as the criteria for successful applications of the models. Specifically, structured and object-oriented models have both conceptual and semantic connections and disconnections. With respect to basic elements, class diagrams have strong conceptual and semantic linkage with both the entity–relationship model and the data flow model. The notion of entity types in the entity–relationship model is the same as that of classes in the class model. Entities are the same as objects except that objects have behavior while entities do not. Attributes in the entity–relationship model are the same as that in class diagrams. The concept of functions in the data flow model is the same as, or at least closely related to, the concept of methods (or behaviors, functions, responsibilities) in class diagrams.

Therefore, the understanding of the structured models will improve the understanding of class diagrams, at least conceptually.

With respect to relationships, class diagrams have strong conceptual and semantic linkage with the entity–relationship model. Both the entity–relationship and class diagrams use the same concept of association and generalization. They also use association and generalization in the same way to represent relationships among entities (objects). Due to the need for data normalization, the entity–relationship model does not have the concept of aggregation (or composition) (Silberschatz *et al.*, 1999) as in class diagrams. However, it often represents aggregation (composition) as associations between strong and weak entities. In other words, it has the concept implicitly and represents a domain construct in the same way as in class diagrams.

With respect to relationships, class diagrams have no conceptual linkage with the data flow model. The data flow model uses data flows to connect functions, data stores, and external entities and functional decompositions to connect functions at different levels. A data flow represents either an input to be processed by a function or an output produced by a function. A functional decomposition represents the delegation of the responsibilities of a high-level function to a low-level one. A class diagram uses the concept of association, generalization, and aggregation to represent physical or logical connections between objects. Therefore, the concepts and the semantics of the relationships used in the data flow model and class diagrams are very different.

Regarding modeling goals, class diagrams overlap with but differ from both the entity–relationship and the data flow models. The goal of capturing data requirements in the entity–relationship model is the same as that in class diagrams. The goal of capturing functional requirements in the data flow model is the same as that of class diagrams. In both entity–relationship and class diagrams, we need to capture the relationships between entities or objects. However, their goals are different. In the entity–relationship model, the goal is to capture data navigation, i.e., tracing the data for one entity to the data for a related entity. In class diagramming, the goal is to capture the collaboration of objects, i.e., the methods (functions) of one object support the methods (functions) of other objects.

The data flow model captures the informational collaboration of functions by using data flows. It shows how functions work together by receiving and sending inputs and outputs to realize some cooperative functionality that is bigger than the sum of all its parts. In contrast, collaboration in object-oriented techniques is defined as a society of objects that work together to provide a cooperative behavior. It has two aspects: A structural part that specifies the classes that work together to carry out the named collaboration, and a behavioral part that specifies the dynamics of how objects interact (Booch *et al.*, 1999). A class diagram captures the structural part of collaboration using relationships like association, generalization, and aggregation. The collaboration model (sequence diagrams or communication diagrams) captures the behavioral part of collaboration as messages passed among objects. A message is like a data flow except that messages between objects represent function calls while data flows represent data inputs and outputs between functions. Therefore, capturing collaboration requirements in the data flow model is like capturing the behavioral part of collaboration in object-oriented development. It is different from capturing the structure part of collaboration in class diagrams. Of course, the behavioral part of collaboration is conditioned on the structural part of collaboration; two objects can send messages to each other only when their hosting classes are connected in a class diagram.

The data flow model has a second legacy. The highest data flow diagram, called the *context diagram*, is a close relative to the use case diagram in the object-oriented analysis (Whitten *et al.*, 2001).

Besides the goal of capturing data and functional requirements, the application of each model has some additional goals or criteria, which can be used to judge the quality of modeling tasks. In the entity–relationship model, the most important criterion is the efficiency of data storage and data processing. The fewer the missing values and the less the redundancy, the better an entity–relationship diagram. In data flow modeling, the additional goals include cohesion and loose coupling of functions (Gibson and Hughes 1994). Cohesion measures how much the functions that are connected to each other support a central purpose. A cohesive section of functions does not rely on other sections of functions for help. Coupling, on the other hand, measures the interdependence of the functions that are connected. Since program modules must be as independent as possible to be reusable, functions should be loosely coupled to each other. Interestingly, the notion of cohesion and coupling also applies to class diagrams (Bahrami 1999; Liberty 1997). Here, coupling measures the strength of association between objects (Bahrami 1999). A class is cohesive if all its attributes and operations relate to the same area of concern (Liberty 1997).

Coupling deals with interactions among objects, while cohesion deals with interactions within a single object. One goal in class diagramming is to maximize object cohesiveness in order to improve coupling, because only a minimal amount of essential information needs to be passed between objects (Bahrami 1999).

Even though both models use a similar concept of cohesion and coupling, end results can be very different. A class diagram requires spreading knowledge (attributes and methods) horizontally among all classes, with each class specializing in one area of concern. Any other class that must accomplish a related task will then delegate the task to the class most responsible (Liberty 1997). In contrast, the data flow model advocates spreading knowledge vertically so that responsibility is centralized in one or two manager functions, and they delegate partial responsibilities to a rabble of worker functions. If this design philosophy is taken in class diagramming, it is often a natural inclination that one creates a small number of omniscient manager classes and a rabble of worker classes that are deeply coupled to the manager classes. As a matter of fact, those C++ programmers who are experienced in a procedural language like C tend to create global manager classes that are essentially global functions in class clothing (Liberty 1997). The data flow modeling philosophy undermines the delegation of responsibility essential to clean and robust object-oriented design. Due to the deep coupling, whenever a manager class is redesigned, the effects ripple uncontrollably and destructively throughout worker classes. The manager classes tend to become large and unwieldy, and it is difficult to reuse them or even derive from them as they bring so much baggage and overhead (Liberty 1997).

Liberty (1997) uses Adam Smith's division of labor as a metaphor to illustrate the principle of delegation in object-orientation. A company might have a lawyer, a developer, and a graphic artist. Each one has a narrow, cohesive set of responsibilities and expertise. The company encompasses a lot of expertise, but it is spread evenly across the people involved. When one adds a new responsibility, one assigns it to the person with the most knowledge. If we were adding the responsibility of ensuring that we have not violated a copyright, we might assign it to a lawyer who is the most knowledgeable about the law. If the lawyer needs to determine which algorithm is to be used in a part of the project, he might delegate that responsibility to the developer, who, again, has the most knowledge in the area. If the lawyer needs to determine the authenticity of a graphic work, he might delegate the responsibility to the graphic artist.

In sum, the three models have conceptual and semantic connections and disconnections. From structured to object-oriented models, some concepts are evolutionary while some are revolutionary. Class diagrams extend the concepts of entities, entity types, and attributes in the entity–relationship model and the concept of functions in the data flow model. The structure of a class diagram is like that of an entity–relationship diagram. Their concepts of relationships are also very similar. However, a relationship in the entity–relationship model represents data navigation while in a class diagram it is the delegation of responsibilities or the structure of collaboration. The structure of a class diagram is different from that of a data flow diagram. However, they both represent collaboration. The difference is that a data flow diagram represents the behavioral part of collaboration while a class diagram represents the structural part of collaboration. Another revolutionary difference between class diagrams and data flow diagrams is in how one uses the notion of cohesion and coupling. The concepts are the same in both models. However, they underline two different systems design philosophies.

Cognitive Connections

In entity–relationship modeling, the cognitive tasks include identifying entities and attributes from data requirements, grouping entities into entity types, and discerning relationships based on the data navigation requirements as follows:

- *Data → Entities*: This activity identifies entities. There are two different approaches to such an identification. First, entities are the business objects that one needs to keep data for. Therefore, the identification of entities boils down to the search for business objects such as customers, employees, accounts, orders, etc. This approach is often referred to as the top-down one. In contrast, the second approach is a bottom-up one. Here, one is given the data to be recorded (e.g., customer name, order date, order quantity, discount, etc.), and the cognitive task is to discern the corresponding data carriers or containers (entities) holding the data.
- *Data → Attributes*: This activity identifies common attributes for a group of entities. There are also two different approaches. First, one can identify attributes by exhaustively searching for the properties of the entities. Second, one can identify attributes based on the presence

of certain data items to be recorded, such as order quantity and order date.

- *Entity Grouping or Classification*: This task groups entities that share common properties and relationships into an entity type. It also includes factorizing common attributes and relationships in order to identify sub-entity types.
- *Data → Relationships*: If an attribute simultaneously depends on two or more entities, one can recognize the existence of relationships between the objects. This task identifies them along with a special entity type called *gerund* based on the presence of such an attribute.
- *Data Navigation → Relationships*: The key to data modeling is to ensure data navigation (National Research Council 1999). Given an entity, if there is a need to trace a related entity, then there must be a relationship between the two entities.

The data flow modeling involves the cognitive tasks of identifying functions from a system's responsibilities, identifying data flows based on information collaboration requirements, and progressively decomposing high-level functions into more detailed sub-functions as follows:

- *Data Activity → Functions*: This task identifies the functions to be performed by the system. We can do so by following a top-down approach, where the responsibilities of the entire system are identified first and then a set of functions are created to assume the responsibilities. We can also do so by following a bottom-up approach, where data to be managed are identified first and then the data actions to be performed are inferred.
- *Information Collaboration → Data Flows*: This task identifies the input and output data flows for a function. For any function to perform its action, it must have enough input data. Due to the separation of data and functions, a function must get data from other functions, data stores, and/or external entities. Similarly, a function cannot absorb its output data. It must send them to other functions, data stores, and/or external entities.
- *Functional Decomposition*: This task vertically decomposes a function into a set of collaborative sub-functions such that the sum of the sub-functions is equal to the parent function. This process was called *progressive refinement* by Fertuck (1995), and in this one uses a zoom

lens to reveal details of a function that are not visible initially. It is viewed as the delegation of responsibilities, where the parent function plays the role of a manager that calls and coordinates its sub functions to perform their tasks (Gibson and Hughes 1994). It can be viewed as a logical aggregation process, where the parent function is comprised of and therefore is decomposed into two or more logically different data activities (Hoffer *et al.*, 1999). Regardless of which perspective a systems analyst may take, cognitively, functional decomposition involves solving a logical puzzle or an under-determined equation. It involves solving a hierarchical set-covering problem, where a set is covered by two or more subsets, which in turn can be further covered by finer subsets. There is infinite number of solutions. However, the goal is to come up with a solution in which sub functions are coherent and loosely coupled to achieve reusability and the functions are efficient so as to support the mission of the system.

In class diagramming, one needs to identify objects, attributes, and methods based on data and functional requirements, group objects into classes, and discern object relationships based on functional collaboration requirements as follows:

- *Data → Objects*: This activity identifies domain objects in object-oriented analysis. As in entity–relationship modeling, there are two different approaches to such identification. First, one searches for the business objects such as customers, employees, accounts, orders, etc. that he or she needs to keep data for. Second, given the data to be recorded (e.g., customer name, order date, order quantity, or discount), one can also discern the corresponding data carriers or containers behind the data. Such as a carrier or container is an object.
- *Functions → Objects*: In class diagramming, all functions are embedded inside classes. Like the abstraction from data to objects, the task here is to discern the performer of a function given the requirement to capture the function.
- *Data → Attributes*: This activity identifies common attributes for a group of objects. This task is the same as its counterpart in entity–relationship modeling. First, one can identify attributes by exhaustively searching for the properties of the objects. Second, one can identify the attributes according to the presence of data items such as customer name and order date.

- *Responsibilities → Methods*: This task abstracts real-world responsibilities into object methods. To a certain extent, this task is like the abstraction from data requirements to object attributes. However, there are some differences. First, a typical high-level responsibility must be carried out by two or more objects collaboratively. Due to the need for data encapsulation, a method in Object A usually cannot directly access the data in Object B. Object A must request the service from Object B to execute one or more of its methods. Therefore, a real-world functionality may have to be decomposed and abstracted into multiple methods contained in multiple objects. Second, it is not straightforward to decide which class should contain which function. Methodologists suggest that a function should be carried out by the object that has the best knowledge or data (Coad and Nicola 1993; Liberty 1997). However, such a decision often requires a judgmental call.

- *Data → Relationships*: If an attribute simultaneously depends on or describes two or more objects, one can recognize the existence of relationships between the objects. This task identifies them along with an *association object* based on the presence of such data. It is cognitively the same as a gerund in entity–relationship diagrams.

- *Data Navigation → Relationships*: Based on the belief that a class diagram is an extension of an entity–relationship diagram, there have been attempts to use class diagrams for database design (Post 1999). In such applications, it is important to ensure data navigation by using relationships (National Research Council 1999). That is, given an object, if there is a need to trace a related object, then there must be a relationship between the two objects. Note that such an application of class diagrams still falls within the paradigm of structured design, where data and functions are separated. In object-oriented techniques, direct access to data in different objects is prohibited. Whenever an object needs to get data from a related object, it calls the methods of the other object by sending messages. The need for data navigation in structured techniques is replaced by the functional delegation between objects.

- *Responsibilities → Relationships*: This task identifies relationships among objects to capture the structure of delegation among objects. An important goal in class diagramming is to create cohesive and loosely coupled classes so that each class is responsible for one area of concern, and any other class that must accomplish a related task will then delegate to the class most responsible. This design criterion states that a typical system responsibility must be carried out by one or more objects in collaboration. The behavioral part of collaboration is captured by function calls (or messages) on collaboration diagrams. The structured part of collaboration is represented by relationships among objects in a class diagram. The two parts of collaboration are interwoven; a function call cannot happen between two objects if they have no relationship in the class diagram. An effective class modeling technique, called *CRC Cards*, is often employed to determine the fundamental associations among classes in object-oriented design projects (Fowler and Scott 1997). A class-responsibility-collaboration (CRC) card is nothing more than a 3×5 index card on which one writes the name of each class, its responsibilities, and the names of other classes with which it must collaborate to get its work done. In this way, the relationships among collaborative classes are identified.

- *Entities Grouping*: This task groups objects that share common properties and relationships into a class. It also includes factorizing common attributes and relationships to identify super- and subclasses.

Empirical Evidence

Empirical studies have been focused on whether existing experience in structured development helps or hinders the migration to objection-orientation, both in terms of performance and cognitive effort. Among them, the primary independent variable is prior procedural modeling experience, except for Grandon and Liu's study (2001) where both data modeling and process modeling experience were considered. Between the two dependent variables that symbolize whether it is easy to achieve migration, performance is typically measured by design quality while cognitive effort is measured variably from study to study. For example, Adelson and Soloway (1985) studied dissimilarities in novice and expert cognitive process in the context of software design. Vessey and Conger (1994) applied process tracing and protocol analysis to qualitatively capture differences in cognitive processing. Morris *et al.* (2000) used SMW, a hypothetical construct that represents the cost incurred by a human operator to achieve a level of performance (Hart and Staveland 1988). Grandon and Liu (2001) employed both time to solution and perceived attention level, which captures how much mental resource is devoted to a cognitive task (Kanfer and Ackerman 1989; Kanfer *et al.*, 1994).

Unfortunately, a general conclusion concerning whether previous experience helps or hinders performance in the same or a new domain cannot be derived from the existing literature of empirical studies since their findings are inconclusive. For example, Boehm-Davis and Ross (1992) found that previous experience was not the major contributor to completion of the program design even when subjects had prior experience in the same domain. In their study, they compared the solution design obtained from professional programmers with years of experience in a specific methodology with less-experienced professional programmers in the same methodology. There was no correlation between the percentage of completion and years of programming experience.

Related studies that have considered the role of previous experience in applying object-oriented approaches include those by Agarwal *et al.* (1996), Morris *et al.* (1999), and Lee and Pennington (1994). The findings of these studies are varied. For example, Agarwal *et al.* (1996) studied the effects of previous experience and task characteristics on performance in systems analysis and design. To explore this relationship, they conducted an experiment in which two groups of subjects applied the object-oriented methodology to two types of tasks: one process-oriented and the other object-oriented. The experienced group had significant knowledge in process-oriented approaches while the other group had no formal experience. The results showed that analysts experienced in process-oriented approaches did not perform better than inexperienced modelers on the process-oriented task. As opposed to what was expected, the experienced group performed significantly better than the inexperienced group on the object-oriented task.

Lee and Pennington (1994) examined differences in cognitive activities and final designs among object-oriented and procedure-oriented expert designers using both object-oriented and procedural design approaches, and among expert and novice object-oriented designers, when novices had extensive procedural experience. When analyzing final designs, the researchers found that object-oriented designers decomposed their design into objects corresponding to the real-world domain entities while procedural designers decomposed their designs according to actions on data structures.

Morris *et al.* (1999) examined whether experience in using procedural methods helps or hinders performance using object-oriented analysis and compared procedure- and object-oriented analysis methods on the SMW (subjective mental workload). To accomplish this, they conducted an experiment using two groups of individuals: the first group was comprised of procedurally experienced subjects and the other one was comprised of novices. The measures of performance they used to assess the influence of previous experience were SMW, solution quality, and time to solution. The results indicated that SMW was higher for those procedurally experienced subjects using an object-oriented tool to solve the problem. However, when those procedurally experienced subjects had to solve the task using a procedural tool — data flow diagram — the SMW was considerably lower. The quality of the solution, on the other hand, was lower for the experienced group compared with the inexperienced group when they solved the task using an object-oriented tool. The quality was higher when the experienced group had to solve the task using a procedural tool. In addition, it was hypothesized that time to solution should be less for the experienced group when they had to solve a problem using data flow diagram. However, the data obtained did not support this hypothesis.

Grandon and Liu (2001) explicitly consider two different types of prior experience — data modeling and procedure modeling — and their effect on performance and cognitive effort in class diagramming. They conducted an experiment using four groups of subjects, with each group having a different mixture of subjects of both experience types. They found a very significant main effect of data-modeling experience in the sense that subjects with data modeling experience performed much better than those without. They also found a significant interaction effect between data modeling and process modeling: for subjects with data modeling experience, their experience in procedure modeling positively influenced their performance of class diagramming. However, for subjects without data modeling experience, the impact of procedure modeling experience was the opposite. Moreover, Grandon and Liu (2001) found mixed support for their hypothesis that subjects with data modeling and/or procedure modeling experience would need less effort than novices in using the object-oriented method.

Summary Notes

Requirements modeling involves both data modeling and process modeling. Data modeling is concerned with capturing end-user data requirements, which are eventually converted into database design specifications.

Process modeling is concerned with capturing business process requirements, which are eventually converted into application specifications. In this sense, both data modeling and process modeling are two integral components of any systems analysis and design task.

In structured development, data modeling and process modeling are performed separately. In object-oriented development, however, both data and processes are captured by a unified model, such as a class diagram. Traditional data modeling and process modeling have many different conceptual connections with class diagramming. As was observed, class diagramming consists of abstract elements and relationships from both the data model and the process model. A class diagram inherits constructs like entities (objects), attributes (data members), entity types (classes), and relationships such as association and generalization from a data model. It inherits the concept of functions (methods) from a process model. In terms of modeling goals, class diagramming needs to capture both data requirements and navigations, and responsibility requirements and collaborations, which are, respectively, in the realms of data modeling and process modeling. Therefore, prior experience in data modeling and process modeling should help in improving the performance of object modeling.

Structured development and object-oriented development have significant differences. First, the concepts of objects that encapsulate both data and responsibilities involve a higher level of abstraction than functions and entities. Second, even though class diagramming inherits the concept of relationships from data modeling, a relationship represents the collaboration of objects in a class diagram while it is for data navigation in a data model. Third, both class diagramming and process modeling involve the concepts of collaboration, functional decomposition, cohesion, and coupling. However, process modeling is concerned with the collaboration of functions and a vertical decomposition of functions, while class diagramming is concerned with the collaboration of objects and a horizontal distribution of responsibilities.

In terms of cognitive tasks, class diagramming involves tasks that can be performed by knowledge in data modeling and process modeling independently. Such tasks include identifying objects from data, attributes from data, relationships from data navigations, methods from responsibilities, and grouping objects into classes. It also performs tasks that integrate concepts and skills across data modeling and process modeling. Such tasks include identifying objects through functions, relationships though functions, and methods through data.

The migration from structured to object-oriented development involves both evolutionary and revolutionary factors. It is the cognitive theories along with empirical tests that may eventually judge whether the migration is revolutionary or evolutionary. The conceptual and cognitive analysis detailed in this chapter would have shed light on how to apply cognitive theories and conduct experimental studies.

In empirical studies, the impact of prior design experience is still a paradox to be addressed. Research findings on the nature of the migration have been sparse and equivocal. One study found that the main and interaction effects of data modeling and process modeling experience can explain 61% of the variance of the performance of applying object-oriented development (Grandon and Liu 2001). If this result can be further replicated in other experimental settings, it may be tempting to conclude that the migration from structured development to object-oriented development is evolutionary.

Bibliography

Adelson, B. and Soloway, E. 1985. The role of domain experience in software design. *IEEE Transactions on Software Engineering.* **11**(11): 1351–1360.

Agarwal, R., Sinha, A. *et al.* 1996. The role of prior experience and task characteristics in object-oriented modeling: An empirical study. *International Journal of Human and Computer Studies.* **45**: 639–667.

Alencar, F., Jaelson Castro, *et al.* 2000. From Early Requirements Modeled by the i* Technique to Later Requirements Modeled in Precise UML. In Anais do III Workshop em Engenharia de Requisitos. Rio de Janeiro, Brazil.

Bahrami, A. 1999. *Object-Oriented Systems Development Using the Unified Modeling Language.* McGraw-Hill: Boston, MA.

Boehm-Davis, D. and Ross, L. 1992. Program design methodologies and the software development process. *International Journal of Man-Machine Studies.* **36**: 1–19.

Booch, G. 1991. *Object Oriented Design with Applications.* The Benjamin/ Cummings Publishing Company, Inc. San Francisco, CA.

Booch, G., Rumbaugh, J. *et al.* 1999. *The Unified Modeling Language User Guide.* Addison-Wesley: Reading, MA.

Chen, P. 1977. *The Entity-Relationship Approach to Logical Database Design.* Wellesley, MA, Q.E.D. Information Sciences, Inc.

Coad P. and Nicola J. 1993. *Object-Oriented Programming.* Yourdon Press: Englewood Cliffs, NJ.

Deming, W. E. 1986. *Out of the Crisis.* Massachusetts Institute of Technology Center for Advanced Engineering Study: Cambridge, MA.

Eaton, T. and Gatian, A. 1996. Organizational impacts of moving to object-oriented technology. *Journal of Systems Management.* **47**(2): 18–26.

Erich Gamma, Richard Helm, *et al.* 1995. *Design Patterns: Elements of Reusable Object-Oriented Software.* Addison-Wesley: Reading, MA.

Fertuck L. 1995. *Systems Analysis and Design with Modern Methods*. McGraw-Hill: Boston, MA.

Fichman, R. and Kemerer, C. 1992. Object-oriented and conventional analysis and design methodologies. *Computer.* October, 22–39.

Fowler, M. and Scott, K. 1997. *UML Distilled. Applying the Standard Object Modeling Language.* Addison-Wesley: Reading, MA.

Garceau, L., Jancura, E. *et al.* 1993. Object-oriented analysis and design: A new approach to systems development. *Journal of Systems Management.* **44**(1): 25–33.

Geoffrey A. Moore. 1991. *Crossing the Chasm: Marketing and Selling High Tech Products to Mainstream Customers*, Harper Business: New York.

Gibson, M. and Hughes, C. 1994. *System Analysis and Design. A Comprehensive Methodology with Case.* Boyd & Fraser: MA, San Francisco, CA.

Graham, I. 1994. *Object Oriented Methods.* Addison-Wesley: Reading, MA.

Grandon E. E. and Liu, L. 2001. An empirical study of how structured modeling experience affect the performance of applying object-oriented analysis methodology. Technical Report. Department of Management, Southern Illinois University, Carbondale, Illinois. A short version is also available in M. Rungtusanatham (ed.), *Proceedings of 2001 Decision Sciences Institute Meeting.* San Francisco, CA.

Hart, S. G. and Staveland, L. E. 1988. Development of NASA-TLX (Task Load Index): Results of empirical and theoretical research. In O. A. Hancock and N. Meshkati (eds.), *Human Mental Workload.* Amsterdam, North-Holland, Elsevier.

Hay, D. 1996. *Data Model Patterns: Conventions of Thought.* Dorset House Publishing, New York, NY.

Hoffer, J., George, J. *et al.* 1999. *Modern Systems Analysis and Design.* Addison-Wesley: Reading, MA.

Howard Podeswa. 2005. *UML for the IT Business Analyst.* Thomson: Boston, MA.

Jacobson I., Booch G. Booch, *et al.* 1999. *The Unified Software Development Process.* Addison-Wesley: Reading, MA.

Jeffries, R., Turner A. A. *et al.* 1980. The processes involved in designing software. In J. R. Anderson (ed.), *Cognitive Skills and Their Acquisition.* Lawrence Erlbaum Associates: Hillsdale, NJ.

Johnson, R. 2000. The ups and downs of object-oriented systems development. Association for Computing Machinery. *Communications of the ACM.* **43**(10): 68–73.

Kanfer, R. and Ackerman, P. 1989. Motivation and cognitive abilities: An integrative/aptitude-treatment interaction approach to skill acquisition. *Journal of Applied Psychology.* **74**(4): 657–690.

Kanfer, R., Ackerman, P. *et al.* 1994. Goal setting, conditions of practice, and task performance: A resource allocation perspective. *Journal of Applied Psychology.* **79**(6): 826–835.

Kent Beck and Cunningham. 1989. A laboratory for teaching object oriented thinking, ACM SIGPLAN Notices.

Korson, T. and McGregor, J. D. 1990. Understanding object-oriented: A unifying paradigm. *Communications of the ACM.* **33**(9): 40–60.

Lee, A. and Pennington, N. 1994. The effects of paradigm on cognitive activities in design. *International Journal of Human Computer Studies.* **40**: 577–601.

Levine, H. G. and Rossmoore, D. 1993. Diagnosing the human threats to information technology implementation: A missing factor in systems analysis illustrated in a case study. *Journal of Management Information Systems.* **10**(2): 55–73.

Liberty, J. 1997. *Object-Oriented Analysis and Design with C++.* Wrox Press Ltd: Birmingham, UK.

Liu, L. and Grandon, E. E. 2002. *Effects of Prior Design Experience on the Perceived Ease of Use of Object-Oriented Analysis Methodology.* Technical Report. Department of Management, University of Akron: Akron, OH.

Michael E. Fagan. 1976. Design and code inspections to reduce errors in program development. *IBM Systems Journal.* **15**(3): 182–211.

Morris, M., Speier, C. *et al.* 1999. An examination of procedural and object-oriented systems analysis methods: Does prior experience help or hinder performance? *Decision Sciences.* **30**(1): 107–135.

National Research Council. 1999. *Funding A Revolution: Government Support for Computing Research.* National Academy Press: Washington, DC.

Newell, A. and Simon, H. 1972. *Human Problem Solving.* Prentice Hall: NJ, Upper Saddle River, New Jersey.

Post, G. V. 1999. *Database Management Systems: Designing and Building Business Applications.* McGraw-Hill: Boston, MA.

Rob, P. and Coronel, C. 2000. *Database Systems Design, Implementation, and Management.* Course Technology. Thomson Learning: Cambridge, MA.

Rosenberg D. 1999. *Use Case Driven Object Modeling with UML: A Practical Approach.* Addison-Wesley: Reading, MA.

Scott, W. 1995. Ambler, The Object Primer: The Application Developer's Guide to Object-Orientation (SIGS: Managing Object Technology), Cambridge University Press, Cambridge, UK.

Silberschatz, A., Korth H. F. *et al.* 1999. *Database System Concepts.* WCB McGraw-Hill: Boston, MA.

Vessey, I. and Conger, S. 1994. Requirements specifications: Learning object, process, and data methodologies. Association for Computing Machinery. *Communications of the ACM.* **37**(5): 102–111.

Watts S. Humphrey. 1989. *Managing the Software Process*. Addison-Wesley: Reading, MA.

Whitten, J., Bentley, L. *et al*. 2001. *Systems Analysis and Design Methods*. McGraw-Hill: New York.

Wiegers, K. E. 2003. *Software Requirements*. Microsoft Press: Redmond, WA.

Wieringa, R. 1998. A survey of structured and object-oriented software specification methods and techniques. *ACM Computer Surveys*. **30**(4): 459–527.

Wilkinson, Nancy, M. 1995. Using CRC Cards: An Informal Approach to Object-Oriented Development, SIGS Books, New York.

Wirth, N. 1975. *Algorithms + Data Structure = Programs*. Prentice-Hall: Englewood Cliffs, NJ.

Yourdon, E. 1994. *Object-Oriented Systems Design. An Integrated Approach*. Yourdon Press: NJ, Englewood Cliffs, NJ.

Yu, E. 1995. *Modeling Strategic Relationships for Process Reengineering*, PhD Thesis, Department of Computer Science, University of Toronto: Canada.

Index

www.ingramcontent.com/pod-product-compliance
Lightning Source LLC
Chambersburg PA
CBHW070745220326
41598CB00026B/3738